Mastering
Instrument
Flying

Other books in the PRACTICAL FLYING SERIES

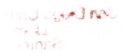

Mastering Instrument Flying

3rd Edition

Henry Sollman with
Sherwood Harris

McGraw-Hill

New York   San Francisco   Washington, D.C.   Auckland   Bogotá
Caracas   Lisbon   London   Madrid   Mexico City   Milan
Montreal   New Delhi   San Juan   Singapore
Sydney   Tokyo   Toronto

Library of Congress Cataloging-in-Publication Data

Sollman, Henry.
    Mastering instrument flying / Henry Sollman, with Sherwood Harris.
—3rd ed.
       p.   cm.
    Includes index.
    ISBN 0-07-059691-3. — ISBN 0-07-059690-5 (p)
    1. Instrument flying   I. Harris, Sherwood. II. Title.
TL711.B6S58   1999
629.132'5214—dc21

            98-52811
            CIP

McGraw-Hill

A Division of The **McGraw·Hill** *Companies*

1 2 3 4 5 6 7 8 9 0 DOC/DOC 9 0 3 2 1 0 9 8

ISBN 0-07-059690-5 (PBK)
ISBN 0-07-059691-3 (HC)

The sponsoring editor for this book was Shelley Ingram Carr, the editing supervisor was Sally Glover, and the production supervisor was Sherri Souffrance. It was set in Times per the PFS design by Michele M. Zito of McGraw-Hill's Professional Group Composition Unit, in Hightstown, NJ.

Printed and bound by R. R. Donnelley & Sons Company.

McGraw-Hill books are available at special quantity discounts to use as premiums and sales promotions, or for use in corporate training programs. For more information, please write to the Director of Special Sales, McGraw-Hill, 11 West 19th Street, New York, NY 10011. Or contact your local bookstore.

This book is printed on recycled, acid-free paper containing a minimum of 50 percent recycled, de-inked fiber.

Contents

CONTENTS

Acknowledgments

My COLLABORATOR ON *MASTERING INSTRUMENT FLYING* SINCE THE beginning has been Sherwood Harris, a veteran pilot who was a senior editor in the books divisions of Readers Digest for many years. His unique combination of talents as a flight instructor, author, and editor made this book possible. He found the words that made my ideas come to life on paper.

I would also like to thank all my former colleagues at Panorama Flight Service, Inc., for their support and assistance, especially my flight instructors. They have used this book in training instrument pilots, essentially field testing it over the past years. They have provided many important insights with their comments and suggestions.

Finally, thanks to all my students down through the years. A good teacher always learns from his students, and mine have taught me very well indeed.

How to use this book

As a flight instructor in the 1960s, I kept running across more and more students who had tried without success for months—sometimes years—to obtain an instrument rating. They would fly countless hours with their instructors, but they never seemed to get anywhere. They couldn't see the light at the end of the tunnel!

So I started developing the concepts for an accelerated instrument course, a 10-day program that would train my students to be safe and confident instrument pilots and give them everything they needed to pass their instrument flight tests with ease. I believe I was the first flight instructor to offer "an instrument rating in 10 days," a course that really worked. The course requires total commitment on the part of the student and instructor. And, if the student intends to complete the course within 10 days, the student must show up with a passing grade on the instrument written test.

I placed three ads in *Trade-a-Plane* for my 10-day instrument course. The response was so great that I had to stop running the ads. I even had people from Europe applying for the course, and, in due course, I enrolled my first foreign students: 2 pilots from Hamburg, Germany.

Since that time, thousands of pilots—American and foreign—have benefited from my system of teaching instrument flying. I have had many imitators, but I believe that no one has duplicated the quality and performance that I have achieved in more than 20 years of training instrument pilots.

I present the basic building blocks of instrument flying as the student actually encounters them in the real world of IFR. We first work on flight planning, weather, and clearances, then proceed to the fundamentals of controlling the airplane by instruments, then to VOR and ADF procedures, then to approaches, and so forth. When an instrument student works with me, we never go up and just bore holes aimlessly in the sky!

Each training flight must have a definite purpose, goal, and competent standards clearly understood and agreed to by both student and instructor. Strict adherence to this

HOW TO USE THIS BOOK

objective will result in the greatest advancement possible and avoid aimless floundering. (Many flight instructors over the years have avoided starvation by frequent training flights for hamburgers to a favorite airport restaurant, resulting in much fun but little learning taking place.)

On *all* instrument training flights, I *always* have the student prepare and file a complete IFR flight plan to a real destination with return. We might not always open the flight plan, and sometimes we might cancel IFR en route to practice basic instrument flying, holding patterns, or instrument approaches to accomplish the objectives of the lesson. But my students *always* prepare and file IFR on *every* flight.

My insistence on planning and filing IFR on every training flight is beneficial. Students quickly become adept at getting good weather briefings, making real go/no-go weather decisions, and choosing appropriate alternate airports. In a surprisingly short time, they become expert at copying clearances and working like professionals with air traffic control (ATC) on instrument departures, en route procedures, and instrument approaches.

Working with ATC on instrument flights is a matter of repetition and routine and is best learned through practice. The sooner the ATC procedures are introduced, the sooner the student becomes competent and confident. And the sooner students become competent and confident in ATC procedures, the sooner they can concentrate on other important aspects of instrument flight. Students who stumble over clearances at the beginning of an instrument flight are just that much more likely to stumble over their dealings with ATC on an instrument approach.

Last, and perhaps not least for many people, going through these routines on each and every training flight gives my students a much better chance for success on the instrument flight test. I am also a designated examiner, and I can tell you that an applicant who has any difficulty with weather, flight plans, or copying clearances has little or no chance of obtaining an instrument rating from me or anyone else. Who wants to share airspace on an IFR day with someone who has not thoroughly mastered these important fundamentals?

Over the more than 50 years that I have been flying and instructing, I have learned many tricks of the trade that help me teach and help my students learn. Many of these are what I call "shortcuts that don't short-change." These special techniques—honed and polished over many years with all kinds of students—are important ingredients in the success of my teaching. I have included all of them in this book.

Finally, I have never allowed my course to stagnate. Aviation is a dynamic, constantly changing field, and no area of aviation changes more rapidly for the general aviation pilot than air traffic control and IFR procedures.

This book is an example of my constantly evolving approach to flight instruction. *Mastering Instrument Flying—3rd Edition* has been thoroughly revised and updated from beginning to end to meet or exceed the instrument flight test standards established by the Federal Aviation Administration. The course was tested and debugged over many months of actual use by the instructors on my staff at Panorama Flight Service at Westchester County Airport, White Plains, NY.

The book is:

A basic course for instrument students. It contains the first syllabus of 20 flight lessons designed specifically to fulfill the requirements of the FAA practical test standards.

Supporting the syllabus are chapters that cover, in detail, the material that should be read and understood prior to each flight lesson. This material has been carefully illustrated to make everything absolutely clear. Flight lessons are cross-referenced for the required reading for that flight lesson.

A set of lesson plans, with explanations, for the instrument flight instructor. The syllabus is designed for completion within two weeks if both instructor and student can devote full time to the course. If this is not possible, there is sufficient reinforcement and review built into the 20 lessons to qualify a student for an instrument rating, with no wasted time or effort, in a less intensive pace.

A handy refresher for the pilot with an instrument rating. Refer to this book to brush up on procedures while preparing for an instrument proficiency check.

Throughout the book the emphasis is on practical problems of instrument flight, rather than theoretical problems. Reference material regarding aerodynamics, weather theory, instrument interpretation, radio signal propagation, and the like, abound. Instrument students, pilots, and instructors should be thoroughly familiar with these sources. The best sources are compiled in Appendix A, "The Instrument Pilot's Professional Library." Also provided is information on how to obtain these publications—a process that is not always as easy as it should be.

Appendix B contains excerpts from Federal Aviation Regulations (FARs) 1, 61, and 91 that deal with instrument training and IFR flight.

Appendix C is a glossary of pilot and controller terms encountered when planning, filing, and flying an IFR flight.

New in this edition

THE WORLD OF INSTRUMENT FLYING HAS BEEN THROUGH AN
electronic revolution since the last edition of *Mastering Instrument Flying* was published in 1994. At that time, electronic developments such as GPS and the Internet were
just barely on the horizon. Now they have moved into the instrument flying mainstream.
The new 3rd edition of *Mastering Instrument Flying* covers these developments and
shows the instrument student how to make the most of them.

But so much more has happened since the last edition! FAR 61 has been thoroughly
revised and has many new requirements relating to general aviation and to instrument instruction and IFR flying in particular. METARs and TAFs have superseded the old
weather reporting codes that we were so comfortable with. The venerable AIM has even
changed its name—it's now the *Aeronautical Information Manual*. The new AIM also
has a lot of new information.

The list of changes goes on and on and has produced revisions to almost every page
of *Mastering Instrument Flying*. We have thoroughly revised, corrected, and updated all
sections for the 3rd edition. All charts and FAR part and paragraph references are as current as practical with publishing deadlines. The charts in our illustrations should not be
used for navigation. As always, fly only with current charts.

1
The psychology of instrument flight

Mastering instrument flying is as much a state of mind as it is training, study, and practice.

FROM THIS MOMENT FORWARD IT IS IMPORTANT THAT YOU ESTABLISH the goal of being the "proud, perfect pilot." Each time you fly, aim to be as perfect as possible in everything you do. Be a nitpicker and be as tough with yourself as you possibly can.

Don't try to master everything all at once. Instead, try to be perfect with just one element and build on the previous skills you have learned. Fly relaxed but push yourself to the very limit. If you find yourself tiring or feeling overburdened, stop and relax.

How do you become a proud, perfect pilot? Let's start with the preflight preparation. Determine what you want to accomplish in your preflight planning and do it over and over again until you can do it perfectly. Concentrate on one thing at a time—the one thing that you want to do perfectly at that moment.

Try to master the planning involved in filling out the flight log. Do everything you possibly can to understand the flight you expect to fly. Be a nitpicker with the details until you can plan a flight perfectly to your satisfaction. Then get an evaluation from your flight instructor and incorporate the instructor's recommendations. Then do it again and again until it is perfect. (In the legal profession, preparation is 90 percent of the law. The same can be said for a perfect instrument flight.)

Plan alternate courses of action in case ATC routes you differently from what you planned. It is fun to try to outwit ATC and understand what goes on in the contollers' minds and why. There must be a reason! Work out a detailed flight log for alternate courses; it will be time well spent.

Considerable attention in your early training should be given to the basics. You can't be a perfect instrument pilot unless you can control heading and altitude. Concentrate on heading control until you can maintain a heading within ±2°; don't worry about altitude.

Then concentrate on altitude until you can maintain altitude within ±20 feet; don't worry about heading. Individually, you will master these basics rather quickly. Then all you have to do is put them together. When you can do this you will have mastered a skill that you will use throughout future flights, whether VFR or IFR. Just think about how much fun it is to be perfect!

2
Mastery of instrument flight in 20 lessons

Mastery of the 20 STEP-BY-STEP LESSONS AND THE FOUR background briefings in the instrument rating syllabus (*see* Chapter 22) will make you proficient in all procedures and maneuvers needed to become a safe and confident instrument pilot. The syllabus also ensures coverage of all the tasks required by the latest *Instrument Rating Practical Test Standards* to pass the FAA instrument pilot (airplane) flight test.

Figure 2-1 is taken from the *Instrument Rating Practical Test Standards*. It is a checklist of the areas of knowledge and skill, or tasks, in which an applicant must demonstrate acceptable performance before an instrument rating will be issued. (Yes, the examiner really must evaluate all tasks on an instrument flight check!)

The *Instrument Rating Practical Test Standards* booklet also has full information on what constitutes acceptable, test-passing performance for each task. This booklet should be part of your "professional library." See Appendix A.

Next to each task in FIG. 2-1 is a checkbox that can be used to record a student's progress through the syllabus. I suggest that the student make one slash from corner to corner through the box when that task has been mastered. The instructor can make another slash going the other way and write in the date when satisfied that the student's

I. PREFLIGHT PREPARATION
- ☐ A. Weather Information
- ☐ B. Cross-Country Flight Planning

II. PREFLIGHT PROCEDURES
- ☐ A. Aircraft Systems Related to IFR Operations
- ☐ B. Aircraft Flight Instruments and Navigation Equipment
- ☐ C. Instrument Cockpit Check

III. AIR TRAFFIC CONTROL CLEARANCES AND PROCEDURES
- ☐ A. Air Traffic Control Clearances
- ☐ B. Compliance with Departure, En Route, and Arrival Procedures and Clearances
- ☐ C. Holding Procedures

IV. FLIGHT BY REFERENCE TO INSTRUMENTS
- ☐ A. Straight-and-Level Flight
- ☐ B. Change of Airspeed
- ☐ C. Constant Airspeed Climbs and Descents
- ☐ D. Rate Climbs and Descents
- ☐ E. Timed Turns to Magnetic Compass Headings
- ☐ F. Steep Turns
- ☐ G. Recovery from Unusual Flight Attitudes

V. NAVIGATION AIDS
- ☐ A. Intercepting and Tracking Navigational Systems and DME Arcs

VI. INSTRUMENT APPROACH PROCEDURES
- ☐ A. Nonprecision Instrument Approach
- ☐ B. Precision ILS Instrument Approach
- ☐ C. Missed Approach
- ☐ D. Circling Approach
- ☐ E. Landing from a Straight-in or Circling Approach

VII. EMERGENCY OPERATIONS
- ☐ A. Loss of Communications
- ☐ B. One Engine Inoperative during Straight-and-Level Flight and Turns (Multiengine)
- ☐ C. One Engine Inoperative—Instrument Approach (Multiengine)
- ☐ D. Loss of Gyro Attitude and/or Heading Indicators

VIII. POSTFLIGHT PROCEDURES
- ☐ A. Checking Instruments and Equipment

Fig. 2-1. *Checklist of tasks that are required for the issuance of an instrument rating, as taken from the FAA's* Instrument Rating Practical Test Standard *booklet.*

performance meets FAA standards for passing the flight test. Not only will this provide a record of the student's progress, but it will also encourage students to analyze their own performance and compare their personal evaluation with the instructor's evaluation.

Take a moment now to note how the material in the Instrument Rating Syllabus is presented in Chapter 22. Each flight lesson and background briefing is numbered in sequence. Each lesson has an introduction.

Assigned reading from this book must be completed prior to each flight lesson. The assigned reading provides the information you must have to understand and perform the pilot tasks prescribed for the flight lesson. If you are not familiar with the assigned reading, it will take you much longer—and cost you a lot more—to complete the flight lesson.

Note also that each flight lesson is preceded by a preflight briefing to cover the objectives of the flight lesson. Likewise, each flight lesson is followed by a postflight briefing to assess the flight and to discuss the next lesson. Flight lessons contain review material and new material; completion standards are also given so that the student's progress can be measured in terms of the levels set in the *Instrument Rating Practical Test Standards*.

No set time is allotted for each of the 20 lessons in the syllabus. A lesson is complete only when the completion standards are met, not when a specified amount of time has been devoted to it. Instructors may repeat lessons or portions of lessons as needed for students to meet the completion standards.

SUPPLEMENTARY EXERCISES

A special feature of the syllabus is maneuvers designated as supplementary exercises in many of the flight lessons. Supplementary exercises provide extra practice to help students cure common problems encountered in instrument training, or to improve proficiency with advanced maneuvers.

Mastery of the supplementary exercises is not necessary to meet the completion standards for a flight lesson, and they are not required on the flight test.

Student note. If you would like to get extra practice on any of the required instrument maneuvers or the supplementary exercises without an instructor on board, you should take advantage of every opportunity to do so. The more you practice the maneuvers, the more competent you will become. But you must fly with an appropriately rated safety pilot aboard.

SAFETY PILOTS

Regulations require a safety pilot to watch out for other traffic, clouds, and obstacles (radio towers and hills). According to FAR 91.109 (b), "no person may operate a civil aircraft in simulated instrument flight unless the other control seat is occupied by a safety pilot who possesses at least a private pilot certificate with category and class ratings appropriate to the aircraft being flown."

If you are acting as safety pilot in a Cessna 172, for example, your pilot certificate must have an airplane single-engine land rating; if you are in an Aztec, your pilot certificate must have an airplane multiengine land rating. A safety pilot must also have a current medical certificate because the safety pilot is a "required pilot flight-crewmember" in terms of FAR 61.3 (c).

FLIGHT EXPERIENCE REQUIREMENTS

As students proceed through the syllabus with an instructor, they will work toward the FAA "aeronautical experience" requirements for an applicant for an instrument rating as stated in FAR 61.65, which are:

(1) At least 50 hours of cross-country flying as pilot in command, of which at least 10 hours must be in airplanes for an airplane-instrument rating.
(2) A total of 40 hours of actual or simulated instrument time...to include—
    (i) At least 15 hours of instrument flight training from an authorized instructor in the aircraft category for which the instrument rating is being sought;
    (ii) At least 3 hours of instrument training that is appropriate to the instrument rating being sought from an authorized instructor in preparation for the practical test within the 60 days preceding the day of the test;
    (iii) For an instrument-airplane rating, instrument training on cross-country flight procedures specific to airplanes that includes at least one cross-country flight in an airplane that is performed under IFR, and consists of—
        (A) A distance of at least 250 nautical miles along airways or ATC-directed routing;
        (B) An instrument approach at each airport.
        (C) Three different kinds of approaches with the use of navigation systems.

PCATDS

Since 1997 the FAA has allowed students credit for up to 10 hours out of the required 40 hours of instrument time for instruction in "personal computer-based aviation training devices" (PCATDs). Instrument students can also log up to 20 hours toward the requirement in a flight simulator or flight training device.

Instruction logged toward an instrument rating in a PCATD must be under the supervision of an authorized instructor. The same is also true for time logged toward a certificate or rating in a flight simulator or flight-training device. PCATDs may not be used for any portion of the instrument practical test or for an instrument proficiency check.

I highly recommend the logging of as much time as you can in a PCATD or flight training device. PCATDs, with their dynamic graphics displays and full controls, do a remarkably good job of teaching complicated procedures such as holding patterns and instrument approaches. The procedures can be stopped and resumed at any time to allow the instructor to correct a student's problem or explain a fine point. This can't be done in flight!

Simulated instrument time can be logged only when operating solely by reference to instruments while using a hood or other device that blocks outside visual references. Actual instrument time can be logged only when operating solely by reference to instruments under instrument meteorological conditions (IMC) when neither the horizon nor the ground is visible because of the weather. The student and the instructor may both log actual instrument time simultaneously in IMC, but only the student may log simulated instrument time in VFR conditions because the instructor will be "visual" while acting as safety pilot.

LOGGING PILOT-IN-COMMAND TIME

The student can only log pilot-in-command time while sole manipulator of the controls in simulated instrument flight. To act as pilot in command during actual instrument conditions, the pilot must already possess an instrument rating. FAR 61.3 (e) (3) states: "no person may act as pilot in command of a civil aircraft under IFR, or in weather conditions less than the minimums prescribed for VFR, unless that person holds the appropriate category, class, type (if necessary), and instrument rating for any airplane...being flown." However, the instructor may log pilot-in-command time for the entire training flight.

Instructor note. You should conduct as much flight training as possible in actual IFR conditions. Training flights in actual IFR weather will prepare your students for the real world of instrument flying encountered after receiving an instrument rating. If the weather is too adverse for a satisfactory training flight, carry out the lesson in an FAA-approved instrument ground trainer.

When the weather is VFR, students must use a hood or other approved view limiting device and log flight time as simulated instrument time. Remember, the examiner will expect the candidate for an instrument rating to use an approved view-limiting device throughout the entire check flight.

PREFLIGHT BRIEFINGS AND POSTFLIGHT CRITIQUES

To get the greatest benefit from the syllabus flight lessons, each lesson must include a preflight briefing and a postflight critique.

The preflight briefing will cover the objectives of the lesson, the procedures and maneuvers to be practiced, and standards that the student is expected to achieve.

The postflight critique will clearly establish the elements of the lesson that the student has mastered, as well as those that need further attention. The postflight critique will also include a preview of the next flight lesson and, because every flight lesson will start by filing an IFR flight plan, the instructor will suggest the destination for planning and filing IFR on the next lesson.

Instructor note. All training IFR flight plans must be filed in the instructor's name according to FAR 61.3 (e)(1). The student cannot act as pilot in command while flying an IFR flight plan.

Students will progress much more rapidly (and at considerably less cost!) if they are thoroughly familiar with the assigned reading material prior to each flight lesson and background briefing. The reading assignments are referenced by chapter number and chapter title.

Assigned reading describes in detail what will take place during each flight lesson. The reading material also contains information and background that must be understood in order to successfully complete the flight lessons.

BACKGROUND BRIEFINGS

Students should note the inclusion of four background briefings prior to four chapters that introduce especially important new material:

- Basic instruments
- Approaches
- IFR cross-country flights
- Instrument flight test

Students should prepare for background briefings by studying the assigned reading on their own and writing out answers to all the questions. A few review questions also appear in each background briefing. These review questions are based upon material previously assigned as required reading for the flight lessons, or based upon material covered in VFR training that must be reemphasized in IFR training.

No trick questions are in the background briefings; however, it might take some research to find satisfactory answers. That is exactly what the background briefings are meant to do—make you dig for the information. What you learn by researching will stick with you much longer than material learned by rote.

That's also the reason why no answers are supplied in this book; if the answers were readily available, there would be no incentive to dig out the information on your own. Your instructor will have the answers and will provide them as you discuss the background briefings.

As soon as you answer all the questions in one background briefing, start working immediately on questions in the next briefing. You want to be prepared to go over the briefings with your instructor at the appropriate points in the syllabus, and it does take time to prepare all the answers!

Instructor note. Be prepared also! Be sure you have all the answers under control before you go over the briefings with your students so you can discuss all the questions and answers intelligently. All the reference material is not in this book. Many of the references cited are other readily available publications listed in Appendix A, "The Instrument Pilot's Professional Library."

You must do your homework on the background briefings before your student's lesson. Answers won't be given verbatim in the reference material. You will have to work out satisfactory answers completely on your own for some of the questions.

Encourage students to obtain the necessary publications in Appendix A on their own. But if a student has not yet had time to obtain them, you should be prepared to lend the publications to the student from either your own professional library or from that of the flight school. You might want to prepare handouts for students by photocopying the relevant pages from government publications.

Do not confuse background briefings with preflight briefings and postflight critiques. The latter are conducted at the beginning and end of every flight lesson. Background briefings are separate sessions between flight lessons. They contain too much material to be included at the beginning or end of a flight lesson.

Background briefings are numbered to show when they should be conducted. For example, Background Briefing 1-2 should be completed after Flight Lesson 1 but before Flight Lesson 2.

The student and instructor should schedule ground school time to cover the material in the background briefing prior to the flight lesson containing new material. The ground school sessions should last two or three hours and can be scheduled over two or three days. Students and instructors should go through the briefings together. Instructors should make certain that students clearly understand the material covered by the assigned reading and questions before proceeding with the next flight lesson.

3
Preparing for an instrument flight

PLANNING AN IFR FLIGHT SEEMS TO BE A LARGE ORDER FOR THE beginning instrument student. Unfamiliar aeronautical charts must be mastered, the weather analysis is more complex compared to VFR analysis, and procedures for the departure, en route, and approach phases of the flight might be unclear.

It's only natural to be a little puzzled at first; however, many things can be done to master the process more quickly and make it more interesting. Believe it or not, flight planning can even be fun!

The single most important factor in taking the mystery out of IFR flight planning is to plan and file an IFR flight plan on every training flight, including the very first one.

Even if you are only going out to practice in the local area, you should still plan and file IFR to a destination 50–75 nautical miles from the home airport. You should also work out an IFR flight plan for the return trip, whether you expect to use it or not.

In the beginning, the instructor will suggest two or three nearby destinations. Get your instructor's ideas well ahead of the flight so you can do your planning at home when you have more time and are under less stress.

Plan for all destinations the instructor has suggested, not just one. This will give you more practice, expose you to a variety of situations, and will not cost you a cent.

Students who show up at the airport without their homework finished waste their time and their instructor's time.

You will be surprised how much you can accomplish at home. You can map out the flight route, review departure procedures, and go over the destination approach charts and airport information in great detail. You can get preliminary weather briefings, check the NOTAMs, and even file your flight plan from home. In fact, you can make several calls if the weather is changing rapidly, or there is something you didn't catch clearly on the first call. This is particularly helpful in the beginning of your IFR training.

ATC will often make amendments to your clearance as the IFR flight proceeds. When you carefully work out your flight at home, familiarize yourself with other VORs and airways between your departure and destination airports and pick out all those obscure intersections that ATC might use for clearance limits, rerouting, or holding fixes. That way there will be no unpleasant surprises!

The cockpit is not the place for basic research. A newly rated instrument pilot departed one of the New York airports in actual IFR weather flying a light twin with his family on board. He had only filed and flown IFR a few times before he received his rating. When airborne, he contacted departure control. The controller responded: "New clearance. Ready to copy?" It wasn't a major change, but due to his inexperience, it overloaded him. He lost control of the airplane and it crashed, killing all on board. If ever an accident could have been prevented by more thorough training, this was it.

GPS EN ROUTE NAVIGATION

En route IFR navigation by GPS (Global Positioning System) requires panel-mounted equipment certified by the FAA as meeting the complex requirements of TSO (Technical Standard Order) C129. In addition, each individual installation must pass an FAA inspection. Handheld GPS units are not approved for IFR.

Furthermore, "aircraft using GPS navigation under IFR must be equipped with an approved and operational alternate means of navigation appropriate to the flight," according to AIM 1-1-22, b 1 (b). For all practical purposes, an IFR fight proceeding by GPS must be backed up by VOR navigation. The FAA is considering transitioning to an all-GPS system of air navigation, but this is still many years away. The *Instrument Rating Practical Test Standards* lists no requirements for competence in en route navigation by GPS. So plan your flights for VOR navigation without regard to GPS, even though you may be conducting your training in an airplane equipped with an approved, panel-mounted GPS system. (For more information on GPS, see Chapters 15 and 21.)

THE FLIGHT LOG

As a VFR pilot, you already have considerable experience planning cross-country flights and navigating with VORs. Your IFR planning is nothing more than an extension of what you have already learned, with more detail in some areas and greater emphasis on other points that are not so crucial to VFR flying. Over the years I developed a flight log form (FIG. 3-1) that covers all information needed for VFR and IFR cross-country flights. I find that if my private pilot students use this log form for the cross-country phase of their training, they make the transition from VFR to IFR flight planning very easily.

Please be my guest; copy the form and use it for your VFR and IFR flying. It gives you a logical, step-by-step way to work your way through all the important elements of flight planning. If you fill out the log properly, nothing will be overlooked.

During the flight itself, the form functions as a running log with places to enter time en route, ETA, actual time of arrival, ground speed between fixes, and fuel consumption. Equally important, it provides a quick reference for all the detailed information you need to conduct the flight without a lot of confusion and fumbling around. Cockpit organization is the key to a good IFR flight.

Let's go through the form step-by-step so you can see how useful it is when planning a flight. We'll be taking a hypothetical IFR cross-country from Westchester County Airport, at White Plains, N.Y., to Broome County Airport, at Binghamton, N.Y., on a typical actual IFR day in late November. I have chosen this example because I have found that this trip to Binghamton contains all the elements of a good training flight. The principles and techniques that apply here are valid everywhere else.

IFR flight planning occurs in two phases. In the first phase you decide your route of flight and fill in your flight log with all the information available before obtaining a weather briefing. In the second phase, get your weather briefing, then fill in the information affected by the weather, such as ground speed and the choice of an alternate airport.

PLANNING THE ROUTE

The first step is plotting the route between your departure and your destination airports, including the appropriate *standard instrument departure* (SID), if there is one. Open up the en route low-altitude chart, or charts, that cover the area—in this case, L-25. Notice which Victor airways make a logical route to the destination; V252 from Huguenot VOR goes directly to Binghamton VOR, so that is the best Victor airway for the en route portion of this flight (FIG. 3-2).

You could also get from Huguenot to Binghamton by departing Huguenot on V162 to intercept V126 to Lake Henry, then taking V149 to Binghamton. This would lengthen the trip unnecessarily so V252 is the better choice. However, on many IFR cross-country flights the best route to your destination might involve two or more Victor airways, and there is nothing wrong with combining several. Also, an airway segment might have several Victor airway numbers. The airway heading southwest from Huguenot on L-25 is numbered V205-252 above the line and 489 below. Which one would you list on your log and flight plan? Either one. It doesn't matter as far as ATC is concerned. Pick the numbered route that will carry you farthest toward your destination.

STANDARD INSTRUMENT DEPARTURES (SIDS)

Note that there is no Victor airway from Westchester Airport to Huguenot. How do you get to Huguenot? Utilize the Westchester's published SID called "Westchester Nine" (FIG. 3-3). If no SID is available at your departure airport, you would simply file "direct" to the first VOR on the airway you intend to use.

FIX		INTERSEC	ROUTE	M.C.	DIST.	G/S	TIME ENR	TIME ARR	DATE	PILOT:
	NAME	NAME / IDENT	VIA	TO	TO	EST	EST	EST	FLIGHT PLAN ROUTE	TAS:
IDENT.	FREQ	FREQ RADIAL	ALT	FROM	REM	ACT	ACT	ACT		RPM:

TAKEOFF:
RWY · W

ATC CLEARS N _____

Field Elev:

ALTIMETER ERROR

LANDING:

Field Elev:
RWY · W

TYPE OF A

TIME ON:

TIME OFF:

TOTAL TIME
ENROUTE:

TACH ON

CLEARANCE LIMIT

WINDS ALOFT

Sta.	A	Dir.	V

VFR WX AT:

AC ID	POSITION	TIME	ALT.	IFR/VFR	EST. NEXT FIX	NEXT FIX	PIREPS

TANK C

Fig. 3-1. *Flight log form for planning and logging IFR cross-country flights.*

PILOT:

			CALL SIGN	FREQ.	ASGND ALT
TAS:	IAS:				
RPM:	MP:				

TAKEOFF: RWY | WIND | KTS | ALT	ATIS		
	CLNC		
	GND		
Field Elev:	TWR		
ALTIMETER ERROR	DEP C		
LANDING:			
Field Elev: RWY | WIND | KTS | ALT			
TYPE OF APPROACH:			
TIME ON:			
TIME OFF:			
TOTAL TIME ENROUTE:	ATIS		
	APP C		
TACH ON | TACH OFF	TWR		
	GND		

WINDS ALOFT

Sta.	A	Dir.	Vel	Temp	A	Dr	VI	T	A	D	V	T

VFR WX AT:

	TANK ON	TANK OFF		TANK ON	TANK OFF

13

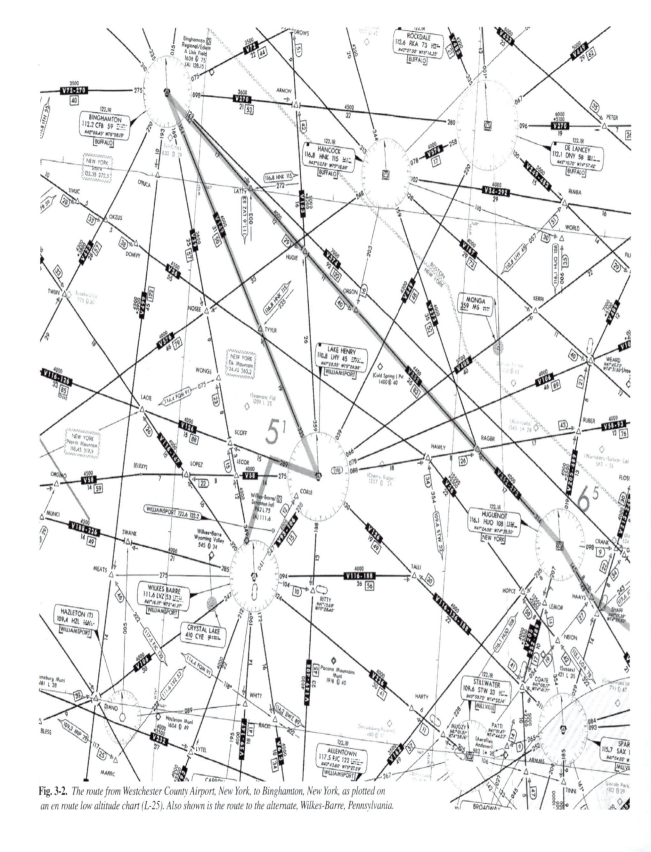

Fig. 3-2. *The route from Westchester County Airport, New York, to Binghamton, New York, as plotted on an en route low altitude chart (L-25). Also shown is the route to the alternate, Wilkes-Barre, Pennsylvania.*

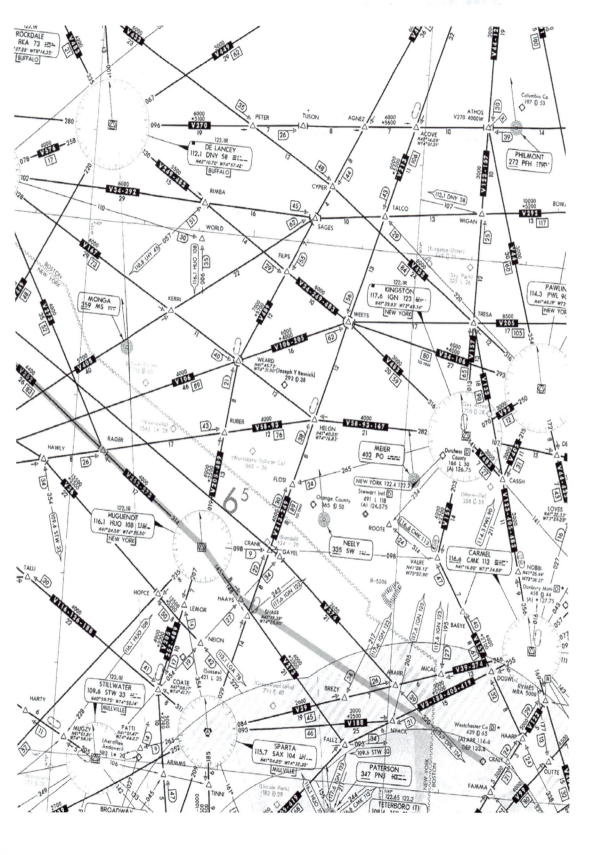

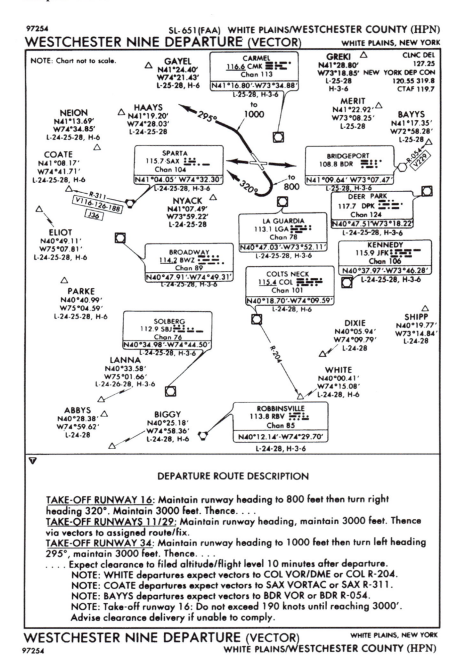

Fig. 3-3. *Westchester Nine Departure, the standard instrument departure (SID) from Westchester County, N.Y., Airport, as depicted on an NOS chart.*

Your IFR flight planning should always include a study of the SIDs available at your departure airport because ATC will expect you to use them. SIDs (and arrival procedure STARs) are now published in the same NOS booklets or packets that contain instrument approach procedure charts. SIDs are also published by Jeppesen and are included as separate sheets along with Jeppesen's approach charts for an airport.

PREFERRED ROUTES AND TECS

Before you enter the planning results on the flight log, check two more items that often can be very helpful. They are Preferred IFR Routes and Tower En Route Control (TEC). (In the beginning of a student's instrument training, I don't emphasize preferred IFR routes and TECs because they can be confusing.) You should know about them and become familiar with them.

Preferred IFR routes are listed toward the back of the *Airport/Facility Directory* (A/FD)—the green book. As the name implies, preferred IFR routes have been established by the FAA to guide pilots in their flight planning and to minimize route changes during the flight. If a preferred route doesn't cover your entire flight, use whatever segments you can. You are more likely to get the route you file for and less likely to have the route amended in the air if you can use all or part of a preferred route.

TEC routes and airports are also listed toward the back of the A/FD. TEC makes it possible to fly from one approach control point to another without entering air route traffic control center (ARTCC) airspace. Designed primarily for instrument flights below 10,000 feet, TEC is usually quicker and simpler than routing through center airspace. But it's available only between the paired airports listed in the A/FD; if your departure airport is not paired with your destination airport, you cannot use TEC. If you can use a TEC route, enter the acronym TEC in the remarks section of the flight plan.

Browse through the preferred routes and the TEC sections of the A/FD. Examine how the information is set up and check to see if your usual departure point fits into either category or both categories. If so, these two special procedures will simplify your planning and save time.

APPROACH PLANNING

After you have worked out the route to the destination, you should study the Standard Terminal Arrival Charts (STARs), if any apply, as well as the instrument approach procedure charts for the destination airport. STARs help you plan your transition from the en route phase to the approach phase.

Look for two things when you study STARs and approach charts. First you need to familiarize yourself with *all* the approaches available at the destination because you must be prepared for the specific approach that ATC assigns. In the beginning of your instrument training, the amount of detail packed on these small charts might be quite mystifying. For example, Binghamton Airport, the destination in our flight-planning example, has no less than five different instrument approach procedures!

No one expects you to be fully cognizant of all the approach procedures on your first few training flights. But you must examine all the approaches for your destination every time you file a flight plan. You will be amazed at how much information you will be able to absorb after you do this a few times.

Don't forget the departure airport; familiarize yourself with its instrument approaches in case a door pops open or a passenger becomes sick or for some other reason you have to return immediately after takeoff.

Airport diagrams

Study the airport diagrams first, just as you would on a VFR cross-country. Note the field elevation. Study the runway diagram and note the number of runways and their headings, lengths, and widths. Will you be able to use all runways? Do mountains or other obstructions affect an instrument approach? How high are the obstructions? Check over the taxiways because getting from your runway to your destination on the field can sometimes be the most complicated part of the trip. (One time, taxiing to the FBO at Montreal took longer than the flight from Ottawa!)

Initial approach fixes (IAFs)

Look at all the VORs, NDBs, marker beacons, intersections, and holding patterns. How will you get from the Victor airway you've been flying to the most likely instrument approach? Ask yourself which approach you will probably be assigned when you are handed off to approach control; consider the prevailing wind. Then pick out the VOR, NDB, or intersection on your inbound route that is closest to the final approach course. Chances are, this fix will be designated as an *initial approach fix* (IAF) on the approach chart.

For our hypothetical flight to Binghamton, the wind is generally northwesterly, so the most likely approach is ILS RWY 34 (FIG. 3-4). We could simply continue on V252 to Binghamton VOR and use that as our destination fix. But that would involve a lot of maneuvering to get back to the final approach course for Runway 34.

A better choice for a destination fix would be Latty intersection. Latty is on both V252 and the final approach course for Runway 34 and Latty appears both on the L-25 chart and the approach chart for ILS RWY 34. In most cases, approach control will give you radar vectors to the final approach course for the approach.

But if there is no radar available, or you have lost radio communications, you will be expected to make the transition from the Victor airway to the final approach course on your own. Beware: your instructor might insist that you make the full approach! So you must plan ahead.

Now we are ready to fill out the flight log. In the "Flight Plan Route" section at the center top portion of your flight log form, write in the abbreviations that will spell out your planned route (FIG. 3-5), in this case—WESTCHESTER 9 SID, HAAYS INTERSECTION, V273 HUGUENOT, V252 LATTY INTERSECTION. This is also entered in Block 8 of the FAA's flight plan form for the "route of flight" (*see* Chapter 5, FIG. 5-11).

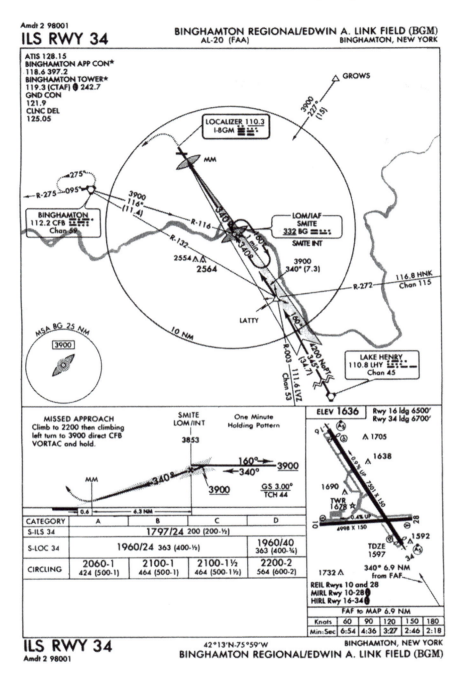

Fig. 3-4. *The ILS 34 approach to Binghamton. Note the location of LATTY intersection, the final fix on the IFR flight to Binghamton discussed in the text.*

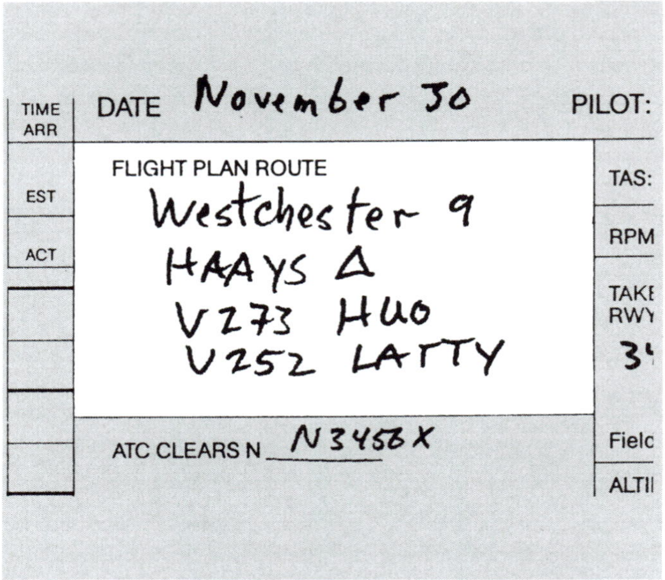

Fig. 3-5. *The flight log form has the flight-plan route block filled in for the flight to Binghamton.*

EN ROUTE FIXES

Now let's fill in more details of your flight. Look at the eight columns in the left one-third of the flight log (FIG. 3-6). No mysteries here, identical to a VFR flight with a few extra details. Spell out the full name of all VORs used for fixes along your route. You might not be familiar with the three-letter identifiers, at least not in the beginning, and the names of some VORs can be very confusing.

For example, who could possibly guess that the identifier HPN stands for Westchester County Airport in White Plains, New York? So to avoid confusion, write out the station name as shown with its identifier and frequency in the boxes below.

If the navigational fix is an intersection, use the column under "intersec," enter the name of the intersection as shown and list the identifier, the frequency, and the radial identifying the intersection. Note that we have entered LATTY intersection here because of its usefulness in locating the final approach course and it is common to both the en route chart and ILS 34 at Binghamton.

The next three columns are for routes, altitudes, magnetic courses, and distances; fill in as shown. The altitude in the third column is what you shall request for the first leg of the flight. The ground speed, time en route, and time of arrival (time arr) columns will have to wait until the weather briefing provides wind information at the planned cruise altitude.

Actual ground speeds, actual times en route, and actual times of arrival will be entered while in flight. This versatile form not only provides a structure for flight planning, but also becomes a very handy log to monitor the flight's progress.

COMMUNICATIONS FREQUENCIES

A few more details can be filled in before the weather briefing. Look up departure and destination airports in the A/FD and enter the frequencies as shown in the three columns at the far right of the form (FIG. 3-7). If the departure airport has an *automatic terminal information service* (ATIS), write this information in at the top of the first column.

The abbreviations for the other entries are: CLNC, clearance delivery; GND, ground control; TWR, tower; and DEPC, departure control. Use the blank boxes to list the en route frequencies as they are assigned.

The column on the far right is reserved for any changes of assigned altitudes. A typical change of en route frequencies might be: "Contact New York Center now, one two eight point five." When you make the frequency change and New York Center advises "Descend and maintain four," you simply write in the new altitude next to the frequency. The sequence of entries in the appropriate boxes would look like this: NY CTR 128.5 40

FIX		INTERSEC	ROUTE	M.C.	DIST.
NAME		NAME	VIA	TO	TO
		IDENT			
IDENT.	FREQ	FREQ RADIAL	ALT	FROM	REM
		HAAYS STW	W-9	325	374
HUO	116.1	109.6 054			
Huguenot			V273	325	8
HUO	116.1			314	
		LATTY HNK	V252	312	66
BGA	112.2	116.8 272			
BGM - Airport			ILS	340	14
I-BGM	110.3				
					125

Fig. 3-6. *Fixes, magnetic courses, and mileages entered for the flight to Binghamton. Note: The initial heading of 325° to HAAYS is an estimate because departure headings from the airport are radar vectors.*

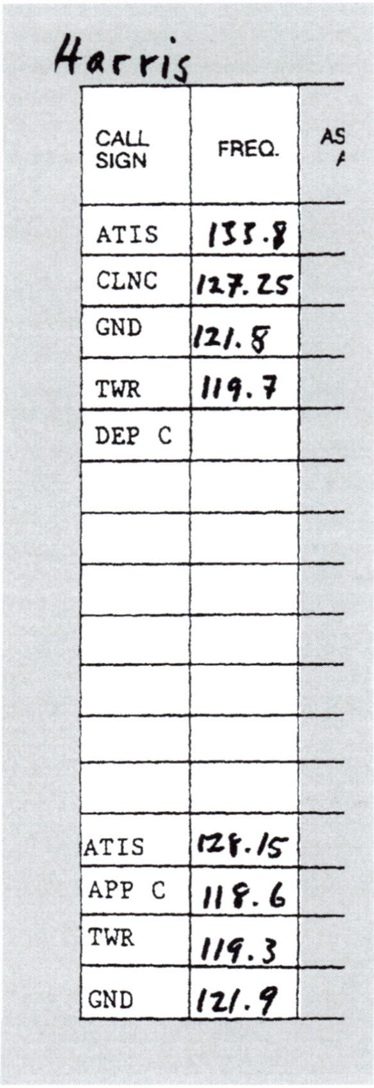

Harris

CALL SIGN	FREQ.	AS
ATIS	133.8	
CLNC	127.25	
GND	121.8	
TWR	119.7	
DEP C		
ATIS	128.15	
APP C	118.6	
TWR	119.3	
GND	121.9	

Fig. 3-7. *The sequence of frequencies for the flight to Binghamton.*

(for 4,000 feet). Once again the flight log form functions as a flight planning guide as well as a running log during your flight.

Toward the bottom of the radio frequency columns are spaces for arrival frequencies, including APP C for approach control.

While working on the right side of the form, fill in the indicated airspeed (IAS) you plan to fly (top of form) and the power setting, to maintain this airspeed. Let's assume that we're making this flight to Binghamton in a Cessna 182. We enter the revolutions per minute (RPM) and manifold pressure—2300 and 23"—that we estimate will maintain 125 knots (FIG. 3-8).

FIELD ELEVATION

When still on the right side of the flight log form, enter the elevation of departure and destination airports (FIG. 3-9). Because you don't know the altitude of takeoff and landing runways you will be assigned, enter the field elevations.

In your VFR flying days, the field elevations you picked off your sectional charts were accurate enough for flying a good pattern and making a good landing. But field elevation is the highest point on an airport's usable runways, and not necessarily the most important elevation for an instrument approach to a selected runway. More precise field elevations are utilized.

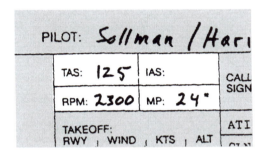

Fig. 3-8. *Performance details for the flight to Binghamton.*

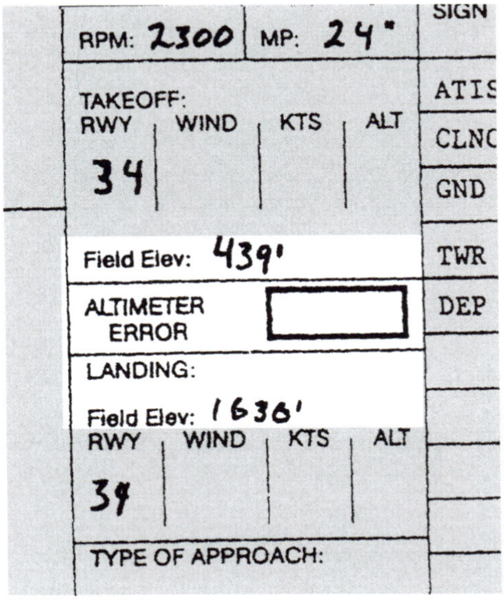

Fig. 3-9. *Field elevations for the flight to Binghamton.*

Take a closer look at the approach chart for the ILS 34 approach at Binghamton (FIG. 3-4). On the airport diagram you will see "TDZE 1597" at the approach end of Runway 34; TDZE stands for *touch down zone elevation*. This is the precise field elevation for the landing area of this instrument approach and it is this TDZE of 1,597 feet on which the landing minimums for landing on Runway 34 are based, not the field elevation of 1630 feet. Keep in mind the fact that no matter which runway you end up using, the TDZE might be substantially different from the field elevation.

The same applies to the departure airport. You will need an equally precise elevation to set the altimeter accurately. At the ramp where I park at Westchester County Airport, the spot elevation is 388 feet, which is 51 feet lower than the published field elevation (highest point) for Westchester!

If an altimeter had the maximum allowable instrument error of 70 feet, this could result in an error totaling 121 feet if you used Westchester's official field elevation. Most precision instrument approaches have a decision height 200 feet above the TDZE. If the aircraft altimeter were off by 121 feet, you could have serious problems on a precision approach.

If you are operating from an unfamiliar field and cannot find the elevation figure for the parking ramp, use the TDZE elevation nearest the run-up area for the takeoff runway.

AIRPORT SERVICES

This completes the information you can enter on the flight log form prior to the weather briefing (FIG. 3-10).

But some unofficial planning items can make the difference between an easy, comfortable journey and a difficult trip. Just because the airport is listed in the A/FD does not mean it is adequate for you.

A lot of important information is not listed in the A/FD. Are there special noise abatement procedures? Will the FBO be open for parking, fuel, and other services when you arrive? Does the airport have more than one FBO? Is there convenient transportation to the office meeting, to a hotel, or to that golf course you've been wanting to play for so long?

Answer these questions before departure. The best way to obtain this information is from somebody on the ground at your destination who knows the area. Determine ahead of time which FBO you'll be using and give them a call. *AOPA'S Airport Directory* is an excellent current source of telephone numbers for FBOs. This annual guide also provides a wealth of information on transportation, lodging, and other services.

All charts, approach plates, and other study materials excerpted from government and other publications contained in this textbook ARE NOT LEGAL FOR NAVIGATION. They are included for illustration and study purposes only! Use current legal charts only!

The materials used in this textbook are intended to prepare the instrument student with the most realistic and practical materials available. Even the purist who went out and bought all these charts would find that within two months they are all obsolete anyway—within a few weeks or a few months at most.

Of course, when I started to fly more than 60 years ago, the most common aeronautical chart was an automobile road map, which was readily available at any gas station at no charge. My, how times have changed.

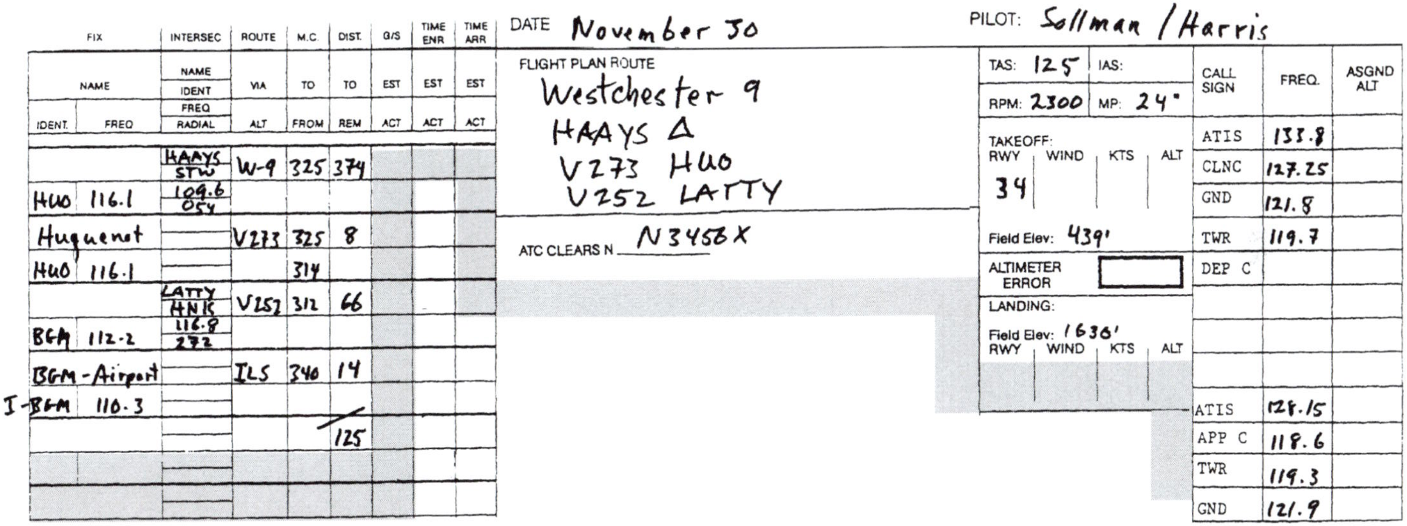
FIX
NAME
IDENT.
FREQ
INTERSEC
NAME
IDENT
FREQ
RADIAL
ROUTE
VIA
ALT
M.C.
TO
FROM
DIST.
TO
REM
G/S
EST
ACT
TIME ENR
EST
ACT
TIME ARR
EST
ACT
HAAYS
STW
109.6
054
W-9 325 374
HUO 116.1
Huguenot
V273 325 8
HUO 116.1
314
LATTY
HNK
116.8
272
V252 312 66
BGM 112.2
BGM-Airport
ILS 340 14
I-BGM 110.3
125
DATE November 30
FLIGHT PLAN ROUTE
Westchester 9
HAAYS Δ
V273 HUO
V252 LATTY
ATC CLEARS N N3458X
PILOT: Sollman / Harris
TAS: 125
IAS:
RPM: 2300
MP: 24"
TAKEOFF:
RWY WIND KTS ALT
34
Field Elev: 439'
ALTIMETER ERROR
LANDING:
Field Elev: 1636'
RWY WIND KTS ALT
CALL SIGN
FREQ.
ASGND ALT
ATIS 133.8
CLNC 127.25
GND 121.8
TWR 119.7
DEP C
ATIS 128.15
APP C 118.6
TWR 119.3
GND 121.9

Fig. 3-10. *Flight log for the flight to Binghamton with all the details that can be filled in prior to the weather briefing.*

4
Weather/whether to fly?

UPON COMPLETION OF THE FIRST PHASE OF FLIGHT PLANNING, YOU must determine if there is anything in the current or forecast weather that might hinder the flight. If anything, weather considerations are simpler for an IFR flight than for a VFR flight. In IFR flying, it's assumed that you'll be in the clouds all the way until breaking out of the overcast during the final approach.

In VFR flying, a good part of the weather analysis is devoted to figuring out where the clouds are, whether to go above, below, or around them, and whether you can safely land at the destination before everything begins closing in. One joy of an instrument rating is setting off on an IFR cross-country when VFR-only pilots are still back at the FBO agonizing over all those VFR weather decisions.

Your decision whether or not to make an IFR flight boils down to five major go/no-go decision factors:

- Thunderstorms
- Turbulence
- Icing
- Fog
- Departure and destination weather minimums

THUNDERSTORMS

Thunderstorms are formed when moisture is combined with rising columns of unstable warm air. The rising warm air cools and the moisture condenses into droplets of rain. Heat is released during condensation, and this additional heat, in turn, increases the speed and power of the rising column of air.

As a thunderstorm begins to tower thousands of feet up into extremely cold temperatures, the rising column of warm air finally cools and a strong column of descending cold air begins to plunge toward the ground outside the core of rising warm air. Rain droplets carried to the top of the thunderstorm are frozen and begin to spill out of the top of the storm as hail.

This pattern is typical of the summertime thunderstorm buildups. But we don't want to fool ourselves with the idea that thunderstorms are just a summertime problem. Thunderstorms can happen in January most anywhere in the country. Anytime you have rising, unstable air and moisture in the atmosphere, thunderstorms can develop.

This is a simplified description of a very complex process; you should learn more about thunderstorms by studying the weather-related publications listed in Appendix A. The FAA book *Aviation Weather* has a particularly good chapter on thunderstorms and good discussions about turbulence, icing, and fog.

What does all this mean to an instrument pilot? The point is that thunderstorms contain powerful columns of rising warm air surrounded by equally powerful columns of descending cold air.

The updraft in a mature thunderstorm cell might exceed 6,000 feet per minute (fpm). Structural limits of most general aviation aircraft might be exceeded when the aircraft passes through the shear between the updraft and downdraft. The vicinity of a thunderstorm must be avoided at all times. The odds are against you in a lightplane.

Hail

Other thunderstorm dangers are not so obvious. When hail spews out of the anvil shape at the top of a thunderstorm, it can come down in clear air as far as 8–10 miles from the storm. A good rule of thumb to avoid getting knocked out of the sky by a shaft of hail is to circumnavigate a towering thunderstorm by 20 miles or more.

A wide circumnavigation will also keep the aircraft outside the *gust front* that rings a mature thunderstorm. The gust front is an area of heavy turbulence caused by the descending currents of cold air reaching the ground and spreading outward. These turbulent, descending currents are called *downbursts*. The gusting cold air currents are a cause of *low-level wind shear* and can transform an otherwise routine instrument approach into a disaster or, at best, a hostile environment that you have to fight all the way.

Embedded thunderstorms

Of particular concern to the IFR pilot are embedded thunderstorms. As the name implies, embedded thunderstorms are hidden by the low-level clouds associated with frontal systems. As a front moves, its wedgelike leading edge forces columns of air aloft, creating columns of rising warm air. The moisture present in the clouds, plus the rising warm air, make for an ideal thunderstorm scenario. Surrounded by clouds, the embedded thunderstorm cannot be spotted visually, and you could penetrate one without warning.

The possibility that you might encounter an embedded thunderstorm can be predicted by meteorologists with fair accuracy. The National Weather Service (NWS) systematically measures air stability (the *lapse rate*), temperature, and humidity at various altitudes, and analyzes these variables along with other factors to develop forecasts.

Microbursts

Also associated with thunderstorms are microbursts—powerful downdrafts caused by descending cold air outside the column of ascending warm air in the core of a storm cell. As these cold downdrafts reach the surface, they produce sudden vertical and horizontal wind shears. The downdrafts can reach 6,000 feet per minute. Horizontal winds near the surface can reach 45 knots, resulting in a 90-knot wind shear from headwind to tailwind for a plane taking off or landing, as shown in FIG. 4-1.

Low-level wind shear detection systems are now in place at many airports. "The early detection of a wind shear/microburst event, and the subsequent warning(s) issued to an aircraft on approach or departure, will alert the pilot/crew to...a situation that could become very dangerous!" says AIM.

While this information is of great value in deciding whether or not to delay a takeoff or a landing, it doesn't help much in planning a flight. AIRMETS and SIGMETS do contain

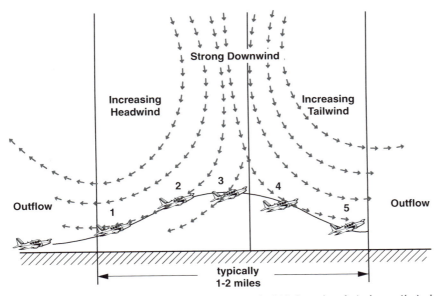

Fig. 4-1. *A microburst during takeoff. The quick shift from headwind to tailwind can reduce performance enough to risk an impact.*

advisories about low-level wind shear. But microbursts cannot be predicted as yet, and they are of such short duration that there is not much point in reporting them—they'll be gone by the time the METAR gets out. Avoid flying in areas of thunderstorm activity and you will stay clear of microbursts.

Thunderstorm forecasts and reports

If thunderstorms are forecast, this will always be stated in the terminal and area forecasts that are reported during a weather briefing.

Thunderstorms are also included in AIRMETs and SIGMETs (Airman's Meteorological Information and Significant Meteorological Information). These advisories are issued to amend area forecasts and to announce weather that is significant or potentially hazardous to flight. The advisories cover thunderstorms, turbulence, high winds, icing, low ceilings, and low visibilities.

AIRMETs concern weather of less severity than weather covered by SIGMETs. AIRMETs and SIGMETs will normally be part of all weather briefings. If not, ask for them. The news they contain—bad or good—is important to making your go/no-go decision.

Ground-based weather radar accurately pinpoints developing storms and tracks them as they grow in size and strength. Air traffic control radar, on the other hand, blocks out weather to a large extent to follow air traffic more precisely. It is important to understand this distinction because air traffic controllers cannot always provide deviation information to get around thunderstorms. Thunderstorms don't always show up clearly on traffic control radar.

Data from specialized weather radar locations is plotted on radar summary charts. The information from these charts can help you during preflight planning to plot a course that is clear of the thunderstorms.

With centralization of flight service stations (FSSs) and the increasing reliance on telephone briefings, many pilots obtain weather information without ever seeing the printed versions of radar summary charts, surface analyses, area and terminal forecasts, hourly weather reports, and other reports and charts provided by the National Weather Service.

Nevertheless, you must know what information these different reports and charts contain. Your knowledge should be sufficient so you can visualize what the weather briefer is reading to you over the telephone and you should know what to ask for if you need special information. There are also many questions about aviation weather services on the instrument written examination.

Purchase a copy of the FAA's book *Aviation Weather Services* (listed in Appendix A) and study the descriptions of forecasts, reports, and charts relevant to lightplane IFR flying.

Locate the FSS serving your area and visit it. (**Instructor note.** Have the student file and fly an IFR cross-country to the FSS airport.) The FSS personnel will show you the material they use for weather briefings and discuss it.

If thunderstorms are forecast for any portion of your flight—or they are developing under the watchful eye of a weather radar—choose a different route or destination or don't make that flight. With all the assistance available there is no excuse for flying into a thunderstorm, whether VFR or IFR. During rapidly increasing thunderstorm activity, a *severe weather avoidance plan* (SWAP) might be in effect to help all aircraft fly routes with more favorable weather conditions.

TURBULENCE

While you can always count on getting a rough ride in turbulence in and around thunderstorms, turbulence can also occur along fast moving weather fronts and around intense high- and low-pressure boundaries.

For the instrument pilot, avoiding turbulence takes on another dimension. It's not just a matter of avoiding severe and extreme turbulence that might cause structural damage to your aircraft. You must also consider the effect of light and moderate turbulence on personal performance as an instrument pilot.

It is much harder to hold headings and altitudes while flying by reference to instruments when the pilot is also buffeted by turbulence. Maximum attention and a lot of work by the pilot are required to maintain heading and altitude in IFR flight in turbulent conditions. Fatigue sets in much more quickly than on a smooth flight.

Fatigue is not just uncomfortable, it is a very real danger. As the AIM points out in the chapter "Medical Facts for Pilots": "Fatigue continues to be one of the most treacherous hazards to flight safety, as it might not be apparent to a pilot until serious errors are made."

Turbulence in clear air presents a problem, too. If you have to contend with turbulence over a long stretch during the VFR portion of an IFR flight, you will be fatigued, tense, and less confident when descending into the clouds to make an instrument approach. Tired, tense, and anxious is not the best way to commence a good instrument approach.

High winds

Turbulence is also associated with high winds. Expect light turbulence when winds of 15–25 knots are forecast or reported; moderate turbulence in winds of 25–50 knots. Base the go/no-go decision on whether or not you can handle winds of up to 50 knots and the associated turbulence. Above 50 knots lies the realm of severe and extreme turbulence. Turbulence in this range can cause structural damage to a lightplane, as well as slam you around unmercifully. Don't try it!

If there is a possibility of light or moderate turbulence during the flight—and particularly if it looks as if the turbulence might last for a long time—decide if you will be able to handle an instrument approach in lousy weather conditions and keep headings and altitudes under control. If there is any doubt, plan a shorter flight, or pick a destination for a VFR approach and landing, or wait for another day.

Terminal and area forecasts contain wind information; the area forecast specifically details turbulence. AIRMETs and SIGMETs detail high winds and turbulence.

A further aid is the surface weather analysis chart that you became so familiar with during your VFR training. If the isobars are tightly packed with a very small amount of space or no space between them, expect high velocity winds and turbulence.

A pilot weather report (PIREP) is even more conclusive evidence of turbulence. If the pilot of an aircraft similar to yours reports turbulence along your route, take it seriously.

Here is someone who has just been through a troublesome or even hazardous experience and is telling it like it is—something no one in a weather station on the ground can ever do.

If a pilot reports turbulence along the route of flight, pick a different route or a different altitude or cancel the flight if you don't think you can handle it. While airborne, don't hesitate to ask the controller to solicit PIREPs on turbulence—or icing or any other conditions that might affect the safety of the flight. Likewise, do not hesitate to report any turbulence as soon as it is safe to do so—fly the airplane first in moderate or extreme turbulence. Other airmen and the controllers will be grateful for the PIREP.

ICING

The situation is different with icing conditions because it is almost impossible to forecast icing with certainty. But the conditions that produce icing are well known and if these conditions are present, assume that the airframe will pick up a load of ice—make other plans, either a different route of flight or alternate transportation.

Aircraft structural icing requires moisture and below-freezing temperatures. But ice might form on the aircraft when flying inside clouds that are above the freezing level. If the freezing level is at or near the ground, icing might occur immediately upon entering clouds after takeoff.

Consider this for a moment. Takeoff and departure are among the busiest and most intense moments of an IFR flight. You are making the transition from visual to instrument flight—sometimes very quickly if the departure ceiling and visibility are low. You are switching from tower to departure control and departure control might well have a clearance amendment. You might not be aware of it, but every human sense is alert to abnormal sounds from the engine, unusual control pressures, and unusual instrument readings—is this any time to be worrying about ice?

The most dangerous form of icing occurs when a mass of rainy, warm air overlies or overrides a cold air mass. As the warmer rain falls through the colder air mass below, the rain becomes cold enough to freeze upon impact with the surfaces of the aircraft. The cold droplets hit, splatter, and coat the surfaces they strike with layers of ice that can build up rapidly.

This rapid buildup can add hundreds of pounds of weight to a small aircraft very quickly. The ice destroys aerodynamic characteristics of wings, control surfaces, and propellers. In extreme cases, 2–3 inches of ice can form on the leading edge of an airfoil in fewer than five minutes. It takes only a half-inch of ice to reduce the lifting power of some aircraft by 50 percent. No doubt about it—icing can be lethal.

Freezing level

If a flight is going to be in the clouds, select an altitude that is beneath the freezing level and avoid areas where warm, moist air overlies colder air. Where do you get information on the freezing level and icing? The area forecast is the first place to look. The National Weather Service "sounds" the atmosphere for temperatures at hundreds of locations throughout the United States and plots freezing levels from these soundings.

Once again the most reliable information comes from PIREPs. If a pilot reports icing along the proposed route of flight, pick a different route or cancel the flight. The pilot isn't guessing or forecasting about this hazard; the pilot can actually see the ice building up on the aircraft. This is much more useful than a forecast, which can only predict conditions favorable for icing.

(If ATC issues a clearance—prior to the flight or while in flight—and you have determined that icing conditions are possible in that area prescribed by the clearance, do not accept the clearance, and explain why by citing, if necessary, the weather information that you obtained and should have written on the flight-planning form.)

FOG

If you are on the ground watching a wall of fog move in, the fog seems to have a sinister, menacing appearance. And with good reason. In a matter of minutes, fog can reduce visibility to zero. In a thick "pea soup" that might be encountered along the coast of New England, fog can be so dense that you literally cannot see where you are walking. This can be very bad news for a pilot, particularly if you are unprepared for it and the fog comes as a surprise.

The best preparation is to understand the conditions that produce fog and then avoid those conditions. Fortunately, this is not very complicated. Fog forms either when air is cooled to its dew point or when the dew point is raised by the presence of additional moisture. Raising the dew point is usually accomplished by the evaporation of water from falling precipitation or by the passage of a body of air over a wet surface.

Fog classifications are: radiation, advection, upslope, precipitation-induced, and ice. Develop a better understanding about fog by studying the weather-related publications recommended in Appendix A.

Temperature/dew point spread

Dew point is the temperature at which air is saturated and the water vapor begins to condense and produce visible moisture; dew point is expressed as a temperature. If the air temperature drops to the dew point temperature, fog forms. This is the most common situation. Along coastal areas, however, moist air moving in from the sea can raise the dew point and produce fog.

Because fog forms so low to the ground, there is nothing like it for closing an instrument runway or an entire airport. Fog can form rapidly, making it especially hazardous. You can start an instrument flight with sufficient ceiling and visibility at the destination to make a comfortable approach, then arrive and find a blanket of fog that makes it impossible to land.

Hourly aviation weather reports will always indicate the presence of fog. The hourly reports also contain temperature and dew point. The difference between the two numbers is the temperature-dew point spread, or simply "the spread" as it is commonly called.

When the spread is about 5°F, be alert for fog. For example, if the temperature is 48°F in the most recent report for the destination airport, and the dew point is 43°F, be prepared for fog on arrival. The possibility of fog also appears in terminal forecasts.

If the destination is socked in with fog, do not depart until the fog begins to lift. Likewise, if the spread is narrowing, according to reports, and drops to 5°F or less, do not depart until the situation begins to improve.

Departure and destination weather minimums

Fog is only one of many reasons for low ceilings and visibilities. Other factors are rain, snow, haze, smoke, low-pressure areas, and frontal systems. Ask yourself whether or not you will be able to land upon arrival and determine—according to regulations—if you need an alternate airport and what the alternate might be. And there is a third item relating to minimums that many pilots overlook: Is there sufficient ceiling and visibility at the *departure* airport to return for an immediate landing, if necessary?

Let's take this third item first. Basically, under FAR 91, you can take off in any weather—including zero-zero—if you're not carrying passengers or cargo for hire. See FAR 91.175 (f) in Appendix B. Read the regulation closely and notice that takeoff minimums apply to FAR Part 121, 125, 127, 129, and 135 operations only. Part 91 operations are not covered by this regulation.

A takeoff in conditions approaching zero-zero is certainly not very smart. What do you do if a door pops open or the engine doesn't sound right? Where can you land?

The departure alternate

Two answers to this question should always be part of flight planning. First, do not depart on an instrument flight if the ceiling and visibility at the departure airport are too low for a safe return in an emergency, possibly requiring an instrument approach.

Or second, if the ceiling and visibility at departure are too low to return, select a departure alternate—a nearby airport with adequate ceiling and visibility for a safe arrival. Establish a route to that field and study the ATC frequencies and details of the approach as if the airport were the final destination. In other words, be completely prepared to land at this departure alternate if necessary.

If departing from an airport that has no instrument approach, select the second solution. Pick out the nearest field with an instrument approach plus a ceiling and visibility that are greater than your "personal minimums" (which are subsequently described in this chapter) and be prepared to make that approach. It's certainly much smarter to have an "out" or alternative when taking off under IFR conditions, no matter what type of airport you are departing from.

DESTINATION MINIMUMS

After analyzing the departure situation, you must next anticipate the weather at the destination airport. Now let's take a closer look at the meaning of minimums. You are familiar with ceilings and visibilities from VFR training; in IFR flying, they take on additional importance.

The *Aeronautical Information Manual* (AIM), published by the FAA, defines ceiling and visibility in the glossary; Appendix C of this book contains pertinent IFR glossary terms. Take time to browse through all definitions in the AIM glossary because it contains a gold mine of information relating to IFR operations, and each nugget is as clear and concise as anything you will ever find.

Ceiling is the height above the earth's surface of the lowest layer of clouds or obscuring phenomena that is reported as broken, overcast, or obscuration, and not classified as thin or partial.

Visibility

The AIM "Pilot/Controller Glossary" lists and defines several visibility classifications in increasing order of precision.

Ground visibility. Prevailing horizontal visibility near the earth's surface as reported by the National Weather Service or an accredited observer.

Prevailing visibility. The greatest horizontal visibility equaled or exceeded throughout at least half the horizon circle, which need not necessarily be continuous.

Runway visibility value (RVV). The visibility determined for a particular runway by a transmissometer. RVV is used in lieu of prevailing visibility in determining minimums for a particular runway.

Runway visual range (RVR). An instrumentally derived value...that represents the horizontal distance a pilot will see down the runway from the approach end...RVR, in contrast to prevailing or RVV, is based on what a pilot in a moving aircraft should see looking down the runway. RVR is used in lieu of RVV and/or prevailing visibility in determining minimums for a particular runway. Which definition is used when flight planning?

Use the most precise visibility measurement available. Weather reports for airports with ILS approaches will have a ground visibility and an RVV or RVR for the runway in use. Visibilities at airports with ADF, VOR, or other nonprecision approaches will usually be given in the less precise prevailing visibility.

Importance of the visibility minimum

Why does the FAA go to such great lengths to define visibility?

The answer is quite clear: "No pilot operating an aircraft...may land that aircraft when the flight visibility is less than the visibility prescribed in the standard instrument approach procedure being used." (FAR 91.175 (d) in Appendix B.) In other words, you cannot land if the visibility is below minimums.

Furthermore, according to another section of the same FAR, you cannot descend below a minimum altitude if the visibility prescribed for the approach is less than minimums. There is a good reason for this because it is possible to be clear of clouds upon reaching altitude minimums, but you might be unable to see far enough down the landing runway to make a safe approach because of fog, rain, snow, or other runway-obscuring condition.

Understand from the beginning that *visibility*, not ceiling, determines whether or not you can initiate an instrument approach.

Let's continue planning the flight to Binghamton, New York. Which ceiling and visibility minimums apply to this flight? Turn first to the approach chart for ILS RWY 34 to Binghamton and look at the landing minimums section (FIG. 4-2).

CATEGORY	A	B	C	D
S-ILS 34	1797/24 200 (200-½)			
S-LOC 34	1960/24 363 (400-½)			1960/40 363 (400-¾)
CIRCLING	2060-1 424 (500-1)	2100-1 464 (500-1)	2100-1½ 464 (500-1½)	2200-2 564 (600-2)

Fig. 4-2. *Landing minimums for Binghamton ILS RWY 34 approach, as shown on the NOS approach chart.*

Approach categories

Notice categories A, B, C, and D. These are "aircraft approach categories" that are based upon actual final approach speeds. Category A is for approach speeds from 0 (for helicopters) to 90 knots. Unless you are training for an instrument rating in a Learjet or something equally exotic, category A will be used on all training flights. (For further information on aircraft approach categories, refer to the first page of any set of NOS approach charts.)

Because you plan to make the ILS 34 approach, use the minimums listed for the straight-in approach (S-ILS 34): 1797/24. Numbers after the minimums have specific information:

- 200: The height above touchdown (HAT) at the ceiling minimum
- 200-½: The ceiling and RVR converted to prevailing visibility in statute miles

The ceiling and visibility minimums for planning this flight are the third set of numbers: 200 feet and one-half mile. Now check the other Binghamton approach charts to determine if any have lower minimums; shifting winds might change arrival runways. There are five instrument approaches to Binghamton. No minimums are lower than those for ILS 34, although the minimums for ILS 16 are close at 300 feet and one-half mile.

To make a go/no-go decision about this flight, look for a ceiling of no less than 200 feet and a visibility of no less than one-half mile. If ceiling and visibility are forecast to be lower than this at the estimated time of arrival (ETA), you will probably not be able to land. It is perfectly legal to file IFR and fly to Binghamton and attempt an approach when the weather is forecast to be below minimums (air carrier pilots are not allowed to do this), but what's the point of making the trip if you can't land?

Well, you might go to Binghamton and try an approach just to see if the weather had improved enough to land. You might get lucky! On the other hand, the weather could be much worse upon arrival and a landing might be impossible. This can happen on even the best-planned flights, especially in winter when weather systems pick up speed and sweep across the country much faster than in the summer.

You must pick an alternate airport for a safe landing if you can't get into Binghamton. This is not just a good idea, it's required by FAR 91.169. You might have to pick an alternate even if Binghamton is forecast to be VFR upon arrival. Many people are fooled by this because they think, "The airport is going to be VFR when I get there so I don't need an alternate." That's not true. To file to an airport IFR without listing an alternate, you need twice the ceiling required for VFR. Review FAR 91.169 for a thorough understanding of its impact.

THE "ONE, TWO, THREE RULE"

If, for one hour before ETA through one hour after ETA, the ceiling and visibility are forecast to be 2,000 feet and 3 miles, according to FAR 91.169 (b) (Appendix B), a pilot must select an alternate airport and include the selection on the flight plan. That is the "one, two, three rule": *one* hour, *two* thousand feet, and *three* miles.

Another rule of thumb to help remember the alternate airport requirements is: You must have VFR conditions plus 1,000 feet for ETA ± 1 hour or you must file an alternate.

SELECTING AN ALTERNATE

How do you pick an alternate airport? Obviously, you can't pick an alternate that is so far away you'll run out of fuel before arrival. And there's not much point in picking an alternate where the weather is so bad it prevents landing.

The FARs are grounded in good common sense on these two points. First, FAR 91.167 (Appendix B) says you must carry enough fuel on an IFR flight to:

- Complete the flight to the first airport of intended landing
- Fly from that airport to the alternate, if one is required, and
- Fly after that (the alternate airport) for 45 minutes at normal cruising speed

You can list an airport as an alternate only if the ceiling and visibility forecast for the alternate at your time of arrival will be at or above 600 feet and 2 miles (if the airport has a precision approach), or 800 feet and 2 miles (if it has only an ADF, VOR, or other nonprecision approach).

There might be a catch to this, however. Some airports might not be authorized for use as alternates, while others available might have *higher* minimum requirements than 600/2 and 800/2 because of local conditions such as hills, towers, or radio towers. How can you find this out?

Turn to the "E" section of your set NOS approach charts and find the listing of "IFR Alternate Minimums." An airport with minimums that deviate from the standard 600/2 and 800/2 will be listed in this section (FIG. 4-3). Note that an airport such as Farmingdale Republic is not allowed for use as an alternate when the control tower is not in operation.

If the airport selected as an alternate is listed in this section, use the minimums in this section. If the choice of an alternate does not appear in this section, use the 600/2 and 800/2 minimums discussed above.

PERSONAL MINIMUMS

Just as most airlines qualify their crews to fly certain minimums, many competent instrument pilots set higher minimums for themselves than the published minimums. This is

98169

INSTRUMENT APPROACH PROCEDURE CHARTS

 IFR ALTERNATE MINIMUMS

(NOT APPLICABLE TO USA/USN/USAF)

Standard alternate minimums for non precision approaches are 800-2 (NDB, VOR, LOC, TACAN, LDA, VORTAC, VOR/DME or ASR); for precision approaches 600-2 (ILS or PAR). Airports within this geographical area that require alternate minimums other than standard or alternate minimums with restrictions are listed below. NA - means alternate minimums are not authorized due to unmonitored facility or absence of weather reporting service. Civil pilots see FAR 91. USA/USN/USAF pilots refer to appropriate regulations.

NAME	ALTERNATE MINIMUMS
ALBANY, NY	
ALBANY INTL	ILS Rwy 1[1]
	ILS Rwy 19[1]
	VOR/DME or GPS Rwy 1[2]
	VOR Rwy 1[3]
	VOR or GPS Rwy 19[2]
	VOR or GPS Rwy 28[2]

[1]ILS, Categories B, C, 700-2; Category D, 800-2¼. LOC, Category D, 800-2¼.
[2]Category D, 800-2½.
[3]Category C, 800-2¼; Category D, 800-2½.

ALLENTOWN, PA
LEHIGH VALLEY INTL .... ILS Rwy 13
ILS, Categories A,B,C, 700-2; Category D, 700-2¼. LOC, Category D, 800-2¼.

ALTOONA, PA
ALTOONA-BLAIR COUNTY .... ILS Rwy 20[1]
VOR or GPS-A[2]
[1]Categories A,B, 900-2;Category C, 900-2½; Category D, 1100-3.
[2]Category D, 1100-3.

BATAVIA, NY
GENESEE COUNTY .... VOR/DME or GPS-A
Category D, 800-2¼.

BRADFORD, PA
BRADFORD
REGIONAL .... VOR/DME or GPS Rwy 14
NA when BFD FSS closed.

CORTLAND, NY
CORTLAND COUNTY-
CHASE FIELD .... VOR or GPS-A
Categories A,B, 1100-2,Categories C,D, 1100-3.

NAME	ALTERNATE MINIMUMS
DUBOIS, PA	
DUBOIS-JEFFERSON COUNTY	ILS Rwy 25
LOC, NA.	

ELMIRA, NY
ELMIRA/CORNING REGIONAL .... ILS Rwy 6
ILS Rwy 24
NDB Rwy 24
Categories A,B, 1300-2; Categories C,D, 1300-3.
NA when control tower closed.

ERIE, PA
ERIE INTL .... ILS Rwy 6[1]
ILS Rwy 24[1]
NDB Rwy 6
NDB Rwy 24
RADAR-1
NA when control tower closed.
[1]ILS, 700-2.

FARMINGDALE, NY
REPUBLIC .... ILS Rwy 14
NDB or GPS Rwy 1
NA when control tower closed.

HARRISBURG, PA
CAPITAL CITY .... ILS Rwy 8
Categories A,B, 900-2; Categories C,D, 900-2¾.
NA when control tower closed.

HARRISBURG INTL .... ILS Rwy 13[1]
ILS Rwy 31[1]
VOR or GPS Rwy 31[2]
[1]ILS, Categories C,D, 700-2. LOC, NA.
[2]Categories A,B, 900-2, Category C, 900-2¾, Category D, 900-3.

NE-2

98169

Fig. 4-3. *If IFR landing minimums are other than 600-2 for precision approaches and 800-2 for nonprecision approaches, they will be listed in this IFR Alternate Minimums section of the NOS instrument approach procedures booklet.*

38

very smart! A brand new instrument pilot might want to start out with a 1,000-foot ceiling and 3 miles visibility (VFR minimums) until gaining more experience and confidence. (It's kind of scary being up there in the clouds without your friendly flight instructor!) This can then be lowered, depending upon frequency of flights, until reaching the lowest minimums available.

If you don't fly very often and are barely meeting currency requirements, it might be prudent to raise personal limits to 500 and 2, for example. Many pilots make it a policy to fly instruments every three or four weeks. Many also sign up for instrument refresher instruction every three months if they feel a little rusty.

To maintain currency as an instrument pilot, regulations require 6 hours of instrument flight and 6 instrument approaches plus "holding procedures" and "intercepting and tracking courses through the use of navigation systems" each 6 months. Does legally current mean that you are competent to fly the published minimums? *Absolutely not.* Set personal minimums with which you are comfortable and with which you feel confident.

WEATHER FACTORS REVIEWED

This brief review should help you focus on necessary elements of a weather briefing:

1. Are there any weather conditions that might make it difficult or impossible to complete the flight as planned: thunderstorms, turbulence, icing, or fog?

2. Are the ceilings and visibilities high enough at departure and destination airports for a safe IFR takeoff and for a safe (and legal) IFR approach and landing? Does the weather meet your personal minimums?

3. Is an alternate needed? If so, what alternate airports can you reach that meet the minimum requirements for ceiling and visibility?

Look again at the flight log form and notice in the lower right-hand corner the words "VFR WX AT" (FIG. 4-4). This item is not found in any FAA publications; it is based upon years of IFR flying and IFR instruction.

VFR WX AT:						
	TANK ON	TANK OFF		TANK ON	TANK OFF	

Fig. 4-4. *Space for noting nearest VFR weather on flight log form.*

Chapter Four

When the chips are down, where can you find VFR conditions? Where can you go for a safe VFR landing if all electrical power—radios, transponder, electrically powered instruments, pitot heat, lights—is lost? Where can you go if an instrument flight becomes horrendous?

Always learn where the nearest VFR conditions are to safely abort the IFR flight and land VFR. A weather briefer will provide this information if requested. If there is no VFR weather at an airport within range, don't go.

"When in doubt, wait it out!"

5
How to get a good weather briefing

IT WASN'T TOO LONG AGO THAT GETTING THE WEATHER MEANT talking directly to a weather briefer, either by telephone or during a personal visit to a Flight Service Station (FSS). Pilots lucky enough to have an FSS located at their airport could go in and examine all the weather maps, forecasts, and reports, then talk to a weather briefer in person.

For most pilots this personal contact is a thing of the past. The FAA has transitioned to a centralized network of Automated Flight Service Stations (AFSSs) that provide much more service than has been available previously from the nonautomated stations. FAA-funded Direct User Access Terminal Service (DUATS) now allows computer users to get complete "official" aviation weather briefings directly, as well as file IFR flight plans. The bottom line, as always, is money. The FAA simply does not have the budget to operate the number of Flight Service Stations it once had, nor can it justify the cost of one-on-one personal briefings for everyone.

COMPUTER WEATHER SERVICES

At the same time that Flight Service Stations were centralized, the explosive growth in the use of personal computers began to produce information for pilots that was undreamed of a few years ago. Nowhere has this been more evident than in the use of computers for weather briefings. The amount and quality of computer-based aviation weather for pilots is truly astounding these days!

If you already have a computer and a modem, you have probably sampled some of the excellent aviation material that is now available on the Internet. If you do not have a computer, you should seriously consider getting one at this point in your flying career. The information awaiting you in the many programs for pilots will enable you to get your instrument rating faster and give you greater confidence in handling a variety of IFR situations.

I recommend beginning with the best computer system you can afford. For good weather briefings, and many other things, your computer should have Windows 95 or better, a hard drive with a capacity of two gigabytes or higher, a CD-ROM drive and sound card, a 3.5-inch disk drive, top-of-the-line color capability, a modem with a capacity of 33.6 bits per second or higher, and the largest color monitor and the best stereo speakers your budget can handle.

A good computer store can show you what will meet your needs in a range of prices, depending on the options. Get a friend with computer experience to help you decide what to get. A computer-knowledgeable friend can also help you set up your system, install programs, connect with the Internet, and download material. Some of these things can seem discouragingly confusing at first.

If you don't wish to buy a computer at this point, don't worry. This chapter will also cover the range of aviation weather services available through a telephone call to Flight Service (1-800-WX-BRIEF) or a personal visit.

THE "BIG PICTURE"

The first step in getting a good weather briefing begins the evening before a flight with a look at the weather on your pick of local TV stations. With the use of satellite photographs and computer-generated displays, TV weather programs have become more and more sophisticated in recent years. The information provided by TV weather programs on fronts, pressure systems, precipitation, and temperatures will give you "the big picture" of what the next day's weather will be. Look for the patterns in the bands of clouds in the satellite photographs and ask yourself what effect these large patterns will have on the flight you are planning.

Is the weather improving or deteriorating? Are low ceilings and visibilities a possibility at departure and destination airports? Will the flight be in the clouds or in the clear? Is there a chance of thunderstorms, turbulence, icing? Don't look for the precise details you need for your go/no-go decision at this point; look instead for trends that can be verified in the detailed weather briefing you will obtain for your specific flight.

Now is the time when a computer is handy for the "big picture." If you are a member of Aircraft Owners and Pilots Association (AOPA) and have an Internet connection, call up <www.aopa.org> and check out the weather pages offered at this Web site. With color maps and up-to-the-minute area and terminal forecasts and reports and much more, the AOPA Web site is a superb resource. Place a bookmark or placemark at this location so you can go quickly to it whenever you wish.

If you are not a member of AOPA, another good weather Web site source is the Weather Channel at <www.weather.com>. This site presents free weather in greater detail than what you see on the Weather Channel's TV report—and you don't have to wait until what you want to see cycles around again after the commercials! A special feature allows you to customize the information by location. Another Web site worth exploring is AccuWeather at <www.accuwx.com>. This site provides free color weather maps with a five-hour time delay. If you take out a subscription to the service, you will get real-time information, plus access to 35,000 weather products. Try the "five hours free" offer and see if it might not be worthwhile for you to subscribe to this excellent service. The Internet offers many other aviation-related sites. So surf the 'net and place bookmarks or placemarks at those sites that best serve your interests. With a little practice you will be surprised at how quickly you can go right to what you want.

Many professional pilots make a habit of checking the "big picture" on the weather *every day*, regardless of flight schedules. For instrument students, this habit will improve your understanding of how weather phenomena and patterns develop and why. These are excellent teaming experiences, and much of the best information is free.

Instructor note. As good as these "big picture" weather sources are, none of them satisfy the requirements of FAR 91.103 (a), which states that before beginning an IFR flight, each pilot must become familiar with "all available information concerning that flight. This information must include...weather reports and forecasts, fuel requirements, alternatives available if the planned flight cannot be completed, and any known traffic delays." The only way to obtain a briefing that will satisfy all these requirements is through Flight Service or DUATS. If a problem develops later, the logs and records kept by these two services can provide proof that you have complied with FAR 91.103 (a).

DUATS

On the morning of a flight, check local TV stations and Internet sites again to see how the weather has changed overnight. Cross-check with local newspaper weather reports and maps. Are you ready to make a "go/no-go" decision? If so, and you have a computer with a modem, your next best move is to get a complete aviation weather briefing by DUATS for your departure, destination, and route of flight.

DUATS can be accessed toll-free 24 hours a day by pilots in the 48 contiguous states with current medical certificates. DUATS provides alphanumeric preflight weather data, NOTAMS, and information on traffic delays, which can be printed out easily for later reference and study. The two DUATS providers also offer free aviation weather graphics.

DUATS will also file IFR flight plans; so it is truly a one-stop service. A record is kept of all briefings, thus DUATS provides solid evidence of compliance with FAR 91.103 (a)—which is not possible with any other computer weather services.

DUATS is provided free of charge by these two commercial services under contract to the FAA:

DTC (Data Transformation Corporation)
1-800-245-3828—Modem access to weather briefings and filing flight plans
1-800-243-3828—Regular help line for customer service and information

GTE (Contel)
1-800-767-9988—Modem access to weather briefings and filing flight plans
1-800-345-3828—Regular help line for customer service and information

TAFS AND METARS

Printing out the coded weather from DUATS provides an additional benefit. The printouts are ideal study guides for mastering TAF and METAR codes. TAFs (Terminal Aerodrome Forecasts) are airport forecasts; METARs (Meteorology Aviation Routines) are hourly and special weather reports by location. If it has an "F," it's a forecast. If it has an "R," it's a report. That's the easy part.

Let's look at the actual METAR and TAF reports as supplied by DUATS for Washington National Airport (DCA) on a rainy winter day. There is a pattern to all those numbers and letters, and once you understand the pattern, you should have no difficulty reading the reports and forecasts. To understand the following METAR, let's break it into the basic blocks in which it is organized:

METAR KDCA 031551Z 33005KT 9SM -RA SCT046 BKN060 OVC080 08/02 A3023 RMK AO2 RAB41 PRESRR SLP235 P0000 T00780017

- METAR KDCA—Type of report and station location. In this case, a METAR reported from Washington National Airport (DCA). (The "K" is an international designator for the 48 contiguous states.)
- 031551Z—Date and time of report: day 3 of the month of the current month, 1551 Zulu time
- 33005KT—Wind from 330 degrees at 05 KTS
- 9SM—Visibility 9 statute miles
- -RA—Light rain
- SCT046—Scattered clouds at 4,600 feet
- BKN060—Broken clouds at 6,000 feet
- OVC080—Overcast clouds at 8,000 feet
- 08/02—Temperature/dew point in degrees Celsius
- A3023—Altimeter setting 30.23

Next comes the remarks section:

RMK AO2 RAB41 PRESRR SLP235 P0000 T00780017

- RMK A02—Remarks, Automated Observation from location type 2 that can discriminate between rain or snow. An Automated Observation with no rain/snow discrimination would be designated AO1

- RAB41—Rain began at 41 after the hour

- PRESSRR—Pressure rising rapidly

- SLP235—Sea level pressure 1023.5 millibars

- P0000—Precipitation less than one hundredth of an inch in the last hour

- T00780017—Temperature 7.8° Celsius, dew point 1.7° Celsius in the last hour, not rounded off

The remarks section of METARs is intended to refine the data reported in the main sections and to provide additional information, such as the status of equipment. This is useful—sometimes critical—information, but it can be hard to decipher. Both AIM, Chapter 7, and *Aviation Weather Services*, Chapters 2 and 4, cover the fine points of METARs and TAFs in considerable detail. If you always make it a habit to look up the codes you don't understand, you will soon see that the same abbreviations and sequences are used over and over again.

Now let's look at the TAFS, the forecasts, for Washington National on the same rainy day covered by the METAR above. TAFs are simpler to decode than METARS; they use the same sequence of groups as METARS—wind, visibility, significant weather, clouds. A TAF usually covers a 24-hour period. As you work through the TAF below, you will see that it is composed of a series of simplified METARS:

TAF KDCA 031130Z 031212 VRB03KT P6SM SCT100 OVC200 TEMPO 12214 BKN100
FM1400 35007KT P6SM SCT070 OVC100
FM1800 01010KT P6SM SCT040 OVC080 PROB40 1822 -RA BKN040
FM2200 02013KT P6SM BKN040 TEMPO 2202 4SM -RA BKN025
FM0200 02014KT 4SM -RA OVC025
FM0600 03015G25KT 3SM RA BR OVC012

- TAF KDCA—Type of report and station location. In this case, a TAF forecast for Washington National Airport (DCA). (The "K" is an international designator for the 48 contiguous states.)

- 031130Z—Date and time the forecast is actually prepared: day 3 of the current month, 1130 Zulu time

- 031212—Date and time of the beginning of the forecast's validity: day 3 of the current month, 1212 Zulu time

- VRB03KT—Wind variable at 03 knots

- P6SM—Visibility greater than 6 statute miles. (The "P" stands for "plus.")

- SCT100—Scattered clouds at 10,000 feet
- OVC200—Overcast clouds at 20,000 feet
- TEMPO 1214 BKN100—Temporarily at 1214 Zulu: broken clouds at 10,000 feet
- FM1400 35007KT P6SM SCT0700 OVC1000—From 1400 Zulu, wind 350 at 07 knots, visibility greater than 6 statute miles, scattered clouds at 7,000 feet, overcast clouds at 10,000 feet
- FM1800 01010KT P6SM SCT040 OVC080 PROB40 1822 -RA BKN040— From 1800 Zulu, wind 010 at 10 knots, visibility greater then 6 statute miles, overcast clouds at 8,000 feet, probability 40% at 1822 Zulu: light rain, broken clouds at 4,000 feet
- FM2200 02013KT P6SM BKN040 TEMPO 2202 4SM -RA BKN025—From 2200 Zulu, wind 020 at 13 knots, visibility greater than 6 statute miles, broken clouds at 4,000 feet, temporarily at 2202 Zulu: visibility 4 stature miles, light rain, broken clouds at 2,500 feet
- FM0200 02014KT 4SM -RA OVC025—From 0200 Zulu, wind 020 at 14 knots, visibility 4 statute miles, light rain, overcast clouds at 2,500 feet
- FM0600 03015G25KT 3SM RA BR OVC012—From 0600 Zulu, wind 030 at 15 knots gusting to 25 knots, visibility 3 statute miles, moderate rain and mist ("BR"), overcast clouds at 1,200 feet

There is much additional aviation weather and information available from DUATS beside METARs and TAFS. I suggest that you print out one full briefing so that you can improve your understanding of how Flight Service reports items as area forecasts (FAs), Winds and Temperatures Aloft forecasts (FDs), AIRMETS, SIGMETS, pilot reports (UAs), radar weather reports (SDs), NOTAMS, and ATC delays and advisories.

FORECAST RELIABILITY

When planning an instrument flight always ask: How good are the weather forecasts?

Pilots should understand the limitations and capabilities of present-day weather forecasting. Don't be lulled into complacency by fancy weather graphics and four- and five-day forecasts! They don't always hold up!

Pilots who understand limitations of observations and forecasts usually make the most effective use of forecasts. The safe pilot continually views aviation with an open mind, understanding that weather is always changing and knowing that the older the forecast, the greater the chance that parts of it will be wrong. The weather-wise pilot looks upon a forecast as professional advice rather than an absolute surety. To have complete faith in weather forecasts is almost as bad as having no faith at all.

According to FAA summaries of recent forecast studies, pilots should consider:

- Up to 12 hours—and even beyond—a forecast of good weather (ceiling 3,000 feet or more, and visibility 3 miles or more) is more likely to be correct than a forecast of conditions below 1,000 feet or less than 1 mile.

- If poor weather is forecast to occur within 3–4 hours, the probability of occurrence is better than 80 percent.
- Forecasts of poor flying conditions during the first few hours of the forecast period are most reliable when there is a distinct weather system, such as a front or a trough; however, there is a general tendency to forecast too little bad weather in such circumstances.
- Weather associated with fast-moving cold fronts and squall lines is the most difficult to forecast accurately.
- Errors occur when attempts are made to forecast a *specific* time that bad weather will occur. Errors are made less frequently when forecasting that bad weather will occur during a *period* of time.
- Surface visibility is more difficult to forecast than ceiling height. Visibility in snow is the most difficult of all visibility forecasts.

Predictable changes

According to FAA studies, forecasters can predict the following at least 75 percent of the time:

- Passage of fast-moving cold fronts or squall lines within ±2 hours, as much as 10 hours in advance.
- Passage of warm fronts or slow-moving cold fronts within ±5 hours, up to 12 hours in advance.
- Rapid lowering of ceilings below 1,000 feet in prewarm front conditions within ±200 feet and within ±4 hours.
- Onset of a thunderstorm 1–2 hours in advance, providing radar is available.
- Time rain or snow will begin, within ±5 hours.

Unpredictable changes

Forecasters cannot predict the following with an accuracy that satisfies present aviation operational requirements:

- Time freezing rain will begin
- Location and occurrence of severe or extreme turbulence
- Location and occurrence of heavy icing
- Location of the initial occurrence of a tornado
- Ceilings of 100 feet or zero before they exist
- Onset of a thunderstorm that has not formed
- Position of a hurricane center to closer than 80 miles for more than 24 hours in advance

1-800-WX-BRIEF

With these sobering thoughts from the FAA about forecast limitations, it is time to call 1-800-WX-BRIEF for a briefing for the flight we have planned from Westchester County to Binghamton, New York. You may get an AFSS that is not in your immediate vicinity, but don't worry. The automated system switches calls to the most available AFSS, and whomever you reach will provide all the information you need wherever you are calling from. Chances are that a live specialist will take your call. If not, you will get a recorded menu of services to choose from. The recorded menu will provide area briefings, hourly observations, forecasts, special announcements, instructions for filing flight plans, and many other items. If you hear the acronyms "PATWAS" or "TIBS," this means you are connected to a "Pilot's Automatic Telephone Weather Answering Service" or a "Telephone Information Briefing Service." Both will provide a menu of services from which you can select the briefing and filing items you want.

It is a typical IFR day in late November in the Northeast. TV weather reports and a newspaper weather map that morning show (FIG. 5-1) that a low-pressure system with plentiful rain has been moving northeastward up the Atlantic Coast. It has been raining off and on at Westchester County Airport and we can see from personal observation that the ceiling is low and visibility is reduced at the airport.

Fig. 5-1. *Newspaper weather map on day of flight discussed in text.*

A go/no-go decision must be made. What weather elements should be considered in making the go/no-go decision? Recall the list from Chapter 4:

- Thunderstorms
- Turbulence
- Icing
- Fog
- Current and forecast weather at departure and destination airports
- Availability of alternate airports
- Nearest VFR
- Personal minimums

Now let's add two items:

- Forecast winds aloft
- NOTAMs

Winds aloft are necessary to determine the estimated time en route (ETE) for each leg of the flight and the total time (TT) en route, just as you did during VFR flight planning.

NOTAMs are a critical item on all flights and are important factors when making the IFR go/no-go decision. It doesn't take much effort to imagine what might happen at the end of a long, tiring IFR flight if you suddenly discovered that a key component of the best instrument approach was "out of service."

FASTER SERVICE

Faster and better service is available by initially telling the briefer:

1. The N number, which immediately identifies you as a pilot, not just someone from the general public calling to find out what the weather is like.
2. The type of airplane; light single-engine, high performance multiengine, and jet airplanes present different briefing problems.
3. "Planning an IFR flight from (departure airport) to (destination airport)."
4. Estimated departure time (in Zulu time).
5. Whether or not you can go IFR (if you have not clarified the point in step 3 and VFR is an option). The briefer doesn't know anything when you call and needs to know whether to provide an IFR briefing or a VFR-only briefing.

The briefer will call up the information for the flight on a computer display and will proceed step-by-step through a briefing appropriate for the flight.

TRANSCRIBING THE WEATHER

A fancy form is unnecessary for a weather briefing. Simply list the categories of information in the proper sequence. The trouble with preprinted weather briefing forms is they leave too much room for unnecessary information and too little room for necessary information.

Below is the briefing sequence; items may be omitted if they are not factors in the proposed flight.

1. **Adverse conditions.** Significant meteorological and aeronautical information that might influence an alteration of the proposed flight: hazardous weather, runway closures, VOR and ILS outages, and the like.
2. **Synopsis.** A brief statement of the type, location, and movement of weather systems, such as fronts and high- and low-pressure areas, that might affect the proposed flight.
3. **Current conditions.** A summary of reported weather conditions applicable to the flight from METARs, PIREPs, and the like.
4. **En route forecasts.** Forecasts in logical order: departure, climbout, en route, descent.
5. **Destination forecasts.** The destination's expected weather plus significant changes before and after the estimated time of arrival (ETA).
6. **Winds aloft.** Forecast winds for the proposed route.
7. **NOTAMS.**
8. **ATC delays.** Any known ATC delays and flow control advisories that might affect the proposed flight.

A simple way to set this up is to list the eight items on the left-hand margin of a blank sheet of paper as a reminder (FIG. 5-2) and leave the rest of the page for notes. Photocopy a supply of lists with the sequence to save time during a weather briefing.

Weather shorthand

Copying a weather briefing verbatim is unnecessary but I strongly suggest recording the vital highlights. Figure 5-3 is a list of easy-to-use "shorthand" weather symbols and letters

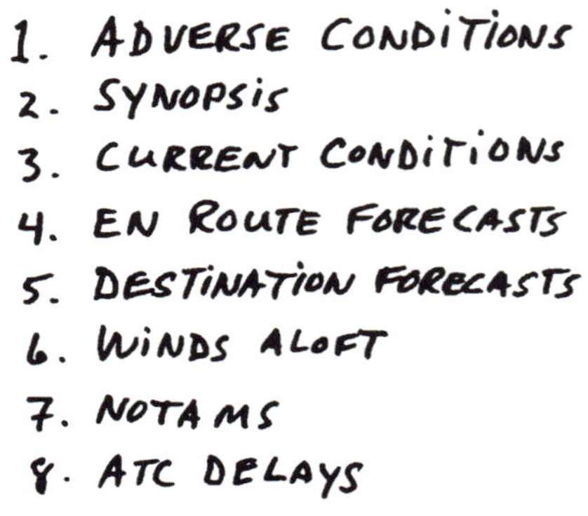

1. ADVERSE CONDITIONS
2. SYNOPSIS
3. CURRENT CONDITIONS
4. EN ROUTE FORECASTS
5. DESTINATION FORECASTS
6. WINDS ALOFT
7. NOTAMS
8. ATC DELAYS

Fig. 5-2. *A quick and simple way to set up a page for jotting down a weather briefing.*

WEATHER PHENOMENA

BR—Mist	**PE**—Ice Pellets
DS—Dust Storm	**PO**—Dust/Sand Swirls
DU—Widespread Dust	**PY**—Spray
DZ—Drizzle	**RA**—Rain
FC—Funnel Cloud	**SA**—Sand
+FC—Tornado/Water Spout	**SG**—Snow Grains
FG—Fog	**SN**—Snow
FU—Smoke	**SQ**—Squal
GR—Hail	**SS**—Sand Storm
GS—Small Hail/Snow Pellets	**UP**—Unknown Precip. (Automated Observations)
HZ—Haze	**VA**—Volcanic Ash
IC—Ice Crystals	

DESCRIPTORS

BC—Patches	**MI**—Shallow
BL—Blowing	**PR**—Partial
DR—Low Drifting	**SH**—Showers
FZ—Supercooled/Freezing	**TS**—Thunderstorm

CLOUD TYPES

CB—Cumulonimbus	**TCU**—Towering Cumulus

ABBREVIATIONS

AO1	Automated Observation without precipitation discriminator (rain/snow)
AO2	Automated Observation with precipitation discriminator (rain/snow)
AMD	Amended Forecast (TAF)
BECMG	Becoming (expected between 2-digit beginning hour and 2-digit ending hour)
BKN	Broken
CLR	Clear at or below 12,000 feet (AWOS/ASOS report)
COR	Correction to the observation
FEW	1 or 2 octas (eights) cloud coverage
FM	From (4-digit beginning time in hours and minutes)
LDG	Landing
M	In temperature field means "minus" or below zero
M	In RVR listing indicates visibility less than lowest reportable sensor value (e.g., M0600)
NO	Not available (e.g., SLPNO, RVRNO)
NSW	No Significant Weather
OVC	Overcast
P in RVR	Indicates visibility greater than highest reportable sensor value (e.g., P6000FT)
P6SM	Visibility greater than 6 SM (TAF only)
PROB40	Probability 40 percent
R	Runway (used in RVR measurement)
RMK	Remark
RY/RWY	Runway
SCT	Scattered
SKC	Sky Clear

Fig. 5-3. *Weather briefing shorthand based on TAF/METAR codes.*

ABBREVIATIONS (*Cont.*)

SLP	Sea Level Pressure (e.g., 1001.3 reported as 013)
SM	Statute mile(s)
SPECI	Special Report
TEMPO	Temporary changes expected (between 2-digit beginning hour and 2-digit ending hour)
TKOF	Takeoff
T01760158, 10142, 20012, and 401120084	In Remarks—examples of temperature information
V	Varies (wind direction and RVR)
VC	Vicinity
VRB	Variable wind direction when speed is less than or equal to 6 knots
VV	Vertical Visibility (Indefinite Ceiling)
WS	Wind Shear (In TAFs, low level and not associated with convective activity)

Fig. 5-3. (*Continued*) Courtesy AOPA

based upon coding used with TAFs and METARs. You'll remember what they mean when encountered on various textual weather reports and forecasts. Eventually a personal short-hand will develop. Don't hesitate to ask the briefer to read the weather slowly, especially in the beginning when the briefing form and the shorthand might be unfamiliar. Plan on having two or three blank forms at hand to write big and still get everything in.

NOVEMBER WEATHER BRIEFING

The following briefing is quoted from an actual briefing at Westchester County Airport for a flight to Binghamton in November. The briefer reads the adverse conditions from the computer screen, then proceeds in sequence through the other items on the briefing checklist.

Adverse conditions

"Covering your route of flight this morning we have SIGMET November Six for occasional severe icing in clouds and precipitation above ten thousand feet.

"We have AIRMET Oscar Four for occasional moderate rime icing from the freezing level to fourteen thousand feet and Oscar Seven for occasional moderate turbulence below ten thousand and moderate to severe low level winds. And here's a NOTAM—Binghamton radar is out of service.

"Westchester at zero niner hundred had wind calm, visibility one-quarter statute mile in light rain and fog, vertical visibility one hundred feet, temperature nine Celsius, dew point eight Celsius.

"Along your route there are scattered to broken clouds below one thousand feet; then two to three thousand broken and four to five thousand broken, variable overcast.

"Binghamton at zero niner hundred had winds from zero three zero at niner knots, visibility twelve statute miles, clouds scattered at six hundred feet, broken at two thousand feet, temperature six Celsius, dew point five Celsius.

"We have a pilot report from seven miles west of Kingston VOR from a BE thirty-three at five thousand feet reporting a base of scattered clouds at one thousand two hundred feet, broken at twelve thousand, visibility five to ten miles. Temperature plus eight, occasional light turbulence."

Forecasts

"Westchester prior to eleven hundred local is forecasting wind from three two zero at seven knots, visibility two statute miles in fog, overcast at seven hundred feet. Occasionally visibility four statute miles in light rain and fog, overcast at one thousand one hundred feet. After eleven hundred local you can expect wind from three one zero at ten knots, visibility greater than six statute miles, overcast at two thousand feet variable three thousand five hundred feet overcast.

"En route, Poughkeepsie is forecasting the same improving conditions.

"Binghamton between ten and twelve hundred local is forecasting wind from two six zero at niner knots, visibility greater than six statute miles, overcast at one thousand five hundred feet, scattered clouds at five thousand feet, with a forty percent probability at eleven-thirty of visibility two statute miles in fog, light rain showers and broken clouds at five hundred feet.

"After twelve hundred Binghamton is forecasting wind from two seven zero at twelve knots, visibility greater than six statute miles, one thousand two hundred scattered variable to broken, two thousand broken."

Winds aloft

"Winds at three, six, and niner thousand feet at Kennedy are: one eight zero at two four, one niner zero at two eight, and two zero zero at three two, with a temperature of plus three.

"At Wilkes-Barre at three, six, and niner thousand feet, the winds are two three zero at six, two zero zero at two seven, and two zero zero at four two, with a temperature of plus one.

"The freezing level at Kennedy is between ten and eleven thousand feet. At Binghamton the freezing level is nine to ten thousand feet."

Alternate

Binghamton is not forecast to be above the "VFR plus 1,000-foot for ETA ±1 hour rule of thumb for requiring an alternate because the ceiling is too low; an alternate airport is required.

"Wilkes-Barre looks good. They are currently reporting wind from two eight zero at thirteen knots, visibility twenty-five statute miles, broken clouds at twenty thousand feet, temperature nine Celsius, dew point five Celsius. After ten hundred and for the rest of the day, Allentown is forecasting wind from two six zero at ten, visibility greater than six statute miles, four thousand five hundred broken variable to scattered.

Not only is Wilkes-Barre a good alternate, but it's also a convenient, easily located airport. Wilkes-Barre is also a good candidate to list on the log as a nearby airport with VFR conditions. Now, all the information needed to make a go/no-go decision has been obtained and written down on the weather briefing form (FIG. 5-4).

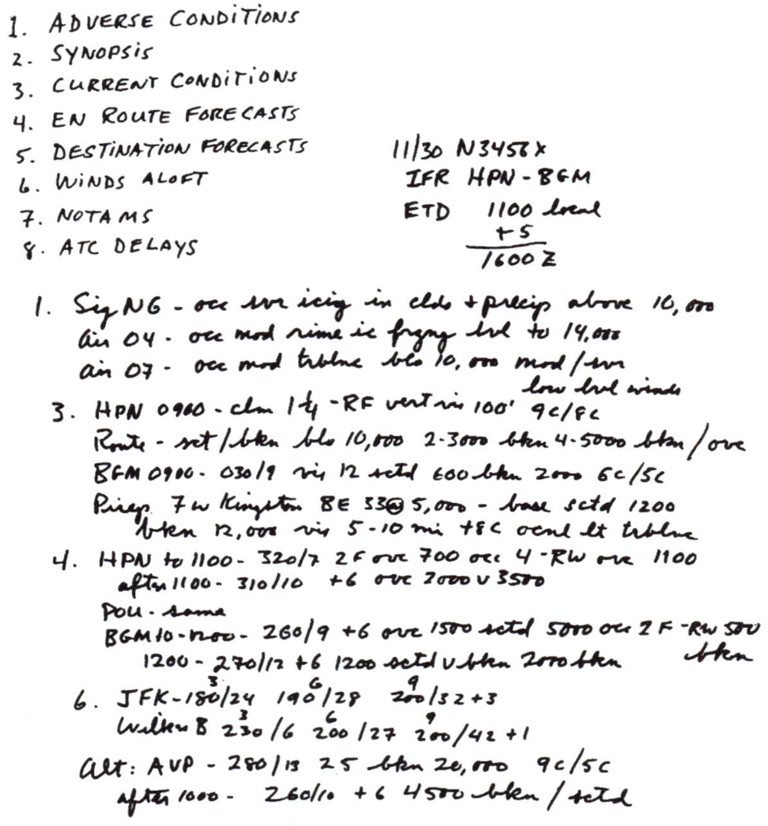

1. ADVERSE CONDITIONS
2. SYNOPSIS
3. CURRENT CONDITIONS
4. EN ROUTE FORECASTS
5. DESTINATION FORECASTS
6. WINDS ALOFT
7. NOTAMS
8. ATC DELAYS

11/30 N3458X
IFR HPN-BGM
ETD 1100 local
+5
1600 Z

1. Sig NG - occ svr icing in clds + precip above 10,000
   Air 04 - occ mod rime ic frzng lvl to 14,000
   Air 07 - occ mod tblnc blo 10,000 mod/svr low lvl winds

3. HPN 0900 - clm 1/4 -RF vert vis 100' 9C/8C
   Route - sct/bkn blo 10,000 2-3000 bkn 4-5000 bkn/ovc
   BGM 0900 - 030/9 vis 12 sctd 600 bkn 2000 6C/5C
   Pirep 7 w Kingston BE 33@ 5,000 - base sctd 1200
   bkn 12,000 vis 5-10 mi +8C ocnl lt tblnc

4. HPN to 1100 - 320/7 2F ovc 700 occ 4-RW ovc 1100
   after 1100 - 310/10 +6 ovc 2000 v 3500
   Poll - same
   BGM 10-noon - 260/9 +6 ovc 1500 sctd 5000 occ 2F -RW 500
   1200 - 270/12 +6 1200 sctd v bkn 2000 bkn   bkn

6. JFK - 180/24 190/28 200/32 +3
   Walker B 230/6 200/27 200/42 +1
   Alt: AVP - 280/13 25 bkn 20,000 9C/5C
   after 1000 - 260/10 +6 4500 bkn/sctd

Fig. 5-4. *Details of the weather for the IFR flight from Westchester County to Binghamton, as copied over the telephone.*

GO OR NO-GO?

What does all this mean? Is it go or no-go?

Return to the checklist of five go/no-go factors and ask whether any will prevent flight as planned:

- Thunderstorms
- Turbulence
- Icing
- Fog
- Departure and destination weather minimums

Thunderstorms. No mention of thunderstorms in SIGMETs, AIRMETs, reports, or forecasts. "Go."

Turbulence. The briefer mentioned an AIRMET for occasional moderate turbulence below 10,000 feet, a PIREP for occasional light turbulence in the vicinity of Kingston, and winds aloft of 32 and 42 knots at 9,000 feet. AIRMET Oscar Seven is a concern with moderate to severe low-level winds.

But turbulence reported in the AIRMET and the PIREP would not damage the airplane. The flight progresses into improving weather, so any turbulence-induced fatigue should not be as important as it might be heading into deteriorating weather.

Surface winds at Westchester and Binghamton should range from calm to 13 knots and that's manageable. "Go."

Icing. Icing is mentioned in SIGMET November 6 and AIRMET Oscar 4. A west-bound (even thousand) cruise altitude of 8,000 feet will stay below the freezing level. "Go."

Fog. Westchester reported dense fog at 0900, but it seems to be dissipating. Neither Binghamton nor Wilkes-Barre, the alternate, have fog. Go, but delay takeoff at Westchester until 1100 so the ceiling and visibility will be sufficient for an instrument approach if an emergency return is necessary.

Minimums. Departure minimums will be a "go" after 1100. Binghamton is forecast to be well above minimums when we arrive. "Go."

Practice this reasoning prior to each instrument flight and it will become easier to weigh the major factors in the go/no-go decision with solid evidence for the decision. Occasionally, the judgment calls will be too close for comfort, or the weather will unpredictably improve one hour and worsen the next hour. We have all seen days like that. For doubtful situations like these, apply the old pilot's rule of thumb: "If in doubt, wait it out."

In this case, it's a "go." Complete the flight log by entering estimated ground speeds, computing the time en route for each leg, and supplying additional details.

ESTIMATED CLIMBOUT TIME

Sometimes instrument students are perplexed about how to compute an accurate time en route to the first fix because that phase of flight involves takeoff, turns, and climbs. I teach students a simple and surprisingly accurate method to figure time to the first fix. Measure the distance from departure to the first fix, estimate winds from the forecast winds aloft reports, compute ground speed and time en route, and add one minute for each 2,000 feet of climb.

This will be very close to the actual time to reach the first fix. If an airplane is a slow climber, such as a Cessna 152, consider adding one minute for each 1,000 feet of climb.

Absolute precision on the first leg is not that important. ATC won't enforce the ±3-minute standard for arrival times because they might amend the clearance with additional turns and changes in altitude. But a reasonably accurate time en route to the first fix will ensure that the time en route for the entire flight will also be reasonably accurate.

WIND AND GROUND SPEED

En route to Binghamton, the distance to the first fix, HAAYS intersection, is a shade farther than 37 nautical miles. Kennedy—the weather observation point nearest our departure airport—was reporting these winds:

180/24 at 3,000 feet

190/28 at 6,000 feet

200/32 at 9,000 feet

The winds at 6,000 feet—190/28—are a good compromise for the climbout to 9,000 feet. Allow 90 knots for the climb speed on a course of 325° and the wind of 190/28 will yield a ground speed of 108 knots. This ground speed computes to 21 minutes en route to HAAYS. A minute for each 2,000 feet of climb adds 4.5 minutes to reach 9,000 feet. Time en route to HAAYS is 21 + 4.5 = 25.5 minutes. Round that off to the nearest even number and enter 26 minutes for the first leg. Use Wilkes-Barre winds at 9,000 feet (200/42) for the remainder of the flight to Binghamton. Ground speed for the approach should be 90 knots.

Calculator options

Many pilots learned to figure wind problems on the E6B computer. This device is perfectly satisfactory for flight-planning calculations. The FAA has approved electronic calculators for use on tests, and there are several good ones that will handle all your flight planning calculations. Other pilots do many calculations mentally, using rules of thumb shown in FIG. 5-5. Select a system and stick to it.

Instructor note. Understand all calculation methods—E6B, electronic calculators, and rules of thumb—because you never know which system a student will prefer to use.

Enter the Kennedy (JFK) and Wilkes-Barre (AVP) winds in the lower right corner of the flight log as shown in FIG. 5-6. While working in this area fill in the VFR WX AT: section for reference, if necessary.

Compute and fill in the estimated ground speeds and times en route as in FIG. 5-7.

The estimated total time en route adds up to 1 hour and 8 minutes, which is written as 1 + 08.

FLIGHT PLAN TO ALTERNATE

Plan a diversion to the alternate, Wilkes-Barre. Weather might be improving, but always be prepared for diversion to an alternate airport. (If you forget to carry an umbrella on a threatening day, it is sure to rain.)

Plan for V149 to Lake Henry VOR (LHY), which is also the initial approach fix (IAF) for the Wilkes-Barre approach that will most likely be in use, ILS 22. Radar vectors to the ILS 22 final approach course are likely, but plan to execute the complete approach in case radio communication is lost or the approach control radar fails.

Identical methods for computing the times for climbout, en route, and descent reveal results shown in FIG. 5-8. Add the times en route from Westchester to Binghamton and

TIME TO CLIMB
1. Estimate distance to reaching altitude. (Include vector, reversal, or circle.)
2. Use filed true airspeed for time/distance calculation.
3. Add one minute for each 2,000 feet of climb.

GROUND SPEED

Crosswind Angle	Effect on True Airspeed
0–15°	Full value of wind speed
30°	.9 wind speed
45°	.7 wind speed
60°	.5 wind speed
75°	.3 wind speed
90°	.1 wind speed

TIME TO FLY

Drop the last digit of the air speed (ground speed, if known); you will fly that many miles in 6 minutes.

Examples:

Speed 150 - 15 miles every 6 minutes.
  Miles to go - 7. Time to fly - 3 minutes.
  Miles to go - 20. Time to fly - 8 minutes.
Speed 90 - 9 miles every 6 minutes.
  Miles to go - 5. Time to fly - 3 minutes.
  Miles to go - 35. Time to fly - 24 minutes.

WIND CORRECTION ANGLE

True Airspeed	Crosswind Angle	Wind Corr. Angle
100 - 120 - 140	90°	½ wind speed
	45°	⅓ wind speed
150 - 180 - 200	90°	⅓ wind speed
	45°	¼ wind speed

Fig. 5-5. *"Rules of thumb" calculations for time to climb, ground speed, time to fly, and wind correction angle.*

thence to Wilkes-Barre plus the 45 minutes required by regulation (FAR 91.167) to obtain a total time for the flight, including alternate (plus 45 minutes), of 2 hours and 32 minutes, well within the fuel range of the aircraft. The completed flight log is shown in FIG. 5-9.

IN-FLIGHT NOTATIONS

Note the ample amount of space remaining for entering in-flight items. Locate where the following in-flight information goes:

- Actual ground speed and actual time en route
- Estimated and actual times of arrival (ETA and ATA)

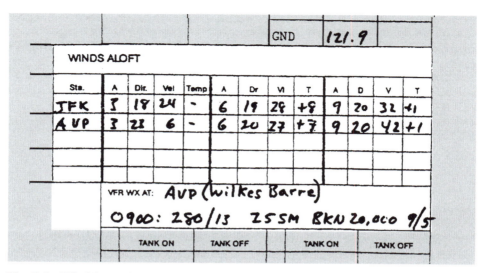

Fig. 5-6. *Filled-in winds aloft forecast and nearest VFR weather blocks on flight log form for IFR flight to Binghamton.*

DIST.	G/S	TIME ENR	TIME ARR
TO	EST	EST	EST
REM	ACT	ACT	ACT
374	108	26	
8	144	4	
66	134	30	
14	109	8	
~~125~~		~~68~~	
		(1+08)	

Fig. 5-7. *Ground speed and estimated time calculations for IFR flight to Binghamton. Note: Actual ground speed and actual times of arrival (ACT) are filled in as the flight progresses.*

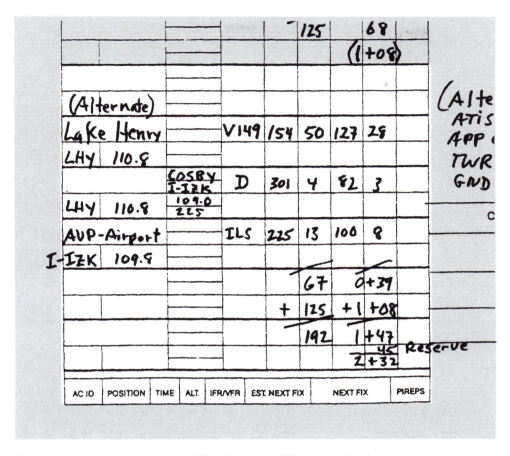

Fig. 5-8. *Planning for alternate, Wilkes-Barre, on IFR flight to Binghamton.*

- Position report sequence (lower left) if needed
- Space for clearances (center) and below that more space for alternate frequencies, ATIS, and clearance limits
- Takeoff and landing runway information, and which approach might be expected
- Time off, time on, total time en route, and tachometer (or Hobbs) reading at the beginning and end of the flight
- Fuel management logs (lower right)

FILING THE FLIGHT PLAN

The completed flight plan is shown in FIG. 5-10.

Use the equipment code in FIG. 5-11; add the code to the briefing card for later reference if a piece of equipment fails, or you are flying a different airplane.

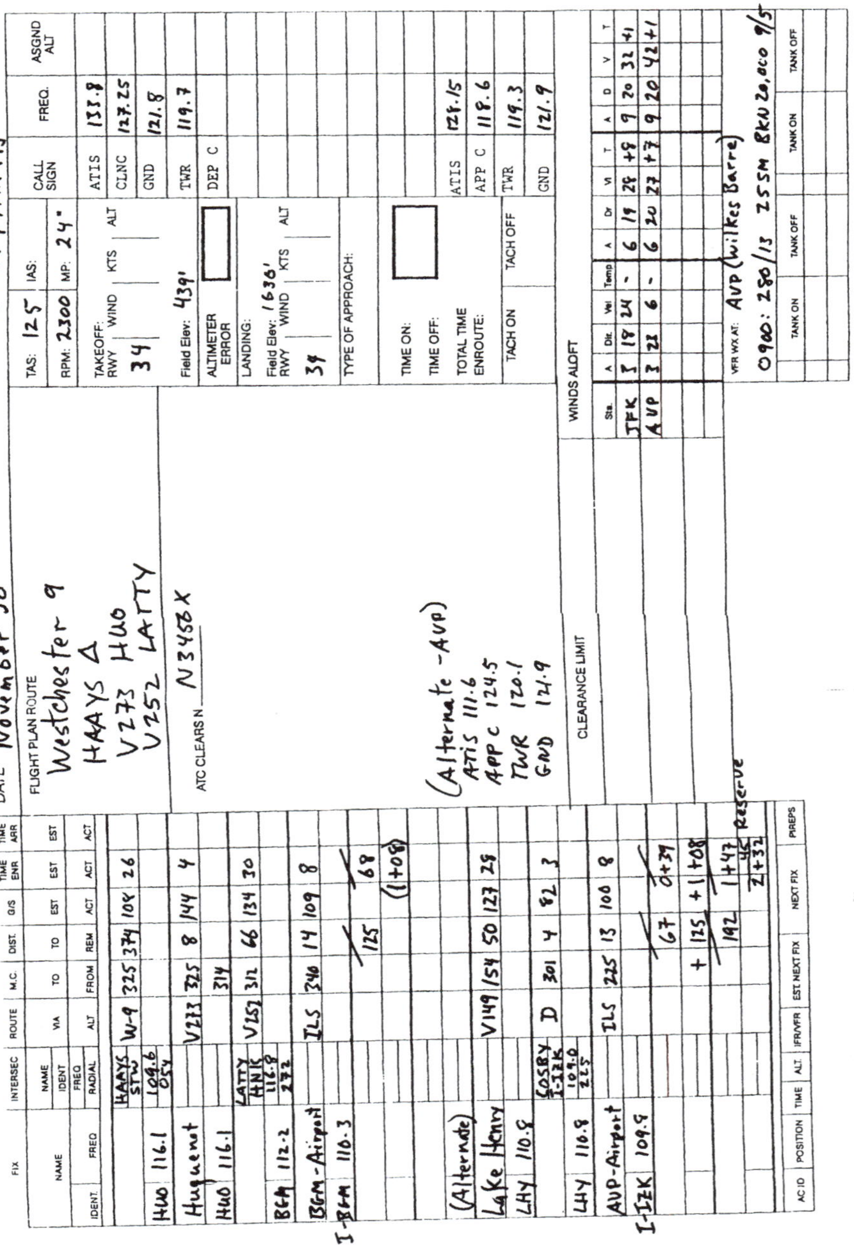

Fig. 5-9. *The complete log for the IFR flight to Binghamton showing all details that can be filled in prior to obtaining a clearance.*

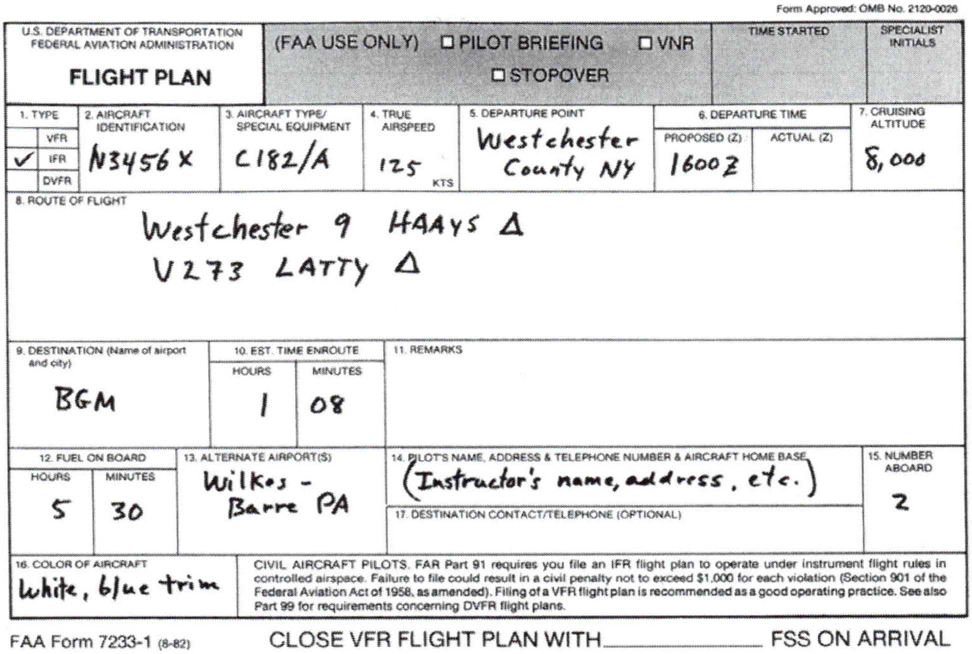

Form Approved: OMB No. 2120-0026

U.S. DEPARTMENT OF TRANSPORTATION
FEDERAL AVIATION ADMINISTRATION
FLIGHT PLAN

(FAA USE ONLY) ☐ PILOT BRIEFING ☐ VNR ☐ STOPOVER | TIME STARTED | SPECIALIST INITIALS

1. TYPE	2. AIRCRAFT IDENTIFICATION	3. AIRCRAFT TYPE/ SPECIAL EQUIPMENT	4. TRUE AIRSPEED	5. DEPARTURE POINT	6. DEPARTURE TIME PROPOSED (Z)	ACTUAL (Z)	7. CRUISING ALTITUDE
VFR / ✓ IFR / DVFR	N3456X	C182/A	125 KTS	Westchester County NY	1600Z		8,000

8. ROUTE OF FLIGHT
Westchester 9 HAAYS Δ
V273 LATTY Δ

9. DESTINATION (Name of airport and city)	10. EST. TIME ENROUTE HOURS	MINUTES	11. REMARKS
BGM	1	08	

12. FUEL ON BOARD HOURS	MINUTES	13. ALTERNATE AIRPORT(S)	14. PILOT'S NAME, ADDRESS & TELEPHONE NUMBER & AIRCRAFT HOME BASE	15. NUMBER ABOARD
5	30	Wilkes-Barre PA	(Instructor's name, address, etc.)	2
			17. DESTINATION CONTACT/TELEPHONE (OPTIONAL)	

16. COLOR OF AIRCRAFT: White, blue trim

CIVIL AIRCRAFT PILOTS. FAR Part 91 requires you file an IFR flight plan to operate under instrument flight rules in controlled airspace. Failure to file could result in a civil penalty not to exceed $1,000 for each violation (Section 901 of the Federal Aviation Act of 1958, as amended). Filing of a VFR flight plan is recommended as a good operating practice. See also Part 99 for requirements concerning DVFR flight plans.

FAA Form 7233-1 (8-82) CLOSE VFR FLIGHT PLAN WITH ________ FSS ON ARRIVAL

Fig. 5-10. *The completed flight plan for the IFR cross-country to Binghamton.*

/X	No transponder.
/T	Transponder with no altitude encoding capability.
/U	Transponder with altitude encoding capability.
/D	DME, but no transponder (per preceding).
/B	DME and transponder, but no altitude encoding capability.
/A	DME and transponder with altitude encoding capability.
/M	TACAN only, but no transponder.
/N	TACAN only and transponder, but with no altitude encoding capability.
/P	TACAN only and transponder with altitude encoding capability.
/R	RNAV and transponder with altitude encoding capability.
/C	RNAV and transponder, but with no altitude encoding capability.
/W	RNAV but no transponder.
/G	Global Positioning System (GPS)/Global Navigation Satellite System (GNSS) equipped aircraft with oceanic, en route, terminal, and GPS approach capability.

Fig. 5-11. *ATC equipment codes for use on flight plan.*

Student note. Use the instructor's name, address, and telephone number because an instrument-rating student cannot legally file an IFR flight plan. The only exception is the instrument flight test.

When filing by telephone, stay on the line for a specialist to copy the flight plan rather than simply recording your message. Request information that has been updated since the previous briefing.

The FAA stated in *FAA Aviation News*: "Live briefers are a good source of unpublished Notices to Airmen concerning important data about airport or runway closures, military flight route training activity, obstructions to flight, outages and shutdowns, etc. The latest pilot reports of weather conditions aloft . . . are also more likely to be available from FSS specialists than from prerecorded messages."

ABBREVIATED BRIEFINGS

There are certain "magic words" that I will introduce from time to time to simplify instrument procedures. One very handy pair is *abbreviated briefing*.

Request an abbreviated briefing for:

- Updated information, such as NOTAMS, to supplement recorded information.
- Updated specifics of a previous full-length briefing.
- One or two specific items. This would be the case, for example, when filing IFR on a "severe clear" day to a nearby airport that you are familiar with. Request a terminal forecast to make sure VFR conditions will continue through the ETA.

Try abbreviated briefings a couple of times during training when weather conditions are favorable. This procedure can save a lot of time when used properly.

OUTLOOK BRIEFINGS

Another good pair of magic words is *outlook briefing*. Request this service for a short, live briefing to supplement other sources regarding a departure time more than six hours away.

Outlook briefings are particularly helpful when it's harder than usual to outguess the weather. There is no way to know when a stalled low pressure system might start moving unless a pilot has access to winds aloft charts, constant pressure charts, and other highly technical information. Even then, the charts might not be much help.

THE ONE-CALL TECHNIQUE

Another method to speed up the weather briefing and filing process is combining the weather briefing and flight plan filing in one telephone call.

Completion of phase one flight planning—elements completed prior to calling flight service—will reveal all details to complete the flight plan form, except estimated time en route and alternate airport.

The weather briefing will quickly establish whether or not an alternate airport is necessary and the briefer will assist with the selection.

Estimated time en route to Binghamton is based upon a true airspeed (TAS) of 125 knots for the Cessna 182. It's a simple matter to bracket this airspeed in increments of 10 knots on either side and make a series of time-speed-distance computations.

Quick estimates

Before calling flight service, compute several figures for time en route in our example using 125 total nautical miles en route. Bracket the 125-knot TAS with headwinds of 30, 20, and 10 knots and tailwinds of 10, 20, and 30 knots to reveal:

125 @ 95 = 1 + 19 (30 knot headwind)
125 @ 105 = 1 + 12 (20 knot headwind)
125 @ 115 = 1 + 05 (10 knot headwind)
125 @ 125 = 1 + 00 (zero wind factor)
125 @ 135 = 0 + 56 (10 knot tailwind)
125 @ 145 = 0 + 52 (20 knot tailwind)
125 @ 155 = 0 + 38 (30 knot tailwind)

When the briefer gives the winds, simply pick the closest estimate of ground speed. A tailwind off the left quarter during this flight probably means a tailwind component of approximately 20 knots, based on the forecast winds for Wilkes-Barre. This produces a time en route of 52 minutes.

Add one minute for every 2,000 feet of climb to the planned altitude of 8,000 feet and the total time en route is 56 minutes. (Simplify the climb estimate by automatically adding 5 minutes for climbs up to 10,000 feet and 10 minutes for climbs above 10,000 feet.)

TOTAL TIME EN ROUTE

Return to the flight log and see the estimated 68 minutes for total time en route. Is the 56-minute quick estimate close enough for filing purposes? Certainly. Time en route establishes when ATC expects the approach to begin in the event of two-way communications failure.

Emergency procedures are subsequently explained in greater detail, but let's briefly consider lost communications. If two-way radio communication is lost, ATC will expect a pilot to carry out the lost communications procedure in FAR 91.185, which pertains to IFR operations during a two-way radio communications failure.

Fly to the destination following the last assigned routing and altitude. According to the FAR: "Begin descent from the en route altitude or flight level upon reaching the fix from which the approach begins [initial approach fix], but not before—

"The expect-approach-clearance time (if received);" or "If no expect-approach-clearance time has been received, at the estimated time of arrival, shown on the flight plan, as amended with ATC." ATC will expect an arrival time as filed. Enter a holding pattern if early. If late, ATC will protect the approach airspace for 30 minutes, then ATC will initiate lost aircraft procedures.

The quick estimate indicates early arrival at Binghamton, which is perfectly all right. The approach would begin out of the holding pattern when the time en route expired.

If the estimated time en route changes either way by more than 5 minutes, notify ATC about a revised ETE or ETA. Not only is this good insurance to cover the possibility of two-way radio communication failure, but it will also help ATC sequence traffic efficiently.

Don't use the one-call procedure if concerned about any of the five go/no-go weather factors. Get a thorough briefing and study the impact of weather on the flight, then make a second call to update the weather information and file the flight plan.

"Cleared as filed"

ATC has its magic words too. If your flight planning has been thorough, you just might hear the welcome words "cleared as filed" when you receive your clearance.

Phase one of a thorough flight plan covers all the details prior to contacting flight service: route, destination, departures, approaches, frequencies, and the big picture of the weather.

Phase two was either a two-call, a one-call, or a recorded communication with flight service to obtain the detailed weather and other information necessary to make a go/no-go decision and to file the flight plan.

Follow this approach to planning an IFR flight from the very first day and the planning becomes easier, faster, and much more interesting. Following this procedure and filing an IFR flight plan for every instrument training flight—even if it's just out into the local area for practice in basic maneuvers—will provide a better chance of passing the instrument checkride.

Save all logs and flight plan forms. It was hard work to get all the information necessary for those flights, so save the information and use it again. Always check the information against the current charts for changes. Use routes and approaches that worked well on previous flights and especially remember any amended clearances received. Amended clearances might suggest better routes or fixes that can be incorporated into future flight planning.

6
Airplane, instrument, and equipment checks

YOU HAVE COMPLETED THE FLIGHT LOG, CHECKED THE WEATHER, and found that the flight can be safely completed. The flight plan is filed and the pilot is prepared but what about the airplane? Is it ready for an IFR flight in actual instrument weather?

Several items on the preflight checklist require special attention for an IFR flight; see FIG. 6-1 for a basic list. Consider adding items for a specific airplane and equipment. (Note: Some of the items are discussed in detail in Chapter 7.)

FUEL QUANTITY

Some checks are obvious. Turn on the electrical power and check fuel gauges to make sure there is enough fuel to fly to the destination plus 45 minutes at normal cruising speed. If an alternate airport is required, verify enough fuel to the destination, then to the alternate plus 45 minutes. This item is first on the checklist so the pilot can continue the checklist while waiting to refuel.

PRE-TAXI CHECK
Before starting:
- ☐ Outside antennas—all secure
- ☐ Fuel—to destination, alternate, plus 45 minutes
- ☐ ATIS—copied
- ☐ Altimeter—set, error noted
- ☐ Airspeed, VSI—both on zero
- ☐ Magnetic compass—shows correct heading
- ☐ Clock—running and set correctly
- ☐ Pitot heat—working
- ☐ Lights—all working
- ☐ VOR—checked within 30 days
- ☐ Charts and logs—sequenced

After starting:
- ☐ Ammeter—checked
- ☐ Suction gauge—normal
- ☐ COM radios—all checked and set in sequence
- ☐ NAV radios—all checked and set for departure
- ☐ Marker beacon lights—checked
- ☐ Heading indicator—set
- ☐ Attitude Indicator—normal
- ☐ Alternate static source (if any)—working
- ☐ Clearance—copied
- ☐ Transponder—code set, checked, turned to standby

TAXI CHECK
- ☐ Heading indicator—responding normally to turns
- ☐ Attitude indicator—normal and stable
- ☐ Turn coordinator—responding normally to turns

PRE-TAKEOFF CHECK
- ☐ Approach charts—emergency return chart on top
- ☐ NAV radios—double-check set for departure
- ☐ COM radios—Departure control on #2
- ☐ Heading indicator—aligned with runway centerline

RUNWAY ITEMS—"STP"
- ☐ Strobes (or rotating beacon)—on
- ☐ Transponder—on
- ☐ Pitot heat—on

Fig. 6-1. *IFR checklist items.*

It's always a good practice to start an instrument flight with full fuel tanks if weight permits, visually confirmed. Nothing is as useless as the "runway behind, the air above, and the fuel in the fuel truck."

During a round-robin trip, at each stop ask "Should I top off again?" If in doubt, top off. There is no reason to worry about running out of fuel in addition to all the other concerns.

ATIS

While the electrical power is on, tune in the ATIS frequency if available at the departure airport and write down the current information. Set the altimeter to the current altimeter setting as reported by ATIS. If there is a difference of more than 75 feet between what the altimeter reads and the ramp elevation where the plane is parked, don't go. There is something seriously wrong with the altimeter.

Reference the altimeter with the ramp elevation where the plane is parked, not the general field elevation. Airports are seldom level. Look at the airport diagram on the approach chart and select the elevation nearest the parking ramp. As noted in the flight-planning chapter, there can be considerable difference between official airport elevation and ramp elevation.

ALTIMETER AND AIRSPEED ERRORS

There is frequent confusion about what action to take with altimeter error. Use the given barometric pressure and note the error on the flight log in the appropriate spot. Fly the altitudes shown on the altimeter and disregard the error until the approach.

Add the error to published minimums, regardless of whether this error is plus or minus. For example, the White Plains decision height on the ILS 16 approach is 639 feet. If the altimeter error is 50 feet, add it to the minimum and use a decision height of 689 feet regardless of whether the error is plus 50 feet or minus 50 feet. Always add the difference to be on the safe side.

Check the airspeed indicator and the vertical speed indicator, which should indicate zero and zero. Most vertical speed indicators have an adjustment screw on the lower left corner of the instrument for calibration to zero.

If there is no adjustment screw, allow for the error when interpreting indications in flight. For example, if the instrument shows a 50 foot-per-minute climb on the ramp, that is level flight in the air.

ELECTRICAL EQUIPMENT

Turn on and check the pitot heat and all outside lights, including landing and taxi lights, even if the flight will occur during daylight hours. Pitot heat is necessary if any part of the flight is in the clouds, especially when climbing through a cloud layer after takeoff. It's a very strange feeling to see airspeed drop toward stalling speed during an instrument climbout because ice is building up in the pitot tube!

Lights might be necessary for visual identification of the airplane by tower controllers during an approach in low visibility.

Instructor note. Many questions and doubts have arisen about the final item of the checklist (FIG. 6-1). Before takeoff the *pitot heat must be* ON, whether the takeoff is into severe clear, cloud, rain, or icing conditions ahead. This is the way I have been teaching other instructors for many years. In addition, if the plane has electric prop heat, it should also be turned on.

It is theorized that a B-727 crashed because the crew missed this item on the checklist, and recently it has surfaced as the most probable underlying cause of the crashes of a number of Malibu airplanes, which had shown no bad traits during flight testing and certification.

Proper pilot training and indoctrination could have prevented some of these accidents and loss of life. Remember the "law of primacy" that you were taught as a student flight instructor? If proper habit patterns had been taught from the beginning, some of these lives might have been saved. When we train pilots, we bear a heavy responsibility to "teach 'em right the first time." We have no way of knowing whether a student will wind up flying a B-747, Citation, Learjet, or P-210. We must train them right! Now!

Although it is not legal to fly in "known ice," anyone who has flown IFR for any length of time has encountered icing conditions where they were least expected—a complete surprise. Then ATC might be in a "bind" and unable to give us relief as fast as we would like, such as a lower or higher altitude to get us out of the ice. This is just the time that the pilot forgets to put on the pitot heat. The power drain is minimal for pitot heat— you can hardly see the ammeter needle move when the pitot heat switch is turned ON and OFF to check. If there is any doubt that the alternator/generator can't handle this added load, then this is not a real instrument airplane. I haven't seen an airplane in the last 10 years that couldn't handle the load, and similarly I haven't heard of many pitot heating elements burning out. So get with it and teach it right.

VOR CHECKS

Verify that the record of VOR checks is current. The VOR check is probably the most overlooked or ignored regulation. There is nothing mysterious about the check nor is the regulation hard to understand. FAR 91.171 says that no person may use VORs on an IFR flight unless the system has been checked within the preceding 30 days and found to be within certain limits.

A record of VOR checks must be kept in the plane. If no log entry is found attesting to a VOR check within the preceding 30 days, it is illegal to make an instrument flight unless the VORs are checked before takeoff. In spite of the legal requirements, it doesn't make sense to depart on an instrument flight without knowing that the VORs are accurate and within limits.

If the receiver and indicator have been checked within the 30-day limit, everything is OK. If not, determine if a VOR check is possible before taking off. VOR checks will be explained later in this chapter. The point here is that checking the status of the VOR systems is an important item on the IFR preflight inspection.

COCKPIT ORGANIZATION

The basic principle of cockpit organization is twofold:

- Organize charts and flight log to avoid fumbling for a vital piece of information
- Check and set up communication and navigation radios ahead of time according to the expected sequence of frequencies

One very real hazard associated with cockpit confusion is vertigo. Sudden head movement while searching for a chart might cause an attack of vertigo. It might be more severe when reaching back to pick up something from a flight case on the rear floor, then snapping around to an upright position.

How to sequence logs and charts

My favorite device for holding charts and logs is a standard $8\frac{1}{2} \times 11$ clipboard. A folding flight desk that rests on the lap, or a kneeboard that straps to a thigh works well.

I place the flight log and flight plan on top. Underneath those I place the en route chart folded so I can see the route and airways or fixes that might be in any amended clearance. Approach charts for the departure airport are placed under the en route chart in case I have to return for an instrument approach shortly after takeoff. Next come approach charts for the destination airport with the expected approach on top. After these come all approaches for the alternate airport.

Radio preparation

Before starting an engine, determine what sequence of frequencies will be used for communications and navigation. Write the frequencies in sequence, starting from the top. A top-to-bottom sequence used consistently will reduce the chances of selecting the wrong radio at the wrong time.

In most cases, the departure communications sequence will be:

- ATIS
- Clearance delivery or pretaxi clearance
- Ground control
- Tower
- Departure control
- Air route traffic control center

Tune the ATIS frequency first on the number one communications radio. Move down and tune clearance delivery. (Clearance delivery is often handled by ground control. Some very large airports might also have a gate control frequency. Airline pilots monitor the gate control frequency for clearance to start engines, thereby saving fuel.)

Listen to transmissions on the first frequency, adjust the volume to a comfortable level, then adjust the squelch sensitivity until the "noise" begins; readjust the squelch until the "noise" stops.

Follow the same procedure with the second radio in the communications sequence.

Presetting frequencies

Some newer navcoms will store up to nine additional frequencies. I once preset nav frequencies to fly all the way from Palm Beach, Florida, to Myrtle Beach, South Carolina, without dialing in a new VOR frequency. The same features are also available on the com side. I frequently preprogram the ATIS, approach, tower, and ground control at the destination airport, which cuts down on the workload in the air.

Use the top-to-bottom approach to check navigation radios as well. I check the VOR receivers' accuracy on every flight. Look in the "VOR Receiver Check" section in the back of the A/FD and see which checks are available at your home airport as well as other frequently used airports (FIG. 6-2). Set up the test frequencies before starting, as you did with the communications radios.

VOR CHECKS WITH VOT

The best method of checking a VOR receiver and indicator is with a *VOR test facility* (VOT). The VOT is a special VOR ground facility that transmits only the 180° radial. Tune in the VOT frequency listed in the A/FD and dial in 180 with the *omni bearing selector* (OBS) knob. The needle should center in the TO position. An easy way to remember the setup and expected indication is "Cessna 182:" The C in Cessna is for center, one-eighty is the radial, and two is the direction (TO). The maximum permissible bearing error is ±4° with the VOT.

Unfortunately, only a limited number of airports in the United States have a VOT. Many are very busy, high-density hubs—such as Kennedy, La Guardia, and Los Angeles—

VOR RECEIVER CHECK
MASSACHUSETTS

VOR RECEIVER CHECK POINTS

Facility Name (Arpt Name)	Freq/Ident	Type Check Pt. Gnd. AB/ALT	Azimuth from Fac. Mag	Dist. from Fac. N.M.	Check Point Description
Gardner (Fitchburg Muni)	110.6/GDM	A/1500	102	13.0	Over intersection of rwys.
Gardner (Metropolitan)	110.6/GDM	A/2000	097	1.9	Over intersection of taxiway and rwy.
Gardner (Orange Muni)	110.6/GDM	A/1500	292	10	Over parachute jump circle.
Gardner (Worcester Regional)	110.6/GDM	A/2000	167	18.8	Over intersection of Rwys 11–29 and 15–33.
Lawrence (Plum Island)	112.5/LWM	A/1500	089	11.8	Over apch end Rwy 10.
Lawrence (Tew-Mac)	112.5/LWM	A/1500	224	9.9	Over apch end Rwy 21.
Marthas Vineyard (Marthas Vineyard)	114.5/MVY	G	216	0.7	On runup block for Rwy 06.
Nantucket (Nantucket Memorial)	116.2/ACK	G	242	1.9	On runup area at apch end Rwy 24.
Providence (Fall River Muni)	115.6/PVD	A/1500	097	15	Over intersection of rwys.
Putnam (Southbridge Muni)	117.4/PUT	A/1700	328	12	Over intersection of taxiway and rwy.

VOR RECEIVER CHECK
VOT TEST FACILITIES (VOT)

Facility Name (Airport Name)	Freq.	Type VOT Facility	Remarks
Laurence G. Hanscon	110.0	G	
Gen. Ed. Lawrence Logan Intl.	111.0	G	
Worcester Regional	108.2	G	

Fig. 6-2. *VOR receiver check information as published in the Airport/Facilities Directory.*

where instrument training flights are not always practical. If you don't have access to a VOT, the next best check is with a VOR located on the field or close enough to receive on the ground. Tune both VOR receivers to this station and note the indicated bearings TO that station. The maximum permissible difference between the two bearings is 4°.

GROUND AND AIRBORNE VOR CHECKS

The A/FD also lists other ground and airborne checkpoints (FIG. 6-2). Again, see what checkpoints are available. One disadvantage to ground checkpoints is that they might be located on the opposite side of the airport. And because airborne checkpoints require a crosscheck between a VOR radial and some point on the ground, such as the end of a runway, they won't be much good on an IFR flight in the clouds!

Another method of checking VOR receivers is by tuning the same VOR in flight and noting the bearings from the station. The maximum permissible variation between the two bearings is 4°. FAR 91.171, which covers VOR checks, also prescribes an airborne procedure for a single VOR receiver and for dual VOR receiver checks.

To legally meet the 30-day requirement, VOR receiver checks must be logged with date, type of check, place, bearing error, and signature of the person making the check.

Take advantage of all opportunities to make airborne checks. Log them properly and keep the log in the airplane to avoid getting stuck on the ground on a routine IFR day simply because the VOR receivers have not had a legal check within the last 30 days.

ILS CHECK

If there is an ILS system at the departure airport, check out the ILS receivers after the VOR receiver check. Tune in the localizer frequency on the number one navigation radio, listen to the identification signal, and observe the needles for correct movement. The needles won't steady up in one position because the airplane is not aligned with the final approach course while at the parking ramp. If you hear a clear identification signal, the red warning flag on the instrument face disappears, and the needles are "alive," the receiver is operating properly.

Check the ILS identifier, then turn the receiver off or turn the volume down to the lowest level and leave the localizer frequency on the number one receiver. You will be ready to make an emergency return shortly after takeoff without fumbling around for the correct frequency. Likewise, set the number-two navigation receiver to the first en route VOR station and it is ready to proceed on course without cockpit confusion.

ADF CHECK

Get a good check of the ADF by tuning the frequency of the outer compass locator at the departure airport and observe the swing of the needle. When the needle steadies it should point toward the locator. With a little trial and error you can soon establish the relative bearing from the ramp to the locator and use this to check the accuracy of the ADF. Use the same procedure if the departure airport has an NDB approach and no ILS.

If there is neither a compass locator nor an NDB near the airport, use a local AM radio station to check the ADF. Try to find a station near the airport that is shown on a sectional chart and plot the bearing to the station and see how well the ADF needle matches that bearing.

TRANSPONDER CHECK

Finally, turn the transponder to the TEST position. If the indicator light comes on and blinks, the transponder has run an internal circuitry check and everything is functioning normally.

Be careful not to switch the transponder to the ON or ALT (mode C) positions because the transponder signal might be sensed by the traffic control radar antenna and fed into the computer to indicate that an IFR flight has commenced. Leave the transponder in the STANDBY position until cleared onto the runway for takeoff and include the transponder (Transponder ON) in the final list of runway items.

TIPS TO REDUCE COCKPIT CONFUSION

The best ways to reduce cockpit confusion are:

- Organize logs and charts on a clipboard with the flight log on top and the other material in the proper sequence. Do this in the planning room before starting out to the airplane.
- Tune the communications and navigation radio frequencies to check before starting the engine. Then use a top-to-bottom sequence for checking out each piece of radio navigation equipment. If frequencies checked are different from departure frequencies, tune in the departure frequencies with the ILS of the departure airport on the number one navigation receiver. Then you can keep one step ahead throughout the flight.

All the checks above may be performed on battery power unless the temperature is below freezing. When it gets that cold, of course the battery has less power and must be conserved. In cold weather it is a good idea to start the engine first, while the battery is still fresh, then perform the radio checks.

GYRO INSTRUMENTS

Other instrument checks can only be made after starting the engine. Prior to taxiing, set the heading indicator to match the heading on the magnetic compass. Next, check the attitude indicator to ensure that it confirms level flight. (Pitch attitude cannot be set on the ground because there is no way of knowing when the airplane is sitting precisely in a level attitude.)

Any drift by either instrument during taxi indicates a malfunction that should be diagnosed and repaired prior to IFR flight.

7
Clearances and communications

COPYING CLEARANCES SEEMS TO BE A GREAT STUMBLING BLOCK IN the minds of many instrument pilots. There is no need for this to happen. If you start out copying clearances the right way, you will soon find that this is one of the easier elements of instrument flight. Because an IFR flight plan is filed for every training flight, you will have to copy IFR clearances from ATC—and read them back correctly—beginning with the first training flight. Based upon experience with hundreds of students, those with the best success in mastering clearance copying are those who are prepared to copy and read back clearances on every flight.

PRACTICE CLEARANCES

These tips will soon have you copying clearances like an airline pilot. But practice, practice, practice is the best way to become competent in copying clearances, especially in the beginning when the jargon is unfamiliar and the controllers seem to set new records for fast talk!

I urge students to buy a multiband portable radio with aviation bands or a hand-held transceiver to practice copying clearances delivered to other flights. The transceiver is a better investment because it can also be used to communicate with ATC if communications are lost while on an IFR flight plan. Other handy uses for a transceiver are subsequently discussed. Read those clearances into a tape recorder then play the tape back and attempt to copy the clearances. It will be fun and it will develop a competence and confidence— within three or four lessons—that is unbelievable.

If you are unable to listen to actual clearances being delivered, ask an instructor to read simple clearances for practice.

Instructor note. Why not make up an audiotape that students can use? A cassette recorder and a few short sessions in a parked aircraft on a busy day at a large airport will produce a tape recording for students to use at home.

CLEARANCE SHORTHAND

It's impossible to copy a clearance in long hand and get it right. Every pilot develops a clearance "shorthand" that works well. Look over the list of simplified clearance shorthand symbols in FIG. 7-1 and practice using them. If a personal clearance shorthand is more comfortable and more in keeping with your personal taste, be my guest and use symbols from the list that are helpful. All that's necessary is an ability to read back the clearance promptly and accurately. If the flight has been planned correctly, there should be little need to refer back to the clearance, and never in an emergency!

On the first few instrument-training flights, tell ATC that you are an instrument student and ask the controller to "please" read the clearance slowly. When controllers realize that they are working with a student, they will read the clearance slower rather than read it over three or four times.

Be frank about expertise; ATC will cooperate in most cases. Controllers get in the habit of talking fast and they assume that any pilot can keep up with them. Conversely, an FSS specialist will quickly request slower speaking if a pilot talks too fast when filing a flight plan.

This advice is not just for students. When an experienced pilot is at an unfamiliar airport expecting an unfamiliar clearance, the pilot should request a slower delivery. ATC will usually cooperate.

Another tip on clearance copying: If the controller reads the clearance a little too fast and parts are missed, keep on copying and leave a blank spot for the missed information. Then read back what you have and ask for a repeat of any section that wasn't clear. Don't give up—keep on writing! The last thing clearance delivery wants to hear is "Please repeat everything after 'seven two Romeo cleared to...'"

HANDLING AMENDED CLEARANCES

Now let's look at what's involved in copying clearances in the air. It's almost impossible to fly IFR anywhere these days without receiving an amended clearance or two—and they always seem to come as a surprise. Be prepared for them and avoid surprises.

Always have a blank spot on the flight log or an extra blank piece of paper handy to reach instantly when the controller issues a clearance.

60	altitude—6,000'
A	airport
<	after passing
>	before reaching
C	ATC clears
X	cross
D	direct
EAC	expect approach clearance
EFC	expect further clearance
CAF	as filed
H	hold
H-W	hold west, etc.
M	maintain (altitude)
O	VOR or VORTAC
RL	report leaving
RP	report passing
RR	report reaching
RV	radar vectors
RY Hdg	runway heading
LT	left turn after takeoff
RT	right turn after takeoff
V	Victor airway
til	until further asking
↑	climb to
↓	descend to
→	intercept

Fig. 7-1. *Clearance copying shorthand.*

Take a pencil and start writing on that piece of paper on the left-hand edge while looking at the attitude indicator because it is the basic reference for control of the airplane. Keep the airplane straight and level.

Learn to copy clearances without looking at the paper by practicing on the ground. Try this experiment: Place a piece of paper on the clipboard, stare off into the distance at an object imagining it to be an attitude indicator, and write "I am a very good IFR pilot" several times, beginning from the left edge. It doesn't have to be perfect, and it's your writing so you'll be able to read it, even if it runs uphill or downhill a little. Try this a couple of times to see how easy it is.

Now go back and play some of those recorded clearances. Copy them also without looking. Airborne clearances seldom cover more than two or three items at a time. Develop a clearance shorthand and practice copying full clearances on the ground without looking. There should be no difficulty copying the shorter airborne clearances.

For example, let's say that an amended clearance while airborne said: "Cessna three four five six Xray cleared direct Haays intersection maintain eight, report passing six." All you would have to write, using the clearance shorthand in FIG. 7-1 is: "D HAAYS M 80 RP 60."

OBTAINING CLEARANCES

After all this practice it is time to get on with the flight. An ATC clearance must be obtained. Complete the preflight and get the cockpit organized before worrying about the clearance, then decide whether or not to start the engine before calling for a clearance. On a cold day it might be wise to start up, then call clearance delivery; engine and instruments can warm up while waiting for the clearance.

On the other hand, the delay might be extensive and the engine might run for a long time prior to taxi. If renting the plane according to the time on the Hobbs meter, this is a needless expense. Instead, call for the clearance then wait for it without the engine running. Today's transistorized radios use very little power and you can listen for up to 15 minutes or so and still have plenty of starting power, unless the battery is weak or the temperature is very cold.

Avoid this dilemma entirely by using a hand-held transceiver to call clearance delivery. Again, the battery-powered transceiver is a backup radio in the event of a lost communications emergency.

When an IFR flight plan is filed, the information goes into an ATC computer. If the computer determines that the flight will not conflict with other traffic, clearance delivery might simply state the magic words: "Cleared as filed."

If there is a conflict, ATC will resolve the conflict by assigning a different altitude, a different route, or by clearance to a fix that is short of the destination. If cleared short of the destination, you will receive an amended clearance to the destination when the traffic conflict has been resolved. The flight plan normally remains in the computer until two hours after the proposed departure time. If the clearance is not requested by that time, the flight plan will be erased from the computer's memory unless an extension is requested.

Clearance on request

Here is a point that many people misunderstand. ATC will not ordinarily get your clearance from the computer until requested. I have seen quite a few students sit on the ramp with their engine running waiting in vain for some message from ATC. It is the IFR pilot's responsibility to inform clearance delivery or ground control when ready to copy the clearance.

The radio communication usually goes something like this:

Pilot: "Cessna three four five six Xray IFR Binghamton with information Romeo (the ATIS)."

ATC: "Cessna three four five six Xray. Clearance on request."

"Clearance on request" means that the clearance delivery controller does not have your clearance immediately available, or that there is a problem with it that must be resolved with ATC before it can be read to you. Sometimes, however, ATC will come right back with the full clearance; always have pencil in hand and be prepared to copy the clearance. In most cases there will be a delay between the time you call clearance delivery and when you actually receive the clearance.

UNACCEPTABLE CLEARANCES

If the clearance is unacceptable, read it back anyway, then explain. Valid reasons include:

- No survival gear for a long section that is over water
- Icing conditions
- Routing will add 50 minutes to the flight and exceed legal fuel reserves

The pilot in command is responsible for the safety of the flight; any compromises with safety are unacceptable.

When operating from an airport with a control tower, ATC will issue a clearance on either a clearance delivery frequency or on ground control. If operating from an airport not served by a control tower (or if the tower is closed) there are several ways to get the clearance.

REMOTE COMMUNICATIONS OUTLETS (RCOS)

Check first to see if the uncontrolled departure airport has a remote communications outlet (RCO). The RCO transmitter and receiver antenna located on the airport is linked by landline to flight service or ATC. You can find RCOs listed in the A/FD in the communications section of the airport listing (FIG. 7-2) Some RCOs are also shown on sectional charts. If an RCO is located at an airport, request and copy a clearance and receive a "release" for takeoff from that RCO.

Check the RCO listings carefully. Sometimes an RCO will receive on 122.1 MHz and transmit on a VOR frequency. Numerous RCOs are found throughout the country at uncontrolled fields, as well as some controlled fields where towers do not operate 24 hours a day. Use an RCO to request and copy a clearance and receive a void time for takeoff.

VOID TIME CLEARANCES

If there is no radio facility to issue an IFR clearance, it must be received by telephone. A clearance issued by telephone is called a *void time clearance* because ATC will always set a time limit after which the clearance is void. When you file IFR by telephone from an uncontrolled field, ask whom to call to pick up a clearance. Flight service will provide a telephone number, which might be that flight service station's number or an ATC telephone number.

Call back 10 minutes prior to the proposed departure time. Flight service or ATC will read the clearance over the telephone and you will read it back. Make sure everything is ready to take off immediately. If unable to take off within the time limit, call ATC back by telephone before the clearance is void and request a later "time window" for release and a new void time. The time window is the time between the release and the void time.

MASSACHUSETTS

NANTUCKET MEM   (ACK)   3 SE   UTC−5(−4DT)   N41°15.18′ W70°03.61′   **NEW YORK**
   48   B   S4   **FUEL** 100LL, JET A   ARFF Index A   **H−3J, L−25D**
   RWY 06−24: H6303X150 (ASPH)   S−75, D−170, DT−280   HIRL CL   0.3%up NE.   **IAP**
     RWY 06: MALSF. VASI(V4L)—GA 3.0°. Thld dsplcd 539′.   **RWY 24:** SSALR. TDZL.
   RWY 15−33: H3999X100 (ASPH)   S−60, D−85, DT−155   MIRL
     RWY 15: REIL. Building.   **RWY 33:** REIL. VASI(V4R)—GA 3.0°TCH 43′.
   RWY 12−30: H3125X50 (ASPH)   S−12
     RWY 12: Trees.   **RWY 30:** Trees.
   AIRPORT REMARKS: Attended continuously. Be aware of hi-speed military jet and heavy helicopter tfc vicinity of Otis
     ANGB. Deer and birds on and invof arpt. Rwy 12−30 VFR/Day use only aircraft under 12,500 lbs. Arpt has noise
     abatement procedures ctc Noise Officer 508−325−6136 for automated facsimile back information. PPR 2 hours
     for unscheduled air carrier ops with more than 30 passenger seat, call arpt manager 508−325−5300. When twr
     clsd ACTIVATE MALSF Rwy 06; SSALR Rwy 24; HIRL Rwy 06−24; MIRL Rwy 15−33 and twy lgts—CTAF. Rwy 24
     SSALR unmonitored when arpt unattended. Twy F prohibited to air carrier acft with more than 30 passenger
     seats. Fee for non-commercial acft parking over 2 hrs or over 6000 lbs. NOTE: See Land and Hold Short
     Operations Section.
   WEATHER DATA SOURCES: ASOS (508) 325−6082. LAWRS.
   COMMUNICATIONS: CTAF 118.3   **ATIS** 126.6 (508−228−5375) (1100−0200Z‡) Oct 1−May 14, (1100−0400Z‡) May
     15−Sept 30.   **UNICOM** 122.95
    BRIDGEPORT FSS (BDR) TF 1−800−WX−BRIEF. NOTAM FILE ACK
    RCO 122.1R 116.2T (BRIDGEPORT FSS)
  Ⓡ **CAPE APP/DEP CON** 126.1 (1100−0400Z‡) May 15−Sept 30, (1100−0300Z‡) Oct 1−May 14.
    BOSTON CENTER APP/DEP CON 128.75 (0400−1100Z‡) May 15−Sept 30, (0300−1100Z‡) Oct 1−May 14.
    TOWER 118.3 May 15−Sep 30 (1100−0300Z‡), Oct 1−May 14 (1130−0130Z‡).
    GND CON 121.7   **CLNC DEL** 128.25
   AIRSPACE: CLASS D svc May 15−Sep 30 1100−0300Z‡, Oct 1−May 14 1130−0130Z‡ other times CLASS G.
   RADIO AIDS TO NAVIGATION: NOTAM FILE ACK.
    (H) VOR/DME 116.2   ACK   Chan 109   N41°16.91′ W70°01.60′   236°2.3 NM to fld. 100/15W.
    WAIVS NDB (LOM) 248   AC   N41°18.68′ W69°59.21′   240° 4.8 NM to fld.
    NDB (HH−ABW) 194   TUK   N41°16.12′ W70°10.80′   115° 5.5 NM to fld.
    ILS/DME 109.1   I−ACK   Chan 28   Rwy 24.   LOM WAIVS NDB. ILS unmonitored when twr clsd.

  NAUSET   N41°41.51′ W69°59.39′   NOTAM FILE BDR.   **NEW YORK**
  NDB (MHW) 279   CQX   at Chatham Muni. NDB unusable 220°−280° byd 20 NM.   **L−25D**

  NEFOR   N41°37.30′ W71°01.06′   NOTAM FILE EWB.
  NDB (LOM) 274   EW   056°4.3 NM to New Bedford Regional.

NEW BEDFORD REGIONAL   (EWB)   2 NW   UTC−5(−4DT)   N41°40.57′ W70°57.42′   **NEW YORK**
   80   B   S4   **FUEL** 80, 100LL, JET A   OX 3, 4   LRA   **H−3J, L−25D, 28I**
   RWY 14−32: H5000X150 (ASPH)   S−33, D−48, DT−95   MIRL   **IAP**
     RWY 14: Bush.   **RWY 32:** REIL. VASI(V4L)—GA 3.0°TCH 52′. Trees.
   RWY 05−23: H4997X150 (ASPH)   S−30, D−108, DT−195   HIRL
     RWY 05: MALSR. Trees.   **RWY 23:** MALSR. VASI(V4L)—GA 3.0° TCH 51′. Thld dsplcd 413′. Trees.
   AIRPORT REMARKS: Attended 1100−0500Z‡. Arpt CLOSED to touch and go ldg and training 0300−1000Z‡ daily. Birds
     and deer on and invof arpt. VASI Rwys 23 and 32 ops 24 hours. When twr clsd ACTIVATE HIRL Rwy 05−23, MIRL
     Rwy 14−32, MALSR Rwy 05 and Rwy 23, REIL Rwy 32—CTAF. Flight Notification Service (ADCUS) available.
     NOTE: See Land and Hold Short Operations Section.
   WEATHER DATA SOURCES: ASOS 126.85 (1200−0300Z‡) (508) 992−0195. LAWRS.
   COMMUNICATIONS: CTAF 118.1   **ATIS** 126.85 508−994−6277. (1200−0300Z‡)   **UNICOM** 122.95
    BRIDGEPORT FSS (BDR) TP 1−800−WX−BRIEF. NOTAM FILE EWB.
  Ⓡ **PROVIDENCE APP/DEP CON** 128.7 (1100−0500Z‡)   **BOSTON CENTER APP/DEP CON** 124.85 (0500−1100Z‡)
    TOWER 118.1 (1200−0300Z‡)   **GND CON** 121.9
   AIRSPACE: CLASS D svc 1200−0300Z‡ other times CLASS G.
   RADIO AIDS TO NAVIGATION: NOTAM FILE PVD.
    PROVIDENCE (H) VORTACW 115.6   PVD   Chan 103   N41°43.46′ W71°25.78′   112° 21.4 NM to fld. 50/14W.
     HIWAS.
    NEFOR NDB (LOM) 274   EW   N41°37.30′ W71°01.06′   056° 4.3 NM to fld.
    ILS/DME 109.7   I−EWB   Chan 34   Rwy 05.   LOM NEFOR NDB. (ILS unmonitored when twr clsd. BC
     unusable beyond 15° either side of LOC centerline and beyond 12 NM). Back course DME unusable beyond 12
     NM.

Fig. 7-2. *Typical listing in the* Airport/Facilities Directory *for a remote communications outlet (RCO).*

TAXI CHECKS

Three important checklist items should be emphasized on an instrument flight.

Turn coordinator. How do the symbolic airplane and the ball move in a turn while taxiing? The airplane's "wings" tilt in the direction of the turn and the ball slides in the opposite direction to the outside of the turn. (On the older turn-and-slip indicator, the needle moves in the direction of the turn.)

The gyro of the turn needle is electrically powered. If the needle doesn't move, or moves erratically, it is either an instrument failure or an electrical failure. In either case, turn back. The airplane is not safe for an instrument flight.

Attitude indicator. The "wing" of the symbolic airplane should remain aligned with the horizon line. If the horizon display behind the symbolic wing pitches up or down or tilts beyond the slight movements seen during taxiing, it is an unreliable instrument. The gyros of the attitude indicator and the heading indicator are vacuum driven.

Cross-check the heading indicator. If it is drifting off heading, the vacuum system has probably failed. Check the suction gauge. If it is not within limits—usually 4.6–5.4" of mercury—turn back because the airplane is not safe for flight. If the suction gauge reads normally and the attitude indicator is not showing a normal display, turn back.

Heading indicator. A failure in the vacuum system will also affect the heading indicator. Headings shown on the heading indicator should change during a taxiing turn, then steady up to correctly match the magnetic compass on long, straight taxiways. Cross-check with the attitude indicator and suction gauge. If they show abnormal indications, it is a failure in the vacuum system. If the attitude indicator and suction gauge are normal and the heading indicator is erratic, it is an instrument failure in the heading indicator.

ROLLING ENGINE RUN-UP

Normally those are the only checks made while the airplane is rolling. You do not want to get distracted while taxiing, particularly at night. However, there is an important exception to this rule when heavy commercial traffic is sharing the taxiways. The pilots of those big jets will certainly get upset if you come to a full stop and block the traffic for an engine run-up and takeoff checklist.

Do the run-up and takeoff checklist while taxiing. But first, discuss this technique with your instructor and practice it a few times with the instructor aboard when the taxi traffic is light.

RUNWAY CHECKS

Six items are on the instrument runway checklist:

1. **Correct Approach Chart.** The departure airport's approach in use in case you have to return shortly after takeoff.
2. **Nav Radios Set.** Nav 1 for approach in use in case you have to return quickly; nav 2 for first airborne fix.

3. **Correct Departure Control Frequency.** Be ready for a change to departure control quickly and without fumbling when tower hands you off.

4. **Lights On.** All ON for night operations, strobes and rotating beacon on for day IFR flights.

5. **Transponder** ALT **or** ON **and Set.** The correct code as given in the clearance.

6. **Pitot Heat** ON. Even on a clear day get in the habit of doing this to avoid inadvertently overlooking this important safeguard when flying into visible moisture.

7. **Check Heading Indicator Against Runway Heading.** Sometimes the heading indicator will drift off while taxiing. Reset if necessary.

Just before adding power for takeoff, do an "STP" check: strobes, transponder, pitot heat.

And finally, note time of takeoff just before adding power. Takeoff time is doubly important on an instrument flight for fuel calculations and for planning the arrival in the event of lost communications.

IFR COMMUNICATIONS

Departure is the pilot's busiest time on an IFR flight: controlling the plane through a wide variety of situations, from parking and taxi through takeoff and into an instrument climb; transitioning between flight using outside visual references and flight by instruments; trying to maintain an efficient climb with turns to comply with the departure clearance.

You must be listening for the airplane's N-number among many others that tower might be working and be ready to respond quickly and correctly when tower has an amended clearance or issues a frequency change.

Reply to every ATC communication promptly. This is especially important during readback of an amended clearance. Read it back immediately; if something puzzles you, figure it out later. ATC wants to hear from you right now. If a clearance is totally incomprehensible, say "stand by for readback" then call back and read back the entire clearance even if only the next section of the instructions is figured out. Research the rest of the clearance later.

STANDARD PHRASEOLOGY

Use concise, standard phraseology and a professional tone of voice when working with ATC and the controllers will be inclined to assist as much as they possibly can. We've all heard transmissions like this: "This is Cessna November one two three four. I'm over (long pause), ah, Hartford. And I'm cruising at four thousand five hundred feet, departed from my home base at, ah (pause) ten o'clock on a VFR flight plan to New Jersey. Request permission to go through the New York TCA. Over."

Do you think ATC will show any enthusiasm for clearing this pilot through one of the country's busiest TCAs? Of course not. Time on the radio is very precious, especially around busy areas such as New York.

ATC is much more likely to grant a request if transmissions are brief and to the point. The more professional you sound, the more cooperation you will receive from ATC. This is especially true with IFR communications. Appendix C contains many standard IFR phrases that should always be used as indicated.

WHO, WHO, WHERE, WHAT

Each time you switch to a new controller the initial call sequence consists of who, who, where, and what:

- Who you are calling
- Who you are (always give the full N number on the first call)
- Where you are located (east ramp, outer marker inbound, etc.)
- What your request or message is

Use the full registration to avoid a possibility of confusing similar numbers, such as 62876 and 67876.

As a general rule abbreviate the call sign after the controller begins doing so, and use the controller's abbreviation. This speeds things up considerably. There is no need to say "November" for flights within the United States. ("N" is the international designation for aircraft registered in the United States.)

For example, if the full identification is N3456X, the initial call would be "Cessna three four five six Xray." If the controller came back with "Five six Xray turn right to zero niner zero," use "five six Xray" until switched to another controller.

If ATIS is available, be sure to copy the ATIS information before calling clearance delivery or ground control for taxi clearance (or approach control for an arrival clearance). Give ground control the code for the ATIS and tell them that you are IFR. Ground control needs to know this to obtain a "release" from departure control or the air route traffic control center. You cannot take off on an IFR flight until the controlling authority for the airspace issues a release to enter that airspace on takeoff.

CALLING GROUND CONTROL

The call to ground control might sound something like this:

"Westchester ground, Cessna three four five six Xray, terminal ramp, IFR Binghamton, information Delta. Ready to taxi."

Delta would be the code for the current ATIS information. It is changed every hour or sooner if there is a significant change in safety information. It is important, especially in IFR weather, to let the ground controller know that you have the latest information. If you don't, the ground controller will request verification. This results in extra transmissions that clutter up the frequency and wastes everybody's time.

Say it on the first transmission and avoid using the phrase "with the numbers" instead of the ATIS code. "With the numbers" tells the ground controller that you either didn't bother to obtain the ATIS information or that you got it and you could not remember the code.

You will remain on ground control until ready for takeoff, then ATC will switch you to tower. On the first call to the tower controller give the full call sign: "Westchester tower, Cessna three four five six Xray, ready for takeoff, IFR." Include "IFR" to alert the tower controller that the flight is IFR.

When switched to departure control after takeoff, the next call might be: "New York Departure, Cessna three four five six Xray, out of one thousand for three thousand [feet]."

Nearly every first contact with a new controller requires acknowledgment of altitude or the altitude passing through to an assigned altitude. This helps the controller verify that the actual altitude is the same as that reported by the Mode C transponder and shown on the controller's radarscope.

WHEN YOU HEAR NOTHING FURTHER

When one facility switches you over to another facility, it's a *handoff*. What if it comes time for a handoff—from tower to departure control, for example—and you receive no further instructions? Wait and hope for something to happen? Call someone? If so, who? And say what? This situation is not uncommon when the traffic is heavy. Sometimes the silence right after takeoff makes it feel like the tower has forgotten you.

If you have taken off and hear nothing further, continue climbing until at least 500 above the ground and established on departure heading. Then call the tower and ask: "Do you want four five Xray to go to departure control?" That will alert the tower to the fact that the handoff has not taken place. Don't just sit there and say "Well, they didn't ask me to go to departure control, so I'm not going to do it." As the pilot-in-command, get things straightened out.

MANAGING FREQUENCIES

Many experienced pilots keep a running list of the various frequencies used during a flight so that it is easy to return to the last frequency in case they or the controller makes a mistake in the next frequency. Another technique is to alternate between comm 1 and comm 2. Others like to work with only comm 1 while en route, leaving comm 2 tuned to the emergency frequency, 121.5 MHz, with the volume adjusted to a comfortable level.

If, for some reason, no one answers after switching to a newly assigned frequency, simply go back to the last assigned frequency, give the abbreviated call sign, and say "unable" with the name and frequency of the facility you couldn't contact. For example the transmission might be:

"Five six Xray unable Boston Center one three three point one."

ATC should respond to an "unable" message with further instructions.

REQUIRED REPORTS

A major difference between VFR and IFR communications is that many IFR situations *require* reports to ATC. Pilots must make the following reports at all times, whether or not in radar contact with ATC:

- When leaving an assigned altitude.
- When changing altitude on a "VFR on top" clearance.
- When unable to climb or descend at a rate of at least 500 feet per minute.
- On commencing a missed approach. The report must include a request for clearance to make another approach attempt or to proceed to another airport.
- When average true airspeed at cruising altitude varies by more than 5 percent or 10 knots.
- Upon reaching a holding fix. The report must include time and altitude.
- When leaving a holding pattern.
- If, in controlled airspace, VOR, ADF, or ILS equipment malfunctions, or there is any impairment of air-to-ground communications (in case ATC is able to receive).
- Any information relating to the safety of the flight.
- Any weather conditions that have not been forecast or when encountering hazardous conditions that have been forecast.

When a flight is not in radar contact (Such as when ATC says "Radar contact lost" or "Radar service terminated" or while flying "VFR on top"), pilots must make these additional reports:

- En route position reports upon reaching all compulsory reporting points. These are indicated on en route charts by a solid triangle. Position reports include:
  - Position
  - Time
  - Altitude
  - ETA and name of next reporting point
  - Name only of next succeeding reporting point

For example: "Cessna three four five six Xray, Carmel, one five, six thousand, Kingston two zero, Albany." You don't need to remember the sequence of items in the position report if the flight log form suggested in Chapter 3 is used. A reminder of the sequence is printed at the bottom of the left-hand page (FIG. 3-1).

- When leaving a final approach fix inbound on final approach.
- When it becomes apparent that a previously submitted estimate is in error in excess of 3 minutes.

CANCELING IFR

Although you will prepare and file IFR flight plans for all training flights in this syllabus, you will probably terminate the IFR portion of many flights to conduct training exercises under VFR. This is a simple procedure. If you are able to carry out the rest of the flight in

VFR conditions and wish to do so, simply tell ATC, "Cancel my IFR flight plan." Your request will be granted immediately—ATC will be glad to have one less airplane to control.

You, rather than ATC, will then become responsible for maintaining safe separation from other aircraft. As you work on the training exercises under the hood, the person in the other seat must act as safety pilot and must be qualified to do so.

There is one more thing to think about when canceling IFR. What kind of airspace are you in? You know you were in controlled airspace of some kind when you were operating IFR under ATC control. But what was it—B, C, D, or E airspace?

Each of these classifications limits the conduct of VFR flights in some respects, so be prepared to have ATC remain in contact with you and issue binding instructions if necessary. As part of your planning routine, make it a practice to determine what type of airspace you will encounter along your route of flight if you cancel IFR, as well as what restrictions apply. Consult AIM for details about VFR operations in each type of airspace.

Familiarity with the ATC system through actually using it builds confidence and competence and it is great practice!

RADIO CONTACT LOST

Anticipate a handoff when you have been with one controller for quite some time and the controller requests an ident. Occasionally you might lose radio contact when flying at low altitudes or in areas of high terrain, or a long distance from the transmitter site. Call for a radio check; if there is no response after two or three attempts, try to reestablish communications through another nearby frequency listed on the IFR charts, such as flight service, approach control, or even a tower. As a last resort use the emergency frequency 121.5 MHz.

Always consider phraseology before transmitting and then speak in the most professional manner possible. Always be brief and to the point. Listen for a break in other transmissions and key the microphone immediately when a break occurs. Quickly give aircraft N-number and a concise message on the first call. That will save precious radio time for everyone on that frequency.

8
Basic instruments

YOUR FIRST INSTRUMENT TRAINING FLIGHT WILL BE AN OVERVIEW of the conduct of an instrument flight—planning, weather, filing, departure, en route, approach, and landing—and it will explore what seems like an overwhelming amount of detail. However, a large amount of this detail is routine, procedural, and bears a close similarity to prior VFR planning.

By plunging right in at the beginning, you will soon find that even such seemingly "impossible" tasks as copying clearances will get easier and easier. This will foster attentive devotion to the basics of flying solely by reference to instruments, and to IFR instrument navigation, approaches, and the other important elements of instrument flight.

FIRST INSTRUMENT FLIGHT

When I take students up on their first instrument flight, I want them to be full participants in all elements of the flight. Of course I help the students when they find that something is a little too difficult to handle or the workload gets ahead of them.

But I want them to go through the motions throughout all phases of the flight—including a precision approach without the hood—so they can at least experience a visual picture of how it's done. It doesn't make any difference if the student makes mistakes because they are part of the learning process.

An amazing number of pilots find that they are able to take on a great share of the work load on this first training flight just by doing what comes naturally. This is especially true if the student has had good instruction in instrument fundamentals for the private pilot certificate.

However, the student must stay within the parameters required for instrument flight—within ±100 feet of assigned altitudes and within the confines of whatever victor airways we use. I'm not looking for expert instrument flight this first time out. The objective on the first training flight is an introduction to the whole process, not a polished performance.

It has always surprised me how poorly the fundamentals of instrument flight are sometimes taught. All too often the student is told to fly straight and level under the hood and get the hang of it. The instructor continually criticizes, reminding the student that the airplane is going off altitude or heading instead of emphasizing how altitude and heading are controlled.

The how-to is as important as the practice. Instructors should be able to show clearly how control is achieved and maintained. With the methods I have developed over the years, control of the airplane can be quickly fine-tuned to a level that is almost unbelievable.

"TWO, TWO, AND TWENTY"

From the very first flight I teach students to maintain heading within ±2°, airspeed within ±2 knots, and altitude within ±20 feet. This is necessary to become a member in good standing of the "2, 2, and 20 Club." This might sound corny, but it makes students think about goals to aim for right from the beginning.

The FAA allows tolerances on the instrument flight test of ±10° on heading, ±10 knots on airspeed, and ±100 feet on altitude. Stop and think about it; these are actually very large deviations—from one extreme to another, up to 20° on heading, 20 knots on airspeed, and 200 feet in altitude.

If a pilot is trained from the beginning to hold 2, 2, and 20, there will be no problem on the flight test. Even more important, the pilot will have better control of the airplane. If your personal maximum altitude deviation is 20 feet, you will use much smaller control adjustments to maintain it, compared to a 100-foot deviation limit. Larger tolerances invite larger corrections, and the larger the correction, the greater the tendency to overcontrol.

OVERCONTROLLING

Remember the problem from VFR training? Overcontrolling occurs when a pilot uses too large a change in attitude to make a correction: too much bank in a turn, nose too high or too low in a climb or descent, for example. Soon after the correction, the pilot corrects again in the other direction to avoid overshooting the heading or altitude. The flight of an overcontrolled airplane is a wobbling, bobbing affair in which it is impossible to hold headings and altitudes with any degree of accuracy.

Overcontrolling is, then, a problem of attitude control. And what is the best way to control attitude, avoid overcontrolling, and achieve the goal of 2, 2, and 20? Flight

Lesson 2 concentrates on the fundamentals of controlling the aircraft by—the answer—reference to instruments.

Instructor note. The student should be under the hood as much as possible from here on. Raise the hood from time to time to help the student compare instrument indications and visual references.

It is helpful to remember that there are only four basic maneuvers in flying: straight and level, turns, climbs, and descents. Every move made with an airplane is based upon these four maneuvers, or combinations of these four, such as a level turn.

ATTITUDE CONTROL

To make the aircraft perform these maneuvers the pilot must maintain or change the attitude. The attitude indicator was specifically developed as the basic reference for maintaining or changing attitude in instrument flight. The other flight instruments are vital, of course, and lifesavers if the attitude indicator does fail. But for normal IFR flying, the attitude indicator is the star player, and the other flight instruments are the supporting cast.

Straight and level

I have found it best to work first on straight and level flight. Most of the time an instrument cross-country is in straight and level flight. And, more often than not, the other basic maneuvers begin and end with a return to straight and level.

This basic maneuver is divided into two elements: altitude control and heading control. It seems to work better to practice altitude control first, then work on heading control.

Choose an odd or even thousand-foot altitude—such as 3,000, 4,000, or 5,000 feet—for practicing straight and level. It's interesting to note in passing that on an IFR flight, most of the time you will be flying at an odd or even thousand foot altitude with the long hand of the altimeter pointing straight up to zero.

Set the power for cruise. Trim the airplane. Trim is important because the closer you can trim the airplane to fly "hands off," the easier it will be to make the small corrections that are so vital to precise instrument flight.

ALTITUDE CONTROL

In level flight, the representative wings on the attitude indicator should superimpose the horizon line to form one line (FIG. 8-1). If it doesn't look like this in level flight, reset the attitude indicator.

Try maintaining altitude within ±20 feet for 20 seconds with reference to the attitude indicator primarily, using the altimeter as a cross-check. If altitude drops, use slight back pressure on the control yoke to pitch the nose up slightly. Raise the wings of the miniature airplane to one-half bar width above the horizon line, keeping the wings level (FIG. 8-2). (Bar width is the thickness of the wings of the miniature airplane represented on the attitude indicator.) When you get back to the desired altitude, relax the back pressure and align the wings with the horizon line.

Fig. 8-1. *In straight and level flight, the miniature airplane on the attitude indicator should be set so that the top of the "wings" make a straight line from left to right with the horizon line.*

Fig. 8-2. *When below a prescribed altitude by 100 feet or less, correct by increasing pitch one-half bar width.*

If altitude increases, apply a slight forward *pressure* on the yoke to pitch the nose down slightly. Lower the miniature airplane to one-half bar width below the horizon line, keeping the wings level (FIG. 8-3). Cross-check with the altimeter. Level off at the desired altitude. Relax the forward pressure on the yoke and align the wings with the horizon line again. *Pressure* (not movement) is the key to smooth control.

Step climbs and descents

The "step" climb and descent is particularly good for developing an automatic reflex for making altitude corrections. The maneuver is very helpful to the student trying to reach the "20" part of the 2, 2, and 20 goal climbs and descents and controlling altitude. Figure 8-4 shows the maneuver from a starting altitude of 3,000 feet.

Instructor note. The step climb and descent is very useful for another reason: It can be introduced early in the syllabus without overloading the student.

After a student has demonstrated an ability to hold some convenient altitude consistently for two or three minutes on a constant heading, practice the step maneuver next—climb 100 feet, level off, and hold that altitude for one minute. The miniature airplane on the attitude indicator is held one-half bar width above the horizon line with back pressure alone (FIG. 8-2). No power changes are needed for a change of less than 100 feet.

After the minute is up, descend to the original altitude, level off, and hold that for one minute. The miniature airplane should be pitched one-half bar width below the horizon line for the descent (FIG. 8-3). Again, there should be no change in the power setting.

One-half bar width on the attitude indicator is a very small adjustment, but with a little practice, it can be easily achieved every time. One-half bar width is also an excellent rule of thumb: For altitude changes of 100 feet or less, use no more than a one-half bar width correction. Remember: *No power changes.*

The remainder of the exercise consists of one more descent and one more climb performed the same way—establishing pitch with the attitude indicator and making no power changes. The wings should remain level according to the attitude indicator.

The exercise should be continued until altitude can be controlled within ±20 feet of what is desired. Students who master step climbs and descents usually have little difficulty with altitude changes of more than 100 feet.

CONTROL, PRIMARY, AND SUPPORT INSTRUMENTS

The attitude indicator provides pitch information during the step maneuver; it is the central control instrument or simply the *control instrument.*

Fig. 8-3. *When above a prescribed altitude by 100 feet or less, correct by decreasing pitch one-half bar width.*

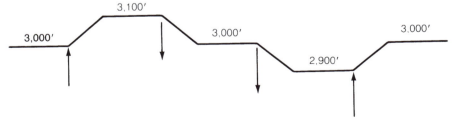

Fig. 8-4. *Step climbs and descents for practicing altitude adjustments.*

Primary information on the quality of control comes from the altimeter. It shows a direct, almost instantaneous reading on whether or not the quality is good enough to stay within ±20 feet.

Here's a good way to determine the primary instrument: It indicates the most pertinent information about how well you are doing and *does not move* when flying precisely. The altimeter is the *primary instrument* in maintaining level flight. Does it meet the test? You bet it does. It indicates the most pertinent information regarding altitude. And if you're doing a good job, the altimeter will not move.

The vertical speed indicator (VSI) has neither a control nor a primary role maintaining level flight. But it is an additional source of information on the rate of climb or descent and may be used for cross-checking. The VSI is thus a *support instrument* during straight and level flight.

Figure 8-5 shows the primary, support, and control instruments for all phases of instrument flight. Note how the instruments just discussed are used to control pitch in level flight. I will classify instruments as control, primary, and support when discussing other instrument flight maneuvers. Refer to this chart from time to time to better understand the roles these instruments play during each phase of flight. The primary instruments will have added importance later during partial panel practice because they become control instruments during simulated failure of the vacuum-driven attitude and heading indicators.

If you can maintain altitude within ±20 feet for 20 seconds, we'll try it for 40 seconds, then carry it on to the next step, one minute. When you get a good grasp of altitude control with this method I will cover up the altimeter, wait 30 seconds or so, then uncover it to see how far up or down the airplane has drifted. With a little practice you will be amazed at how well you can fly without looking at the altimeter. This also serves as a painless little introduction to partial panel work, which we will take up on later flights.

Instructor note. Once again it is a good idea to raise the hood every now and then so the student can compare what is seen on the instruments with the real horizon and other visual references. For example, when the nose is pitched up on the attitude indicator, raise the hood and show that the nose is also pitched up with reference to the actual horizon. This sounds obvious but it's not obvious to the beginning instrument student unless actually confirmed. The beginning instrument student needs all available help at this point.

Before moving on to turns, climbs, and descents, let's work on straight and level until it becomes automatic. Concentrating on one element at a time, you will succeed in learning very quickly to work within tolerances of 2, 2, and 20. The control pressures necessary to maintain altitude in straight and level flight are minimal, the tendency to overcontrol is much reduced, and even the heading will remain relatively constant because wings are level. And best of all, you will soon do this without hardly thinking about it!

Most readers recall getting a bicycle for a birthday or Christmas. Within the first week or two you probably wore the skin off your knees learning to ride by trial and error. Then after two weeks, suddenly the secret of balancing came and you were in full control. You could command that bicycle to turn left or right merely by leaning or shifting weight on the seat of the bicycle. And it wasn't long before you were riding along gleefully in front of friends with arms folded, just leaning left and right to make turns.

	CONTROL	PRIMARY	SUPPORT
STRAIGHT and LEVEL			
Pitch	Attitude Indicator	Altimeter	VSI (Rate of Climb)
Bank	Attitude Indicator	Heading Indicator	Turn Coordinator
Power		Airspeed	RPM/MP
SPEED CHANGES			
Pitch	Attitude Indicator	Altimeter	VSI
Bank	Attitude Indicator	Heading Indicator	Turn Coordinator
Power		Airspeed	RPM/MP
STANDARD RATE TURN			
Pitch	Attitude Indicator	Altimeter	VSI
Bank	Attitude Indicator	Turn Coordinator	Sweep Second Hand
Power		Airspeed	RPM/MP
MINIMUM CONTROLLABLE AIRSPEED			
Pitch	Attitude Indicator	Airspeed	RPM/MP
Bank	Attitude Indicator	Heading Indicator	Turn Coordinator
Power		Altimeter	VSI
CLIMB ENTRY			
Pitch	Attitude Indicator	Attitude Indicator	VSI
Bank	Attitude Indicator	Heading Indicator	Turn Coordinator
Power		RPM/MP	
CONSTANT AIRSPEED CLIMB			
Pitch	Attitude Indicator	Airspeed	VSI
Bank	Attitude Indicator	Heading Indicator	Turn Coordinator
Power		RPM/MP	
CONSTANT RATE CLIMB			
Pitch	Attitude Indicator	VSI	Altimeter, Sweep Second Hand
Bank	Attitude Indicator	Heading Indicator	Turn Coordinator
Power		Airspeed	RPM/MP
LEVEL OFF (to cruise from climb or descent)			
Pitch	Attitude Indicator	Altimeter	VSI
Bank	Attitude Indicator	Heading Indicator	Turn Coordinator
Power		Airspeed	RPM/MP

(Continued on page 92.)

CONTROL   =   Main reference instrument
PRIMARY   =   Key quality instrument*
SUPPORT   =   Back-up or secondary instrument

    Variable Power (cruise, etc.) - "Power to the speed, pitch to the altitude."
    Constant Power (min. controllable airspeed, etc.) - "Pitch to the speed, power to the altitude."
    Power + Attitude = Performance
    *The Primary Instrument is always the instrument that gives the most pertinent information and is not moving when flying precisely.

Fig. 8-5. *Control, primary, and support instruments for all the basic regimes of flight.*

(Continued from page 91.)

CONTROL		PRIMARY	SUPPORT
CONSTANT AIRSPEED DESCENT			
Pitch	Attitude Indicator	Airspeed	VSI
Bank	Attitude Indicator	Heading Indicator	Turn Coordinator
Power		RPM/M	
CONSTANT RATE DESCENT			
Pitch	Attitude Indicator	VSI	Altimeter Sweep Second Hand
Bank	Attitude Indicator	Heading Indicator	Turn Coordinator
Power		Airspeed	RPM/MP

Fig. 8-5. (*Continued*)

Turning had become an automatic reflex—not reacting, but acting. That's the secret of straight and level flight.

Students develop an automatic reflex so that if the long needle of the altimeter is to the left of the big zero, low, apply back pressure to move the miniature airplane one-half bar width above the horizon line. And if the long needle is to the right, apply forward pressure to lower the miniature airplane one bar width below the horizon line.

I practice this with students until they are making corrections automatically for sudden updrafts and downdrafts without pausing to analyze what needs to be done. Then it will be time to move on to heading control. This isn't the end of altitude control. I'll have much more to say when considering climbs and descents to accomplish altitude changes.

HEADING CONTROL

As an expert at maintaining altitude in straight and level flight, it's time to exercise the same finesse to control and maintain heading. Set up straight and level on a convenient heading. With wings level on the attitude indicator, apply control pressure equally and very gently to both rudder pedals. It's not necessary to be heavy-footed; just keep both feet in contact with both pedals and balance the pressure.

Constant heading with wings level is maintained by small adjustments in the balance of pressure on the rudder pedals. Without constant pressure on the rudder pedals, the airplane will drift off heading if it yaws due to poorly adjusted rigging or constantly changing air currents.

Make minor adjustments in the pressure on the rudder pedals to compensate for heading changes caused by these factors. For heading adjustments of 5° or fewer use rudder pressure only, holding wings level; flight will be uncoordinated momentarily, but you will avoid the tendency to overcontrol.

Overcontrolling is an impulse that all students seem to have and they need to work consciously to avoid it. Think about it. If you want to make a 2° heading change and you roll in 20° of bank, what happens? Suddenly the heading has changed 20° or 30°, and you are powerfully tempted to roll into a steep bank in the opposite direction.

As the plane swings back and forth through the sky, control deteriorates and precise flight becomes impossible. When you were a student pilot, your instructor probably took over at this point and steadied the plane so the lesson could continue. But now you are expected to reestablish control without a helping hand. Or much better still, use very small control inputs to avoid overcontrolling altogether.

If the plane tends to drift left with wings level, increase the pressure on the right pedal. If it drifts right, increase the pressure on the left pedal. In other words, "step on the ball" to make minor corrections with rudder pedals. As learned in VFR training, apply rudder pressure in the direction the ball has moved to return to coordinated flight.

If the heading has drifted off more than 5°, make a coordinated turn to adjust the heading using an angle of bank that is *no more than half* the desired degrees of heading change. If you need to adjust the heading 10°, for example, put in an angle of bank of no more than 5°. This will help avoid overcontrolling.

Airplane rigging and shifting air currents are common causes of heading drift, but the major culprit is banking. The attitude indicator, once again, is the control instrument for keeping the wings level. If you maintain a carefully adjusted pressure on the rudder pedals, the wings will stay level almost automatically. The heading indicator is the primary instrument that gives a direct, almost instantaneous quality reading of efforts to maintain level flight because every change in bank will result in a heading change.

The supporting instrument for level flight is the turn coordinator. The turn needle is extremely sensitive and will often show evidence of a bank that is almost imperceptible on the attitude indicator. The ball must remain centered with rudder pressure in level flight or else the airplane will yaw and drift off heading.

SCAN

An efficient instrument scanning technique will develop while practicing altitude and heading control in straight and level flight by means of control, primary, and support instruments—it happens almost automatically. The scan develops even further to include the support instruments.

Students often ask what is the best way to scan an instrument panel: up and down, left to right, clockwise or counterclockwise? It doesn't really make much difference as long as you adhere to two very important principles:

- Don't fixate on one instrument
- Always return to the attitude indicator, the control instrument, after checking each of the other instruments

Teaching scan

I have found that an ideal method of teaching the scan is *attitude, heading, altitude.* Repeat these three words while flying to help guide your scan.

Look first at the attitude indicator to determine if any attitude corrections are required. Next, check the heading indicator to see if any drift in heading has occurred. Glance at the altimeter to determine if any correction in altitude control is required. Go back to the attitude indicator and make any adjustments dictated by the glance at heading and altitude.

Repeat this process continually throughout the flight. On every 10th scan, include the VOR or the ADF, depending on which is used for navigation. Every couple of minutes, include all the instruments in your scan, including engine gauges. Adopt a system and stick with it.

DISTRACTIONS

The human eye is constructed to immediately respond to any movement it picks up. This is a well-known trait from ancient forebears who had to respond quickly to movement of any kind in their environment; movement either indicated the presence of something that they would eat or the presence of something that would eat them.

The implication for the modern instrument pilot is that we descendants still tend to respond with greatest interest to something that moves. In bumpy weather our attention tends to become fixed on the oscillations of the turn needle or the rapid up-and-down movement of the VSI needle instead of the attitude indicator. Our attention is riveted on the wanderings of the VOR needle near a station while neglecting heading and altitude.

Make an effort to tear your eyes away from an instrument that is showing rapidly changing indications and methodically scan from one instrument to another in whatever sequence is most comfortable.

The control instrument for all phases of instrument flight is the attitude indicator (FIG. 8-5). Center the scan pattern on the attitude indicator to maintain control of the airplane on instruments. When the eyes move away from the flight instruments—tuning a new communication or navigation frequency—always return to the attitude indicator to resume the scan.

You will quickly see small changes in attitude develop and make the small automatic adjustments of yoke and rudder pedals that are so important in smoothly controlling the airplane.

SUPPORT INSTRUMENTS

Now, let's look at the role of support instruments, which serve two purposes. First, they might be required to assume the role of *control* instruments in case of mechanical failure. If the gyro in the attitude indicator fails, the airspeed indicator can now substitute for the attitude indicator, supported by the VSI. (At a certain power setting in level flight you know what the airspeed should be—a 5-knot increase in airspeed would indicate a slight descent and vice versa.) The ADF needle (tuned to a strong standard broadcast station) in combination with the magnetic compass could substitute for the heading indicator, and so on.

The second purpose of the support instruments is to stabilize flight and minimize overcontrolling and to detect instrument malfunctions. If the airspeed remains steady on the indicator, for example, and the VSI and altimeter both show a rapid descent, you can conclude that the airspeed indicator is malfunctioning.

AIRSPEED CONTROL

Achievement of precise altitude and heading for longer and longer periods in straight and level flight leads to precise airspeed control. Recall from VFR flying that to increase airspeed in straight and level flight it is necessary to increase power—to decrease airspeed, reduce power. The same principles apply to IFR flying, of course, but the principles are applied more precisely than before.

Precision will demand very small changes in power settings—increments of 100 RPM or 1" of manifold pressure in a high-performance airplane, unless a specific power setting is required. Light control pressure makes it easier to hold altitude and heading.

Likewise, small power changes will also make it easier to control the airplane within the goal of 2, 2, and 20. Large changes in power, or "throttle jockeying," is a form of overcontrolling and must be eliminated or else precise instrument flight cannot be attained.

A power change of 100 RPM or 1" of manifold pressure yields a change in airspeed of $7^{1}/_{2}$ knots. Strangely enough, this rough rule of thumb can be applied to any propeller-driven airplane you are likely to fly, Cessna 152, Cherokee Arrow, Beech Baron, or whatever. For example, if cruising at 100 knots and you want to reduce the airspeed to 95 knots, the 5-knot change would equate to a reduction of 75 RPM or $^{3}/_{4}$" manifold pressure. To increase or decrease airspeed by 10 knots, you would increase or decrease RPMs by 125 or change manifold pressure by $1^{1}/_{4}$". Remember that this rule of thumb is not set in concrete. Minor adjustments are almost always necessary. It is something to start with. (I will offer many rules of thumb throughout this book; they are valuable guides to precision IFR flight and you won't have to waste time reinventing them—don't waste time trying to rediscover old knowledge.)

AIRSPEED TRANSITIONS

Let's take our discussion of airspeed control one step further and talk about airspeed transitions—how to change from one airspeed to another with facility and precision.

First, establish a specific speed for each flight condition for the type of airplane flown. For a Cessna 172 in level flight, for example, consider 110 knots as normal cruise, 90 knots as slow cruise, and 70 knots as slow flight.

Next, establish the power settings to maintain these selected speeds in a Cessna 172. Maintain 110 knots with 2450 RPM, 90 knots with 2100 RPM, and 70 knots with 1900 RPM.

To slow down from 110 to 90 knots, for example, reduce power to the setting established for that speed, 2100 RPM.

Maintain level flight by slowly raising the nose of the miniature airplane on the attitude indicator, with wings level. Increase back pressure on the yoke as the airspeed bleeds off, and trim out the pressure as the desired airspeed approaches.

Cross-check with the primary instruments: altimeter for the best information on pitch, heading indicator for bank, and airspeed for power. Include the support instruments in your scan: VSI for pitch, turn coordinator for bank, and tachometer for power.

To speed up and resume 110 knots, reverse the procedure. Increase power to 2450 RPM and maintain level flight by slowly lowering the nose with reference to the horizon line on the attitude indicator. Increase forward pressure on the yoke, then trim out the pressure near 110 knots. Cross-check the same primary and support instruments.

Power changes induce airspeed changes, while adjusting attitude changes altitude. In straight and level cruising flight, it's "power to the speed, pitch to the altitude." You'll hear instructors say this over and over as you make airspeed changes in level flight.

POWER

There are only two basic power conditions for flight: variable (adjustable) and not variable (either by choice or by accident).

Variable power is used in level flight to control airspeed (power to the speed, pitch to the altitude). Variable power is also used in climbs and descents at a specified rate, such as 500 feet per minute, and on the glide path of an ILS approach. You will see how this works later when these maneuvers are analyzed.

The conditions under which power is not variable occur when using full throttle during takeoff, when the throttle is closed or the engine fails, or when the power is in transit during a transition from one maneuver to another, such as intercepting a glide slope.

In these situations remember "pitch to the speed, power to the altitude" from VFR training and flight at minimum controllable airspeed.

INSTRUMENTS THAT LIE—WHEN AND WHY

Murphy had a great gift for simplifying technical problems in an unforgettable way. The first law: "If a part can be installed wrong, it will be." Here is Murphy's law as it applies to instrument flying: "If an instrument can fail, it will."

Here, in summary form, is a list of instruments and problems to prepare for.

Attitude indicator. This instrument is driven by gyros powered by the vacuum system. The indicator fails when there is a failure in the vacuum system, the result, usually, of an engine-driven vacuum pump failure. Failure of the attitude indicator shows up gradually. The instrument doesn't just roll over and die but begins to drift off slowly at first as its gyros wind down. For a good idea of what this looks like, watch the way an attitude indicator behaves after engine shutdown at the end of a flight.

If you have trouble maintaining straight and level, stabilized climbs and descents, or smooth standard rate turns, suspect failure of the attitude indicator. Cover it up to prevent scan distraction and switch to partial panel operation, which is explained in Chapter 12.

The attitude indicator might also show erroneous information if it has been set incorrectly. There is only one situation in which the attitude indicator can be set correctly. And that is straight and level unaccelerated flight. That is the only time the miniature airplane can be accurately matched to the horizon line.

If you reset the attitude indicator on the flight line, it will be incorrect due to the nose-up attitude of the parked airplane. Other errors will occur if a reset is attempted in turns, climbs, and descents.

Heading indicator. Like the attitude indicator, the heading indicator is driven by the vacuum system. It will fail if the vacuum system fails, and it can also fail when the vacuum system is operating normally. Like the attitude indicator, a failure of the heading indicator is rarely dramatic. If you suspect a failure, cover the instrument and switch to partial panel operation.

Make sure that the heading indicator shows the same heading as the magnetic compass. Students make three common mistakes when setting the heading indicator.

First, they fail to reset the heading indicator when lined up on the runway centerline for takeoff. This is the best time to reset a heading indicator because you know the runway heading. If it's an instrument runway, the heading is often shown to the last degree on an approach chart instead of rounded to the nearest 10° interval.

Second, because of precession the heading indicator slowly drifts off. It should be reset every 15 minutes, or after maneuvers that involve a lot of turns in a short time, such as holding patterns.

Third, the heading indicator cannot be reset accurately unless the airplane is in straight and level, unaccelerated flight. The magnetic compass is the culprit here. It is accurate only when stabilized in straight and level, unaccelerated flight. (Chapter 12 details how to cope with magnetic compass errors.) Meanwhile, resist the temptation to reset the heading indicator in a turn. I have seen students make errors of as much as 30° while trying to match heading indicator and magnetic compass in a turn.

Magnetic compass. Chapter 12 has more about the magnetic compass, but let's review a few points learned in VFR training:

- In calculating headings, account for *variation* due to the earth's magnetic field, and *deviation* due to magnetic influences on a specific compass because of its location in the airplane.

- Turning toward the north, the compass *lags* behind the turn due to dip error, turning toward the south, the compass *leads* the turn.

- On easterly and westerly headings, acceleration produces an indication to the north, deceleration produces an indication to the south. Remember ANDS (Acceleration North Deceleration South).

Altimeter. The altimeter will read erroneously if not set to the correct barometric pressure at all times. It will also read erroneously if the static port is clogged. Insects, ice, and dirt can clog a static port. The problem becomes apparent when airborne and the altimeter needles don't move.

An alternate static source aboard an airplane can restore the altimeter to normal operation if the static port has become clogged: however, the altimeter will read *higher* than normal.

If you do not have an alternate static source aboard, create one by breaking the glass face of the VSI. This vents the pitot-static system to the cabin, the same as an alternate static source. Again, the altimeter will read higher than normal. (Breaking the glass face of the VSI usually damages the needle and renders the instrument inoperative.)

Airspeed indicator. A blockage of the pitot tube will render the airspeed indicator useless. As is the case with the altimeter, you won't know this until airborne. Pitot heat will prevent ice from clogging the pitot tube. That's why I recommend turning on the pitot heat before takeoff on every instrument flight. When ice has clogged the tube, pitot heat might melt it too slowly. (This is developing good habit patterns.)

Insects love to nest in the pitot tube; use the pitot tube cover. Insects can clog those tiny air passages in just a few minutes.

Blockage of the static source also causes erroneous airspeed readings. An alternate static source will produce an indicated airspeed a knot or two faster than normal.

Vertical speed indicator (VSI). The most important thing to remember about the VSI is that it only gives an accurate reading when the needle has been stabilized for 7 seconds or longer. If the needle is moving, forget it.

Another quirk of the VSI is that when you first raise the nose of the airplane to begin a climb, the VSI needle initially shows a descent. The reverse is true in the beginning of a descent when the needle will momentarily show a climb.

The needle of the VSI should point to zero when the aircraft is sitting on the ground. If not, the needle can usually be zeroed by turning a small screw at the lower left corner of the instrument case. If this adjustment cannot be made, add or subtract the error for an accurate reading in flight.

Turn coordinator. The turn coordinator is powered by electricity; it will continue to operate even if there is a failure of the vacuum-powered attitude indicator and heading indicator. If the turn coordinator fails, the needle won't move. It will remain fixed in an upright position. That's why it's important to check the movement of the turn coordinator while taxiing out. It's pretty rare, but I have also seen the ball of the turn coordinator get stuck in the tube.

Fuel gauges. Here's another Murphy's law: "On land, air, and sea, the second half of the tank always empties faster than the first half."

Never trust any fuel gauges. There is no way to judge how accurate they are and most of them are fairly crude. Always note takeoff time and the time en route from each major position fix, then calculate fuel consumption based upon airplane performance figures. Ask "What if my fuel gauges failed completely? Am I keeping track of fuel calculations well enough independently of the gauges to know exactly how much more flying time I have left?" (It is *time* in your tanks.)

Oil pressure and temperature gauges. Engine instruments should be scanned every few minutes. The main concerns are low oil pressure and rising oil temperature. When these symptoms appear, a serious problem is developing in the engine oil system. Land the airplane as soon as possible. Don't stop to think about whether or not the gauges are functioning properly.

Low oil pressure with no rising oil temperature indicates either an instrument error or an incorrectly set pressure relief valve. Keep an eye on this situation. As long as the indications remain stable, the flight can be continued. But if the oil pressure drops and the oil temperature rises, land.

High oil pressure with normal oil temperature usually means that the pressure relief valve has been set incorrectly. As long as engine indications remain normal there is no reason to discontinue the flight.

9
Turns, climbs, and descents

A S A VFR PILOT YOU BECAME ACCUSTOMED TO FLYING STABLE, medium-banked turns. The basic turn used for instrument flying—a standard rate turn—is much milder, requiring about 15° of bank in a Cessna 172 at cruising speed. It is easy to execute, doesn't require a lot of attention during the turn, and altitude control is less demanding than a steeper turn.

Overcontrolling is a common problem. The habit of setting up that steeper VFR medium-banked turn could be so deeply ingrained that you might bank too steeply on the first practice standard rate turns. If you bank too steeply, the rate of turn increases rapidly. Then you have to reduce the angle of bank to slow everything down. And by that time the altitude is way off. Take it easy!

Standard rate turns are precision turns that produce a heading change of exactly 3° per second. If the maneuver starts out on a heading of 360°, then becomes a standard rate turn to the right, the heading in 30 seconds will be 090°, in one minute 180°, and so forth.

Enter the turn by using the attitude indicator to establish the angle of bank for a turn rate of 3° per second. The formula for the degree of bank is:

Airspeed/10 + 5 = bank angle for a standard rate turn.

Example: Airspeed is 125 knots. Divide by 10 and add 5 to the result, which equals a 17.5° bank angle for a standard rate turn.

TURN COORDINATOR AND CLOCK

The turn coordinator is the primary, or quality, instrument. In a standard rate turn, the symbolic airplane tilts and its wings line up with left and right benchmarks (FIG. 9-1). The ball remains centered throughout the turn with proper rudder input. (In standard rate turns with an older turn and slip indicator, the needle is displaced one needle width in the direction of the turn.)

The sweep-second hand of the clock may be included in the scan during standard rate turns. It is an invaluable support instrument because it shows whether you're turning faster or slower than 3° per second. No matter where the second hand is when you commence a turn, it should make half a sweep for every 90° of turn, and a complete sweep for every 180° of turn.

If you have turned 90° and the second hand is more than halfway around the clock face, you know the rate of turn is too slow; speed up the rate of turn by increasing the angle of bank slightly. Likewise, if you have turned 90° and the second hand has not

Fig. 9-1. *Standard rate turn to the left at 100 knots. The wings of the miniature airplane match the benchmark on the turn coordinator. But the angle of bank falls between marks on the attitude indicator because the airspeed is 100 knots.*

reached halfway, the rate of turn is too fast. Decrease the rate of turn by decreasing the angle of bank slightly. Be careful, however, not to fixate on the clock. That steadily advancing sweep-second hand is a powerful attention-getter. Make a conscious effort to direct attention away from the clock and return to your normal scan, resuming, as usual, with the attitude indicator.

Practice level standard rate turns in both directions for a full 360°, then try turns of 180° and 90°. To hit the rollout heading exactly, anticipate the target heading by one-half the degrees of bank, just like VFR turns to a heading. If the angle of bank in the turn is 20°, for example, begin the roll out 10° before reaching the target heading. A 10° angle of bank requires a lead of 5°.

TRIM IN A TURN

A level turn requires back pressure on the control wheel to maintain altitude. Students frequently ask me whether or not they should apply nose-up trim during a turn to reduce the back pressure needed to keep the nose up. This is an individual matter and there is no strict rule for trimming. I think you will find that a little extra back pressure on the yoke during the turn will work very well in counteracting the tendency of the nose to descend. But it's really up to you; find the method that works best for you and stick with it.

Cross-check attitude indicator, altimeter, and VSI to make sure you are using the correct amount of back pressure to maintain constant altitude. In a full 360° turn, which requires two minutes, it *is* helpful to add a quarter of a turn of nose-up trim. Remember that this trim will have to be removed after completing the turn or the nose will rise.

OBOE PATTERN

Practice rolling smoothly out of a standard rate turn in one direction, without pausing, directly into the opposite direction. Here I would like to introduce the Oboe pattern, which is designed to help perfect standard rate turn technique, plus rolling from one direction to another (FIG. 9-2).

Start from straight and level flight on a cardinal heading with the sweep-second hand of the clock approaching 12 o'clock. Begin a standard rate turn to the left and continue the turn for 360° (two minutes). Include the clock in the scan so that you can adjust the rate of turn according to whether it is slow or fast passing the 6 and 12 o'clock positions.

The original cardinal heading becomes the "reversal" heading. Use the "one-half the angle of bank" rule of thumb to anticipate the reversal and turn to a standard rate turn in the opposite direction.

Don't pause in straight and level flight at the reversal heading; roll smoothly into the turn in the opposite direction. The nose will tend to pitch up during the maneuver. Prepare for this and apply forward pressure on the yoke. The attitude indicator is a big help during the turn reversal; just keep the dot between the miniature wings aimed right at the horizon line and your altitude will remain constant. Think in terms of riveting that dot on the horizon and rolling around it during the reversal.

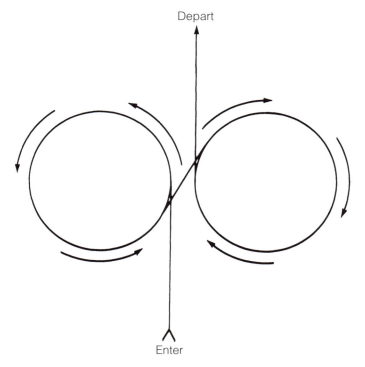

Depart

Enter

Fig. 9-2. *Oboe pattern. Complete the first 360° turn and roll right into the second 360° turn without pausing when wings are level.*

I think you will enjoy flying the Oboe pattern, especially if you feel a bump from wake turbulence indicating a perfect circle as you come around to the original cardinal heading. Do the pattern with both left and right entries. The Oboe will reappear when it is time to practice climbing and descending turns and partial panel procedures.

PATTERN A

It's time to assemble all the elements covered so far: straight and level, speed changes, and standard rate turns. A good exercise for this is Pattern A, shown in FIG. 9-3. It also contains all the maneuvers required for a full-scale instrument approach, except for the descents. You will fly this pattern, or portions of it, on every instrument approach. Make a copy of Pattern A and attach it to your clipboard as a ready reference.

Set up the exercise in straight and level flight at normal cruise speed. Start the pattern on 360° the first few times you try it. As soon as you are comfortable with the pattern, vary the initial headings and start on 090°, 180°, or 270°. This will be good practice to learn which rollout headings will be with different initial headings.

Start timing at the beginning of the exercise with the sweep-second hand at 12 o'clock. Time each leg consecutively; each new leg starts when the time for the old leg has expired and control pressure is applied to adjust for the new leg.

The Oboe pattern and Pattern A are excellent exercises to practice under the hood without an instructor, but with a safety pilot in the right seat (Appendix B: FAR 91.109 (b)(1)). Don't continue practicing these patterns if problems develop. It might be frustrating and you might unknowingly develop bad habits in scanning or procedures. As any instructor will tell you, it takes a lot of extra time and effort to break bad habits. Try again another day with an instructor.

MINIMUM CONTROLLABLE AIRSPEED

I like to introduce minimum controllable airspeed under the hood early in the training syllabus. This surprises a lot of students. They are just beginning to master straight and level flight under the hood and I ask them to try flight at minimum controllable airspeed!

Introducing minimum controllable airspeed early serves several important purposes. First, it will improve your handling ability on instruments because you will learn to control

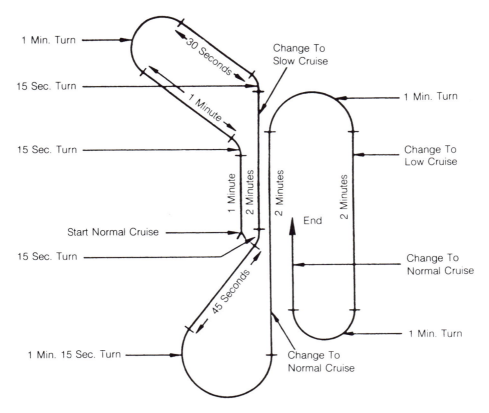

Fig. 9-3. *Pattern A practice maneuver.*

the aircraft throughout a wide range of pitch and power changes. And you will learn to anticipate the major changes in control pressures that accompany these transitions. A student who learns to handle the plane well in minimum controllable airspeed under the hood, always finds it easier to reach the goal of 2, 2, 20 when flying straight and level and other fundamental maneuvers.

Second, the student learns to extend and retract flaps (and landing gear if so equipped) while under the hood. With very little practice extending and retracting flaps and gear on instruments becomes an automatic reflex. This is very important on instrument approaches. The last thing you want to happen while descending on final approach is to break your scan while fumbling with flaps and landing gear.

Simulation of missed approach

The third reason for practicing minimum controllable airspeed under the hood also has to do with instrument approaches. Imagine this situation: You have flown a perfect approach down to the ceiling minimum and the runway is not visible; it's time to execute a missed approach. The airplane is low and slow, the nose is pitched up, and flaps are extended. To stop the descent, you must add full power, level off, and start to retract the flaps.

The airplane might be as low as 200 feet above the runway. You must be in absolute, positive control of the airplane to increase airspeed rapidly and establish a climb to avoid obstructions and put some distance between yourself and the ground. The transitions during recovery from minimum controllable airspeed are identical to those during a missed approach: full power, increase speed, retract flaps, maintain altitude.

Missed approaches will be introduced and practiced in later lessons. Minimum controllable airspeed provides several fundamental building blocks that must be mastered early in the course to become a precise instrument pilot who can handle all maneuvers with absolute safety. The best way to master minimum controllable airspeed on instruments is to think that it is an extension of the second fundamental item, speed changes, except that the changes are carried to greater limits.

Entering the maneuver

Set up the maneuver from straight and level flight on a convenient cardinal heading. Start reducing power; for the Cessna 172 or Piper Cherokee 180, you will find that reducing power to 1500 RPM will be adequate. As the nose starts to feel heavy, and the airspeed slows to the flap extension range, apply the first notch of flaps, or 10°. As airspeed decreases further, lower the flaps another notch. Adjust the trim as necessary throughout the maneuver. As the nose gets heavy again, extend full flaps. Do not attempt to drop full flaps all at once because the changes in attitude and control pressures will be so abrupt and heavy that you will have difficulty controlling the airplane by reference to instruments.

As minimum controllable airspeed approaches, the instructor will request power applications (perhaps full power) to maintain altitude at the lowest possible speed without stalling. Aim for the airspeed at which altitude control becomes difficult. That will be the minimum controllable airspeed for this exercise and it will not place the aircraft in an imminent stall situation.

Another important lesson to be learned from minimum controllable airspeed is when flying with full power, the power is "constant" and not "variable." Recall and implement the memory aid "pitch to the speed, power to the altitude," as discussed in Chapter 8 in the subsection on power. In other words, maintain airspeed by adjusting nose attitude and maintain altitude by adjusting power.

Once again, the attitude indicator is the instrument used to control the airplane. The airspeed indicator is the primary, or "quality" instrument for pitch. The heading indicator is primary for bank and the altimeter is primary for power (FIG. 8-5). Include VSI, turn coordinator, and tachometer or manifold pressure in your scan as supporting instruments for pitch, bank, and power respectively.

Instructor note. The student must do clearing turns prior to minimum controllable airspeed. Make sure there are no airplanes in the area because the high nose-up attitude in this maneuver will limit your vision as safety pilot. My recommendation is that in the beginning the student should do standard rate turns left and right 90° prior to reducing power. As the student gains more experience, the clearing turns may be made while reducing power and lowering flaps. This is good practice because in the real world of IFR there will be many times during approaches when the plane will have to be slowed and flaps extended while in a turn.

CONSTANT AIRSPEED CLIMBS

In VFR flying you have already become accustomed to constant airspeed climbs. After takeoff, climb at a speed that will give the best rate of climb, usually 80 knots for a Cessna 172 or Cherokee 180. In IFR flying, just as in VFR, the most important thing is to pick the best rate of climb speed (V_y) and stick to it.

Full power is used for climbs, so climbs are situations in which power is "constant," not "variable" and you will "pitch to the speed." Climb speed is thus established by setting the pitch attitude. You will set and control the pitch attitude by reference to the attitude indicator. Establish the climb by pitching up to that first index line in the blue "sky" section of the attitude indicator (FIG. 9-4). If entering the climb from straight and level, add full power and simultaneously increase pitch.

Pitch adjustments

Make adjustments from that first reference line, as necessary. If airspeed is too high and climb performance is poor, correct by a slight additional pitch up adjustment to reduce the airspeed; if airspeed is too low, make a slight pitch down adjustment. Hold the pitch constant on the attitude indicator and the airspeed will remain constant.

Pitch adjustments on the attitude indicator—like all adjustments in instrument flight—are small and deliberate, not more than a quarter or half a bar width at most. Bar width is the thickness of the miniature airplane "wings" on the attitude indicator. The smaller the adjustments, the smoother the flight.

Fig. 9-4. *Establish a climb by pitching up to the first index line above horizon line on the attitude indicator.*

Note that when entering a climb, the attitude indicator serves a triple purpose: the control instrument and the primary instrument for pitch and the control instrument for bank. The main concern in climb entry is establishing the correct pitch; the attitude indicator gives the best information on the "quality" of pitch, as well as serving as the reference by which you control the airplane in pitch and bank.

When the climb is stabilized at the best rate of climb speed, the airspeed indicator becomes the "quality" instrument for pitch. The primary and support instruments for both climb entry and stabilized climb are the same, as seen in FIG. 8-5 in the previous chapter.

DEPARTURE CLIMBOUT

Be prepared on the first instrument training flight to climb out on instruments after take-off. Under ATC clearance, adhere to standard instrument climbs, turns, and descents. Make full use of this excellent opportunity for practicing instrument climbs and you won't have to use valuable time later in the flight to hone climb skills.

Let's see how this works in the real world of IFR. The clearance from ATC contains climb instructions; ATC frequently requires a step or two in the climbout before reaching cruise altitude for better traffic separation. A typical clearance might be "Maintain three thousand feet, expect further clearance to five thousand feet in ten minutes."

Two climbs are in this clearance. First is the initial climb to 3,000 feet after takeoff. Then there is a stretch of level flight until ATC calls back 10 minutes later with further clearance to climb to 5,000 feet. How, in practical terms, do you handle this type of climb clearance?

As a rule of thumb, climb at the highest practical rate up to the last 1,000 feet before the assigned altitude. Then reduce the rate of climb to a constant 500 feet per minute for the last 1,000 feet to avoid overshooting the assigned altitude when leveling off.

In the previous example, you would take off and climb to 2,000 feet with full power at the best rate of climb airspeed. Then you would adjust the attitude to produce a 500 foot-per-minute climb from 2,000 to 3,000 feet. Then level off until ATC clearance to re-sume the climb to 5,000 feet, approximately 10 minutes later.

CONSTANT RATE CLIMBS

Upon reaching 2,000 feet you would switch from a stabilized *constant speed* climb to a stabilized *constant rate* climb at the rate of 500 feet per minute. (You will find this less complicated in flight than it sounds in print.) Most single-engine planes can't manage much more than a 500 foot-per-minute climb at full throttle anyway. To attain a constant rate climb, simply reduce pitch slightly, leaving the throttle at full power.

The VSI is the primary instrument for "quality" information on the rate of climb (FIG. 8-5 in the previous chapter).

You will have to wait approximately 7 seconds for the VSI to stabilize at the new pitch. In constant rate climbs and descents the attitude indicator continues to be the instrument by which you control the airplane. The VSI becomes the primary instrument for attitude (FIG. 8-5 in the previous chapter).

If the VSI stabilizes at more than 500 feet per minute, lower the nose to decrease the rate of climb; if less than 500 feet per minute, raise the nose to increase the rate of climb. Always remember that you cannot get an accurate indication from the VSI until it has stabilized for 7 seconds.

CHASING THE NEEDLE

When I give instrument checkrides I can tell very quickly if candidate pilots do not understand the limits of the VSI because they "chase the needle." If the needle is rising fast, they push forward on the yoke to slow it down; if it is descending fast they pull back. This produces a form of porpoising, which is a sure tip-off that the candidate has developed neither a good scan nor an understanding of instrument control. Instead, the pilot's attention remains fixed on one instrument too long.

Instructor note. A student who develops the VSI needle chasing symptom has an instrument fixation problem. Return to level flight and teach the student to forcefully shift his or her focus from one instrument to another, starting and stopping each scan cycle with the attitude indicator.

CLIMB LEVEL OFF

Leveling off from a climb is a simple procedure, but it's the source of a common problem for beginning instrument students. That problem is overshooting or undershooting the target altitude. To avoid this, anticipate reaching the target altitude by 10 percent of the rate of climb. If the VSI shows a rate of climb of 600 feet per minute, for example, begin leveling off 60 feet before the target altitude.

Lower the miniature airplane to the horizon line on the attitude indicator when reaching the desired altitude and allow the airspeed to build up to cruise before reducing power. Exert forward pressure on the yoke to prevent ballooning above the assigned altitude as airspeed builds up, and trim out the excess control pressure as it becomes heavier.

It is very important to have a predetermined idea of what the cruise power setting should be. You can automatically reduce power to that specific cruise power setting in one motion upon reaching the cruise speed. Any further adjustments of power and trim will then be relatively minor.

Include the sweep-second hand of the clock in your scan. No matter when you start the constant rate climb, you know that the airplane should gain 500 feet by the time the sweep-second hand returns to its starting position. The altitude gain should be 250 feet when the sweep-second hand is 30 seconds from its starting position.

DESCENTS

Overshooting the target altitude when descending is a more critical matter. Most precision instrument approaches go down to 200 feet above the runway, where there's no room for error. You must be able to control descents so there is never a question of coming too close to the ground or obstacles in the airport vicinity.

It's worth noting that airspeed control during descents becomes very important when making VOR and other nonprecision approaches. Understand that the *missed approach point* (MAP) on a nonprecision approach is frequently based upon how long it takes at a given airspeed to fly from the final approach fix to the airport.

If you can't control airspeed on a descending final approach course, you might as well throw the stopwatch away. Elapsed time on that final leg to the airport will be meaningless if the airspeed on which it is calculated is not constant. If the time is off, the airplane might end up off course. If there are hills or obstacles around the airport, poor timing due to poor airspeed control might be disastrous.

CONSTANT SPEED DESCENTS

In a constant speed descent use "pitch to the airspeed" like the constant speed climb. While in straight and level flight, reduce power to slow cruise. Set 1900 RPM in a Cessna 172 or Cherokee 180, approximately 90 knots for a comfortable and efficient descent speed.

"Pitch to the airspeed" again. Hold the nose up in level flight until speed bleeds off to 90 knots, then gently allow the nose to pitch down to one line below the horizon line on the attitude indicator (FIG. 9-5). That should produce a 90-knot descent at about 500 feet per minute. (It might vary slightly with different types of airplanes.)

Fig. 9-5. *Establish a descent by pitching down to first index line below horizon line on attitude indicator.*

"Pull the plug" if a faster descent is necessary. Reduce power to 1500 RPM, slow to 90 knots, then lower the nose to maintain 90 knots. This will produce a rate of descent of about 1,000 feet per minute at a somewhat steeper angle.

Some high-performance airplanes can descend 2,000 feet per minute by reducing power and lowering the landing gear and flaps. The *dirty* descent (gear and flaps down) can be made without reducing power excessively. This procedure avoids the shock of sudden overcooling, which can damage the engine.

CONSTANT RATE DESCENTS

Let it be known here and forever carved in stone that a change of 100 RPM or 1" of manifold pressure will produce a change of 100 feet per minute (fpm) in the rate of descent. Suppose you are in a 400-feet-per-minute descent with power at 1900 RPM, and airspeed 90 knots. How do you change the rate of descent to 500 feet per minute? Simply reduce power by 100 RPM to 1,800. Power adjustments of 100 RPM or 1" in descents will produce changes of 100 feet per minute in almost every plane that you are likely to fly.

This rule of 100 RPM or 1" = 100 fpm will become vitally important later on as you learn to fly the ILS approach with its very sensitive electronic glide slope. But the rule also applies to normal descents, and it will help avoid overshooting that critical altitude.

ATC will expect a descent at 1,000 fpm until reaching 1,000 feet above the target altitude. Then you are expected to reduce the rate of descent to 500 fpm. In other words,

- A change to a lower altitude begins with a *constant speed* descent at 90 knots and approximately 1,000 fpm
- At 1,000 feet above the target altitude, switch to a *constant rate* descent at 500 fpm

This is similar, of course, to the situation in a climb. The first part of a climb is at a *constant speed*—the best rate of climb speed. At 1,000 feet below the target altitude, the descent becomes *constant rate* at 500 fpm. The first part of a descent is at a *constant speed*. At 1,000 feet above target altitude, change to a *constant rate* descent of 500 fpm.

Refer to FIG. 8-5 in the previous chapter again and see that the instruments have the same control, primary, and support roles in a descent as they do in a climb.

Again, the sweep-second hand of the clock should be included in the scan to ascertain whether you are ahead or behind in the climb.

DESCENT LEVEL OFF

The target descent altitude should be anticipated by 10 percent of the rate of descent—the same as in a climb. If descending at a rate of 500 fpm, begin to level off 50 feet prior to reaching the target altitude.

Instrument students sometimes fixate on the target altitude and don't start the transition until they reach it. That is a sure way to overshoot the altitude and, on an instrument approach to go below the minimums for that approach. Don't do it! "Busting the minimums" is hazardous to your health when close to the ground. And on an instrument checkride, busting minimums is an automatic failure, no matter how brilliant the rest of the checkride might have been.

To level off, simultaneously increase power to the cruise setting and raise the nose and set it on the horizon bar of the attitude indicator. Adjust trim as the speed builds up to prevent ballooning above the level-off altitude. Scan the attitude indicator and the primary and support instruments as shown in FIG. 8-5 in the previous chapter.

APPROACH DESCENTS

Before leaving the subject of descents, I want to cover two slightly modified descents used on instrument approaches. (I will spend more time on these two descents in the approach phase of the training syllabus, but you should understand what they are at this point and see how closely they relate to standard descents.)

The first is a constant airspeed descent with the addition of 10° of flaps. Let's call them approach flaps because the setting might be more or less than 10° on different airplanes. Most airplanes handle better at slow speeds with a notch of flaps down. This is the configuration used on most instrument approaches on the final approach leg just before landing.

Start from cruise speed, reduce power, and extend approach flaps. Allow the airspeed to bleed off while in level flight, then "pitch to the airspeed." Set the nose of the miniature airplane on the first black line below the horizon on the attitude indicator. Adjust power for 500 fpm, remembering that 100 RPM or 1" = 100 fpm, or "power to the altitude."

If flying a high performance airplane with retractable landing gear, reduce power, set approach flaps, and allow the speed to decrease to the desired descent airspeed while remaining in level flight. When you reach the descent speed, extend the gear; the nose dips automatically to just about the right pitch for a descent when the gear is lowered.

HIGH-SPEED FINAL

More and more these days ATC might say: "Keep your speed up on final." This is frequently followed by something interesting such as like "727 overtaking." If you cannot comply with this request, do not hesitate to tell ATC right away. But it is best to cooperate with this request whenever possible, for obvious reasons. ATC won't let the separation between you and that big jet get too narrow. If a jet is behind you coming in 10 or 20 knots faster and chewing up the distance in between, guess who is most likely to be ordered to go around, you or the big jet? ATC might request a 90° left or right turn to let the jet pass before vectoring you back for the approach.

Practice constant rate descents at cruise speed, or faster, as well as at the normal descent speed. Some pilots automatically make high-speed final approaches whenever they fly into airports with a lot of jet traffic. This certainly makes it easier for ATC.

To set up a high-speed descent, lower the nose of the miniature airplane to the first black line below the horizon line on the attitude indicator. Reduce power 500 RPM to set up a 500-fpm descent. When stabilized, "pitch to the airspeed, power to the altitude" to maintain cruise airspeed at a 500-fpm descent. Leave flaps and landing gear up during a high-speed approach. If jets are on the approach behind you, you know the runway will be plenty long enough to slow down and extend flaps and gear when the runway is in sight and landing is assured.

VERTICAL S

When students file and depart IFR on every flight, they will usually have more than enough opportunity to practice climbs and descents in the real world of IFR. So there isn't much point in practicing additional vertical maneuvers. An instructor has to avoid a natural tendency to teach mechanics of the maneuver rather than the goal of the maneuver.

The Vertical S (FIG. 9-6) and its variations, the S-1 and S-2, are excellent exercises for an instrument student to practice with a safety pilot. The Vertical S consists of climbs to 500, 400, 300, and 200 feet with reversals at the top of each climb and descents back to the original altitude before climbing to the next altitude in the series. The Vertical S can also be a series of descents as shown in FIG. 9-6.

The Vertical S-1 is a combination of the Vertical S and a standard rate turn. Make a standard rate turn each time you return to the original altitude. Alternate turns to the left and to the right.

The Vertical S-2 differs from the S-1 in that the direction of turn is reversed with each reversal of vertical direction.

PATTERN B

The Vertical S, S-1, and S-2 are recommended maneuvers in the FAA's *Instrument Flying Handbook* (*see* Appendix A, "Instrument Pilot's Professional Library"). However, I have found that Pattern B is much more effective in teaching students how to combine the fundamentals of instrument flight: straight and level, speed changes, standard rate turns, climbs, and descents. It's an excellent maneuver for "putting it all together."

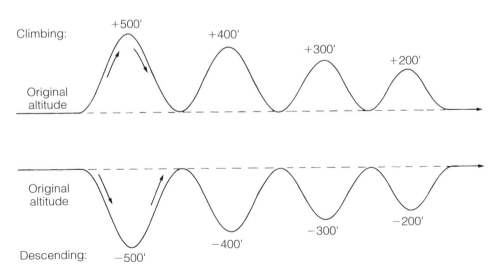

Fig. 9-6. *Vertical S practice maneuver.*

The turns and straight stretches in Pattern B (FIG. 9-7) are the same as those in Pattern A (FIG. 9-3). But B adds speed changes and includes a descent and an emergency pull-up to simulate an approach and missed approach.

Roll out on headings regardless of time passage. The turn to the final leg is a descending standard rate turn. Note that a prelanding checklist is included, then a little later you extend $1/4$ flaps as if commencing the final "approach." If you are flying an airplane with retractable gear, also lower the landing gear at this point.

At the emergency pull-up, don't forget to retract approach flaps and landing gear, if so equipped. Does something seem familiar here? Right! It's the recovery from minimum controllable airspeed. The pieces indeed come together at this point. Maybe not perfectly, but the goal of 2, 2, and 20 is in sight.

PATTERN C

Don't worry, Pattern C isn't required! But you will feel a great sense of achievement if you can do it. It has been called a basic airwork "graduation exercise." (FIG. 9-8.)

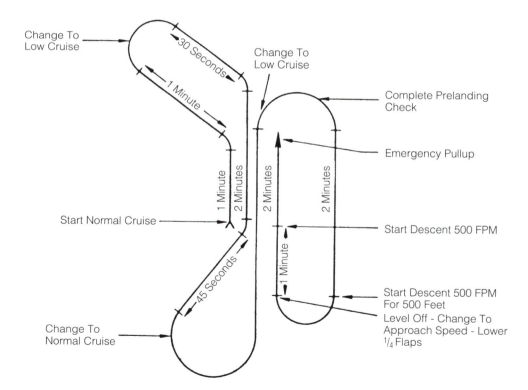

Fig. 9-7. *Pattern B practice maneuver.*

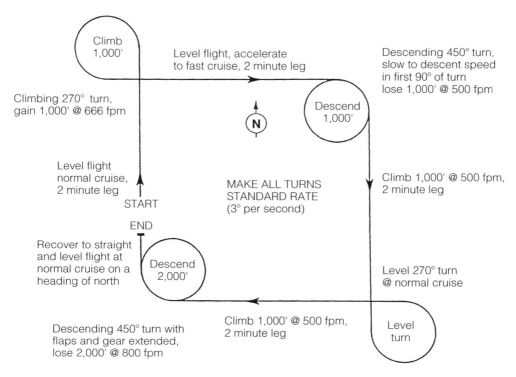

Fig. 9-8. *Pattern C practice maneuver.*

If you can fly C with its nonstandard climbs and descents and maintain 2, 2, and 20, you will have certainly mastered the fundamentals of attitude instrument flight. Patterns B and C are good exercises to practice with a safety pilot. Break off practice if the pattern work is not going well, otherwise you might unconsciously develop bad habits. Work with an instructor on whatever is causing the problem before any bad habits have a chance to take hold.

Master Pattern C on full panel, then try it on a partial panel. It is a sure cure for overconfidence; it is also instant insanity. Some dedicated instrument students have done this. I think they were former military pilots who had partial-panel Pattern Cs inflicted on them by sadistic military instructors. Civilian instructors, of course, would never pull a stunt like that. Flying is supposed to be fun!

10
VOR procedures

DESPITE THE GREAT PROMISE OF GPS, VHF OMNIDIRECTIONAL RANGE stations (VORs) will remain the heart of the airway system for many years. The radials from a VOR are highways in the sky. VOR radials form most intersections, and VORs are used for more instrument approaches than any other type of facility.

VOR use requires positive identification of the facility. Obviously it can be fatal if you fail to make a positive identification and use the wrong frequency on an instrument approach. So get in the habit of automatically turning up the volume and checking the identification for every VOR.

HEADING INDICATOR ERRORS

Another point to emphasize is the necessity of periodically checking the heading indicator and readjusting it to match the magnetic compass. Maintaining an accurate course is difficult if not impossible if the heading indicator has drifted off.

The gyro of the heading indicator might precess a small amount due to bearing friction. The turns and reversals during instrument departures, approaches, and training maneuvers will produce additional precession errors. A heading indicator that precesses up to 3° every 15 minutes is within acceptable tolerances. I recommend checking the heading indicator at least every 15 minutes and prior to intercepting the final approach course on all approaches. Here are several other times the heading indicator must be checked or reset:

- When you line up on the centerline of the active runway for takeoff. This is the best opportunity to set the heading indicator with greatest precision: The gyro is up to speed, the plane is stable, and you know the runway magnetic course.

- When you begin every approach, even after a missed approach. This item must be on the approach checklist.

- On leaving a holding pattern and after practicing Patterns A, B, and C, holding, or any similar maneuvers that require numerous turns.

- After practicing unusual attitudes. These maneuvers might cause the heading indicator to wander off considerably or "tumble" because of the extra bearing friction produced by the maneuvers.

VOR PROFICIENCY

In the beginning of an instrument student's training, VOR skills must be determined. Does the student understand the basic principles of VOR orientation, and intercepting, tracking, and bracketing bearings and radials? Have any bad habits crept in since obtaining the private certificate? For example, does the student tend to get fixated on the course deviation indicator (CDI) needle and neglect other instrument indications? Remember, the goal remains 2, 2, and 20 in VOR work as well as other phases of instrument flight.

Don't spend expensive flight time to determine VOR proficiency. Make a quick pencil-and-paper check of the basic principles of VOR orientation by completing the VOR diagnostic exercise in FIG. 10-1.

Begin at the top line with the omni bearing selector (OBS) set at 030°. Then pick the correct VOR presentation for each lettered position. For example, with the OBS set at 030°, the most appropriate display for the A position is number 5. Proceed across through the G position, then drop down and complete the 090° line the same way, and so on.

Answers are in FIG. 10-2 on the next page. If any answer is wrong, review the exercise with your instructor. Proceed to the exercise in FIG. 10-3 for more VOR fundamentals.

16-POINT ORIENTATION

I developed this exercise many years ago. It works very well when teaching private pilot students exactly what happens around a VOR. The 16-point Orientation Exercise is a teaching exercise and a good diagnostic exercise because it will quickly reveal whether or not a student understands the basic principles of VOR. If not, the exercise can be repeated to bring the student up to par in short order.

Flying from west to east, with a VOR station to the south, one setting of the OBS will reveal 16 lines of position with precision while flying around the VOR station.

Most students are puzzled when I describe this exercise the first time. Sixteen lines of position from a single OBS setting? How can this be? It's really very simple when you understand what's going on, and when you understand, you will have mastered the basic principles of VOR work.

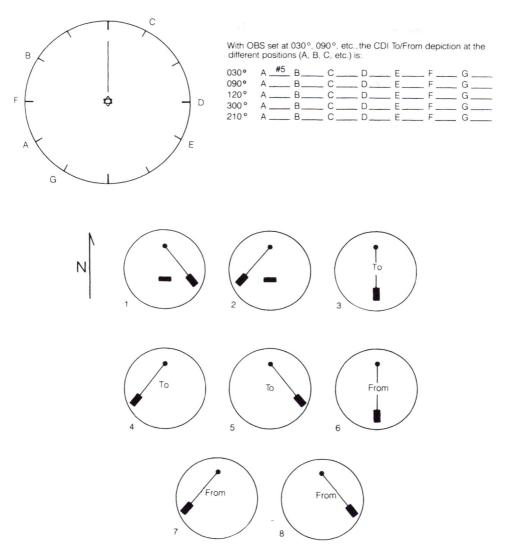

With OBS set at 030°, 090°, etc., the CDI To/From depiction at the different positions (A, B, C, etc.) is:

030° A __#5__ B ____ C ____ D ____ E ____ F ____ G ____
090° A ____ B ____ C ____ D ____ E ____ F ____ G ____
120° A ____ B ____ C ____ D ____ E ____ F ____ G ____
300° A ____ B ____ C ____ D ____ E ____ F ____ G ____
210° A ____ B ____ C ____ D ____ E ____ F ____ G ____

Fig. 10-1. *VOR diagnostic test. Answers are on the next page.*

Enter the pattern on a heading of 090° (FIG. 10-3) with a convenient VOR station to the southeast. Set the OBS to 090°. As you begin the exercise, the TO-FROM indicator will show TO.

The 360° radial is the boundary between TO and FROM. The red flag will start to appear approximately at the 350° radial. When the red flag indicates OFF, you will be passing the 360° radial. That's the first precise line of position. When the red flag disappears, you will be approximately on the 010° radial. So the first leg of the exercise gives at least one precise line of position at 360° plus the lines of position 350° and 010° with lesser accuracy.

With OBS Set at:	CDI/To-From Depiction:						
030°	A #5	B #1	C #6	D #7	E #2	F #5	G #3
090°	A #4	B #5	C #8	D #6	E #7	F #3	G #4
120°	A #4	B #3	C #1	D #8	E #6	F #4	G #2
300°	A #8	B #6	C #2	D #4	E #3	F #8	G #1
210°	A #7	B #2	C #3	D #5	E #1	F #8	G #7

Fig. 10-2. *Answers to VOR diagnostic test on previous page.*

Continue on the 090° heading another 2 minutes or so, then turn right to a heading of 180°. The CDI needle will soon come alive and start moving in from right to left. When it reaches the outermost dot, you will be on the 080° radial. When the needle reaches the edge of the bull's-eye in the center, you will be at the 085° radial, and when the needle centers, you will be at the 090° radial. That's three lines of position.

As you continue on the 180° heading, the needle will pass the other edge of the bull's eye at 095°, and the last dot on the left at 100°, adding two more lines of position.

You have produced three lines of position on the first leg and five on the second leg for a total of eight. If you continue the pattern as shown in FIG. 10-3, you will add three more lines of position on the heading of 270°, and five more on the 360° heading for a grand total of 16 for the full exercise—all without resetting the OBS.

If further practice is needed, enter the maneuver at other cardinal headings.

Instructor note. Have your student anticipate and call out the radials while flying around the VOR station. This will help the student visualize what is going on and it will also give you a good indication of whether or not he or she understands what is happening.

VOR TIME/DISTANCE CHECK

Another excellent exercise in teaching the fundamentals of VOR orientation is the VOR time/distance check, or poor man's DME. Use this procedure to estimate out how long it will take to reach a VOR station in no-wind conditions. To tell the truth, I have never heard of a situation where someone has had to use the time/distance check to determine the time to a VOR on an actual flight; however, practicing the time/distance check will sharpen your VOR skills and, as we shall see later, introduce you to a similar procedure used in making a DME arc approach.

Turn toward a convenient VOR station and adjust the OBS so the CDI needle centers in the configuration. Note the heading to the station. Next, turn the airplane 80° right or left of the inbound course. Rotate the OBS in the *opposite* direction of the turn to the nearest increment of 10. In other words, if you turn right 80°, turn the OBS *left* (counter-clockwise) to the nearest 10° increment (FIG. 10-4). You are flying a short tangent to an imaginary circle around the station.

Maintain the new heading. When the CDI needle centers, note the time. Continue on the same heading and change the OBS another 10° in the same direction as above. Note the number of seconds it takes for the CDI needle to center again. Divide the number of seconds by 10 to determine the time to the station in minutes. The formula is:

$$\text{Minutes to station} = \frac{\text{Time in seconds}}{\text{Degrees of bearing change}}$$

You can also calculate the distance to the station by using this formula:

$$\text{Distance to station} = \frac{\text{TAS} \times \text{minutes flown}}{\text{Degrees of bearing change}}$$

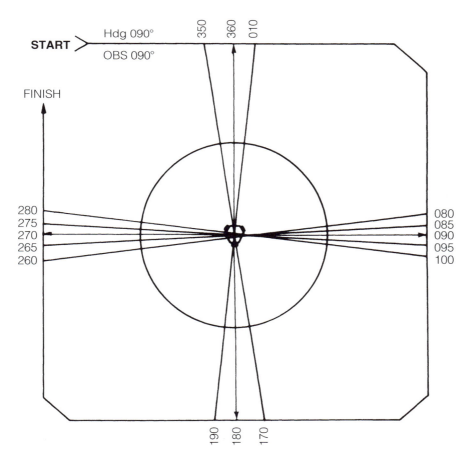

Fig. 10-3. *A 16-point VOR orientation exercise. Begin heading 090° with OBS set on 090. You get 16 lines of position without changing OBS setting.*

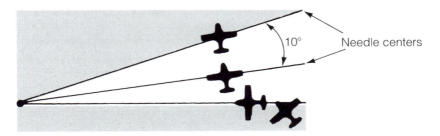

Fig. 10-4. *VOR time-distance check.*

For the second leg, turn 20° *toward* the station and stop the turn. Turn the OBS 10° in the *same* direction as the turn. When the needle centers, note the time and turn another 10° in the same direction. When the needle centers again, note the elapsed time and calculate time to station.

Turn the airplane another 20° toward the station to start a new leg, and repeat the process. Continue these short tangents around the station as many times as you wish. Complete the exercise by selecting an inbound bearing to the station that lies ahead of the last tangent leg you plan to fly. Start turning inbound 10° before reaching the inbound course.

When are you 10° from the inbound course? Simple. Start turning inbound when the needle reaches the outermost dot on the CDI display. This will indicate that you are 10° away from the inbound course.

If you need more practice in either the 16-point orientation exercise or the time/distance check, use a simulator—you can get in more practice in an hour in a simulator because you don't have to copy clearances, take off, and fly to a practice area. Whenever you begin to have difficulty, you can stop the exercise and analyze what's causing the problem. Furthermore, an hour in a simulator costs much less than an hour in an airplane.

INTERCEPTING A BEARING OR RADIAL

The first practical application of VOR work will most likely be clearance to a VOR station after takeoff—cleared "direct" to the first VOR on the clearance or to intercept a specific radial. Clearance via a specific radial involves intercepting that radial and tracking inbound to the station. Two assumptions will simplify this procedure.

First, departure instructions and any radar vectors from ATC will always point in the right direction to make a quick, efficient intercept, unless you are being diverted away from traffic, higher terrain, or other obstacles. This is also true with en route clearance changes that direct you to intercept a specific radial.

Second, clearances always state *radials*, not magnetic courses to the VOR. Because a radial radiates *from* the station, you will fly *toward* the station on the reciprocal of the radial. A quick and easy way to determine the reciprocal is to refer to the heading indicator. Follow the radial from its number on the edge of the dial, through the center, and out in a straight line to the number on the opposite outer edge (FIG. 10-5). Set the assigned radial on the OBS and continue on the assigned heading. When the needle begins to move, the airplane is on the radial that lies 10° before the target radial.

Inbound turn

When you reach this 10° lead radial, turn to intercept the inbound bearing at an angle of 60°. When the needle reaches half-scale deflection, turn an additional 30° and maintain this 30° intercept heading until just before the needle reaches the bull's-eye, the small center circle on the CDI presentation. When the needle touches the bull's-eye, set the OBS to the inbound magnetic course.

Figure 10-6 illustrates how this works. In this situation, you have been assigned a heading of 320° and you have been cleared to the VOR via the 226° radial. As you reach the lead radial—216°—the needle comes alive. Turn right to a heading of 346° to set up a 60° intercept angle.

If you're not sure what heading will produce a 60° angle, refer again to the heading indicator, and count off 60° from the inbound bearing (FIG. 10-5). You will see this is 346°. With a little practice, you will be able to read reciprocals and intercept angles off the heading indicator at a glance.

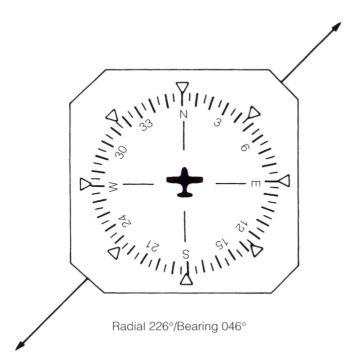

Radial 226°/Bearing 046°

Fig. 10-5. *Use heading indicator to visualize reciprocals.*

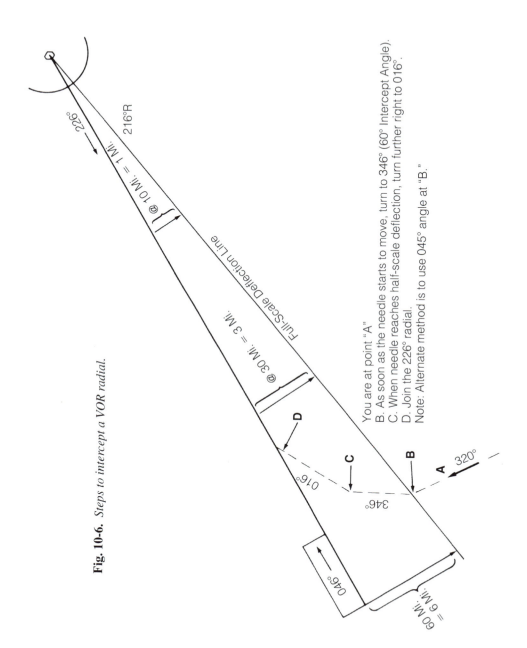

Fig. 10-6. *Steps to intercept a VOR radial.*

You are at point "A"
B. As soon as the needle starts to move, turn to 346° (60° Intercept Angle).
C. When needle reaches half-scale deflection, turn further right to 016°.
D. Join the 226° radial.
Note: Alternate method is to use 045° angle at "B."

When the needle reaches half-scale deflection, turn right again to 016°. Hold 016° until the needle reaches the bull's-eye, then steady up on the inbound heading of 046°, plus or minus whatever wind correction is necessary to hold the needle in the center.

WIND CORRECTIONS

I'm always surprised when instrument students are unable to offer even an educated guess when asked, "Which way is the wind coming from?" Knowledge of the wind should almost be second nature by the time a person receives a private pilot certificate. If not, work on it during instrument training. An instructor should keep asking "which way is the wind?" until you begin anticipating and adjusting for the wind automatically.

Flight planning revealed wind forecasts at various geographical points and altitudes and you know exactly what the wind was at takeoff. It is a simple matter of deciding whether the wind is going to push to the left or to the right departing the airport toward the first VOR fix.

A tailwind will speed interception and a headwind will delay interception. And when turning onto the inbound heading, add a wind correction factor automatically, maybe 2°, 5°, or 10° according to your best estimate. Refine this correction by making adjustments en route toward the station.

There are two other interception techniques. If you believe you are close to the station when intercepting a radial, make the first turn 45° toward the station, rather than 60°. Hold that 45° interception course until the needle is about three quarters of the way from full-scale deflection.

The needle reaches this position about $2\frac{1}{2}°$ from the assigned radial; turn to the inbound heading at this point. Add a correction for the wind when established on the inbound course.

How can you tell if you're close to the station? The more sensitive the needle, the closer you are to the station. When in close, the 45° intercept will put you on the inbound bearing quickly and at a greater distance from the station. The 45° intercept will provide time to adjust the inbound heading for the wind and it also gives you a better chance of being exactly on course over the station.

The second interception procedure is a reinterception technique utilized when off course and the needle is pegged at full deflection. This happens when a strong wind changes abruptly or when you are seriously distracted and drift left or right without correcting the problem.

In either case, make an en route correction to return to the desired radial or bearing. If the wind is from behind, use an intercept angle of 10° or 20° to return to course and avoid overshooting; if a headwind, use an intercept angle of 20° or 30°. The larger angles will get you back on the correct course sooner.

COMMON INTERCEPTION MISTAKES

The needle never centers. This indicates that (1) you turned to the inbound heading too abruptly or too soon or (2) a headwind was much stronger than anticipated. In either case, use the reinterception technique described above in the wind corrections subsection. Turn 10°, 20°, or 30° toward the needle and wait until the needle centers to resume the inbound heading.

The needle passes through the center and moves toward the opposite side. As in the first case, the inbound heading turn might have been faulty. You might have made the turn too slowly or waited too long to start the turn or the tailwind was stronger than anticipated. Use the reinterception technique to center the needle.

Another possibility is that you were so close to the VOR that the width of that 10° arc from full deflection to the center might have been only a few feet. A 45° intercept angle 1 mile from the VOR is almost impossible. The needle will peg with a FROM indication almost as soon as you turn to the inbound course. Ideally, you should have at least 5 miles before you get to the VOR to do a skillful job of intercepting a radial. The only solution is to steady up on the outbound heading and reintercept after the needle has settled down in the FROM position.

A good interception with the needle perfectly centered after the first turn to the in-bound heading is not a matter of luck. With practice, and using the correct techniques, you will learn to judge the wind and turn so that the needle will center every time.

CLEARED DIRECT

In some cases you will be cleared "direct" to the first VOR on the route; ATC means directly to the VOR in a straight line. ATC expects you to establish a course to the station and to stay on that course—with the needle centered, which is the next VOR challenge.

To fly direct to a VOR, turn the OBS knob until the needle centers in the TO position and read the course in the window. That is the course toward the station and that is the course ATC expects you to fly.

You know whether the wind is from the left or right and approximately how strong it is. After establishing the course to the station, fly a trial correction, left or right, of 2°, 5°, or 10° to keep the needle centered. (See Fig. 10-7.)

REFERENCE HEADING

Note the heading. This is a "reference heading" or "holding heading" because it is the heading that holds the airplane on the correct magnetic course with the needle centered. Make small corrections left and right of the reference heading and the needle should hover around the center.

The overall procedure for establishing a reference heading and adjusting it to keep the needle in the center is known as *tracking*. In addition to using tracking to stay on course directly to a VOR, tracking will also maintain a course along a prescribed airway.

Keep in mind that the wind will rarely remain constant for any length of time. So when you are established on your radial or bearing and have worked out the holding heading, you will still have to adjust it a few degrees for slight changes in wind direction and strength. This, too, is part of tracking.

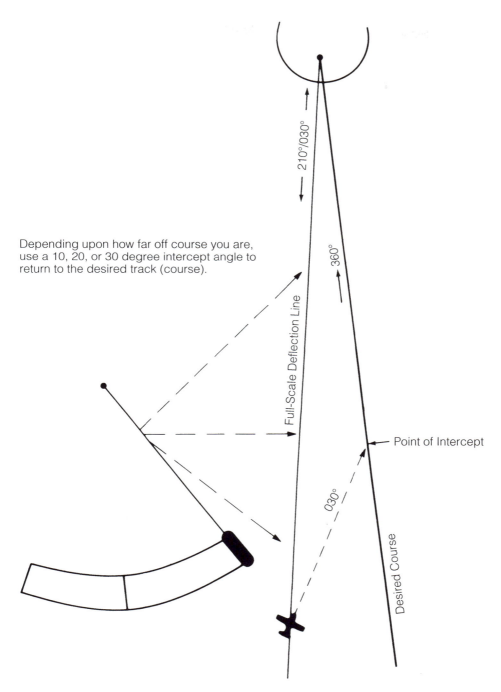

Depending upon how far off course you are, use a 10, 20, or 30 degree intercept angle to return to the desired track (course).

210°/030°

360°

Full-Scale Deflection Line

Point of Intercept

030°

Desired Course

Fig. 10-7. *En route correction to return to a VOR course.*

BRACKETING

In the real world of IFR, you might not always be able to determine wind strength and direction, especially if you fly through a front or other rapidly changing weather conditions. Or you might become momentarily disoriented and unsure of what correction to make to stay on course and keep the needle centered. In either case, *bracketing* will get you back on course quickly and, at the same time, show what the holding heading should be to maintain that course.

Bracketing is a series of smaller and smaller turns from one direction to another across the desired course. Bracketing can be done on inbound and outbound legs. Start by turning to a heading that is the same as the desired magnetic course. Then make a turn 30° toward the needle. Hold the correction and make the needle move back to the center.

When the needle has returned to the center, cut that first 30° correction in half, and turn 15° toward the needle. Make it move back to the center again. When the needle has centered cut the correction in half and turn $7\frac{1}{2}°$ toward the needle. Again, make it move toward the center. You will quickly find a reference heading that will position the needle near the center and stop it from moving. Make minor adjustments left and right of that holding heading to keep the needle centered.

CHASING THE NEEDLE

In describing bracketing, you will note that I was careful with each turn to say "make the needle move." What I meant by this is maintain the correction until achieving the desired result—in this case, returning the needle to the center.

Because the human eye is always quick to pick up motion, there is a great temptation to fixate on a VOR needle as soon as it starts to move and to turn toward it—to "chase" it. If you chase the needle making larger and larger corrections it will be impossible to predict when the needle will stop or reverse direction; pretty soon you will be way off course. If you chase the needle when the airplane is a half a mile from the station, you could make a 45° correction when only 50 feet off course and blow the station passage.

Set in a correction, hold it until the needle moves to the center, then adjust the correction. Don't start taking the correction out as soon as the needle starts to move.

STATION PASSAGE

If you bracket and track the VOR properly, the adjustments to the holding heading will become smaller and smaller near the station. This is very important because the needle gets extremely sensitive closer to the station. If you are still making large heading changes close to the station, the airplane will pass way off to one side or the other. If the VOR that was just missed is the final fix on a VOR approach, execute a missed approach and try again.

Make your greatest efforts several miles out to establish the reference heading for perfect station passage. When you approach within a quarter of a mile of a station (and this is slant range), you enter a zone of confusion where none of the VOR instrument indications will hold steady.

Maintain the reference heading through this zone of confusion and note the time that the TO-FROM indicator flips to FROM. Maintain the reference heading—or turn to a new outbound heading and hold that—until the instrument indications settle down.

Don't chase the needle! As you maintain the reference heading, or turn to a new one, analyze whether that track is to the right or the left of the outbound course and then set up a reinterception or a bracketing procedure to get back on course. Then make minor adjustments to keep the needle centered.

PRACTICE PATTERNS

Patterns A and B are excellent for practicing VOR interception and tracking. Start the patterns over a VOR station, using a VFR altitude (odd or even thousand feet plus 500 feet) to avoid IFR traffic that might also be using the same VOR station. Start the pattern at the station and plan for each straight leg to return over the station.

While developing basic attitude instrument flying skills you will also hone VOR interception and tracking skills. And as noted before these patterns contain all the elements in an instrument approach and when practiced at a VOR, the effort becomes an introduction to VOR approaches.

After your instructor has introduced these patterns, practice them with a safety pilot. As always, discontinue practice if problems begin to crop up.

Instructors, students *and* **safety pilots note:** Your most important responsibility at this stage is collision avoidance. As any experienced pilot can tell you, all the airplanes in the sky at any one time are converging on the VOR, NDB, or airport that you are approaching. Here is the most popular spot for midair collisions, according to FAA statistics. Keep your eyes wide open and searching at all times! Look out!

An additional safety measure I have always practiced is to invite extra observer pilots to join us on training flights whenever possible. Impress upon them the importance of their role by directing them to keep their eyes open and their mouths shut except when they identify a potential collision hazard approaching.

11
Holding patterns

VOR HOLDING PATTERNS ARE INTRODUCED AS EARLY AS FLIGHT lesson 5 for two reasons. First, holding patterns around a VOR station provide excellent practice in VOR interception and tracking close to the station. The needle indications are very sensitive; holding pattern practice will quickly sharpen your tracking skills.

Second, something seems to make holding patterns awe-inspiring and difficult. They are really quite simple when you go about them the right way. Nevertheless, many pilots are wary of holding patterns, so I tackle them early in the course so that students will feel comfortable with them later on.

My technique for teaching holding patterns works very well. Initially, you will learn how to fly the pattern while correcting for wind drift. Only after you are comfortable with the racetrack pattern and have mastered wind correction techniques do I then teach pattern entries.

Figure 11-1 illustrates the basic elements of a standard holding pattern around a VOR. The standard pattern has right turns, the inbound leg is 1 minute long. Nonstandard patterns have left turns. Above 14,000 feet the inbound leg is $1^{1}/_{2}$ minutes.

Pick a nearby VOR and fly inbound on any convenient course with the OBS needle centered. At station passage you will be very busy for a few seconds running an important checklist.

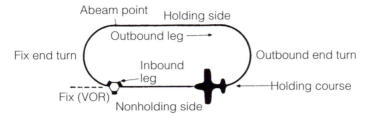

Fig. 11-1. *Elements of a standard holding pattern.*

FIVE TS: TIME, TURN, TWIST, THROTTLE, TALK

Do these at station passage:

1. Note the *time* of arrival at the fix and write the time down on the log or directly on the chart.
2. Start a 180° *turn* to the outbound course with a wind correction as necessary.
3. *Twist* the OBS knob to set new inbound VOR course.
4. *Throttle.* Reduce power to conserve fuel while in the holding pattern; you are not going anywhere.
5. *Talk.* Report reaching the holding fix and altitude ("...Approach, 56 Xray entering hold, level at five."). Always ask the question, "Do I need to report to ATC now?" During an instrument approach the answer to that question will most often be yes. (When in doubt, report.)

These are the five Ts of instrument flying: *time, turn, twist, throttle, talk.* In other words, aviate, navigate, and communicate. You will encounter the five Ts over and over, especially during instrument approaches. Use this mental checklist with every holding pattern and keep using it until it begins popping into your head automatically.

After running through the checklist, the next task is to determine when to start timing the outbound leg. This is easy. After station passage, and while you are making the turn to the outbound leg, the TO-FROM indicator will show FROM. When the airplane is abeam the holding fix heading outbound, the indicator will change to the TO position. Start timing when the indicator changes to TO.

WIND CORRECTIONS

Recall from VOR tracking practice that wind is almost always a factor. How do you correct for the wind in a holding pattern? As you flew inbound to the VOR you kept the needle centered, which should provide a pretty good idea of the wind correction for the inbound leg upon reaching the fix. The wind correction angle on the outbound leg will be *double* and *opposite* the correction on the inbound leg.

If you held a wind correction angle of 4° into the wind on the inbound leg, hold a wind correction of 8° into the wind while flying outbound. The reason for doubling the wind correction angle is to compensate for the effect of the wind on those two 1-minute turns at the fix end and at the outbound end.

Even if you fly for precisely 1 minute on the outbound leg, you will probably find that the inbound leg is not exactly 1 minute. Again, this is because of wind. The correction is simple; adjust the time of the next *outbound* leg to compensate for the difference.

If the inbound leg is only 45 seconds, for example, add 15 seconds to the outbound leg and make it 1 minute and 15 seconds. If the inbound leg is 1 minute 30 seconds, subtract 30 seconds from the outbound leg. The first case compensates for a tailwind on the inbound leg; the second case corrects for a headwind on the inbound leg.

EN ROUTE HOLDING

The wind correction tips are essentially a description of the procedure used for establishing an en route holding pattern. If there is a delay while en route, ATC might simply issue a hold on the present course, at the same altitude, at a convenient fix.

If no holding pattern is shown on the en route chart, ATC will state:

1. What fix to use and where the holding pattern will be located in relation to the fix (north, south, southwest, etc.).

2. What radial to use.

3. Nonstandard instructions, such as left turns and length of legs in miles if DME is utilized.

4. An *expect further clearance* (EFC) time, or the time to *expect approach clearance* (EAC) if held on a segment of an instrument approach; these "expect" times will be given either as a specific clock time, such as 2045, or in minutes. It is not uncommon to hear ATC say "expect further clearance to XYZ in 10 minutes."

   EFCs and EACs are important for two reasons:

   • They tell when ATC expects to issue a clearance to resume the flight. You need to know this to adjust the holding pattern to arrive over the fix at the "expect further clearance time."

   • EFCs and EACs also tell you when to depart a holding pattern in the event of lost communications. If ATC does not issue an EFC or an EAC, be sure to request it.

   A typical holding clearance might be:

"Cessna three four five six Xray, hold southeast of the Huguenot VOR on the one four five degree radial, maintain five thousand, expect further clearance at one two one five."

Many holding patterns are already depicted on en route and approach procedure charts. If the holding pattern is shown, the clearance will be simpler. For example:

"Cessna three four five six Xray, hold as published northeast of SHAFF intersection, maintain five thousand. Expect further clearance at one two one five."

In this case, ATC would expect you to find SHAFF intersection on the L-25 en route chart and fly the depicted pattern northeast of the VOR.

ATC recognizes that holding patterns can be somewhat imprecise because of the wind, the skill of the pilot, and the different airspeeds of different types of aircraft. So they establish a buffer zone on the holding side that is at least double the amount of protected airspace around the pattern.

HOLDING PATTERN ENTRY

When comfortable with the patterns, you will have also automatically mastered the *direct* method of entering holding patterns. The direct entry, simply stated, means that you fly to the station or fix and make a turn directly to the outbound leg, which is exactly what you have been practicing. On IFR flights, you will use direct entries for most holds issued by ATC.

The other two ways to establish holding patterns are the *teardrop* and the *parallel* methods (FIG. 11-2). To enter a standard holding pattern with the teardrop method, cross the holding fix and proceed outbound at an angle of 30° to the holding course for 1 minute, then turn right to intercept the holding course.

For the parallel method, cross the fix and fly outbound parallel to the holding course for 1 minute, turn left, fly direct to the fix, then turn right to the outbound course.

Before going into further detail, some basic points will make holding entry much easier. Direct, teardrop, and parallel entry methods are *recommendations only*. They are *not* required. The FAA's *Practical Test Standards* require only that you use "an entry procedure that ensures the aircraft remains within the holding pattern airspace for a standard, nonstandard, or nonpublished holding pattern."

One very easy—and completely acceptable—method of entering holding is to fly to the holding fix, then turn on the holding side to the outbound heading. Fly the required time (or distance), then turn inbound on the holding side and reintercept the inbound leg.

Think about this for a minute. No matter how you arrive at the holding fix, simply turn to the outbound heading, fly the time specified, turn back toward the fix, and intercept the inbound leg, making all turns on the holding side. There is no longer the need to make all those confusing, distracting calculations about whether your incoming course is greater or less than 70° to the inbound leg of a standard right-hand pattern (opposite in the case of a left-hand pattern). You won't get into trouble as long as you establish an accurate pattern quickly and efficiently and do not violate the protected airspace.

The entry and first time around the holding pattern are free. ATC understands that it is difficult to fly a perfect one-minute racetrack on the first circuit. Flying slightly off course on the first pattern is tolerated by ATC, provided you maintain the assigned altitude. Better tracking is expected by the second circuit.

You should still become proficient in all three holding entries—direct, teardrop, and parallel. They are the most efficient ways of entering holding that have been devised so far, and your ability to use the three methods with precision will help create an atmosphere of competence and confidence on your flight test. So please read on and learn the three methods as you work toward your goal of becoming a "proud, perfect pilot."

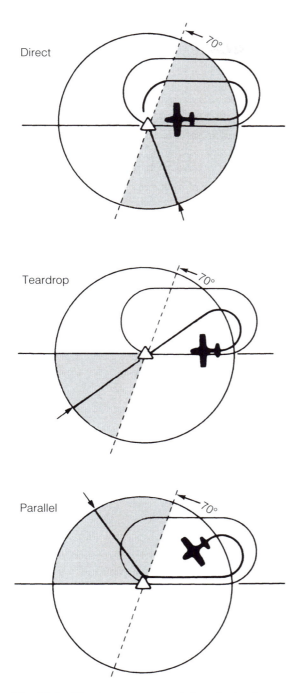

Fig. 11-2. *The three recommended methods of entering a holding pattern.*

CHOOSING THE CORRECT ENTRY

Note the line that has been drawn in FIG. 11-2 at an angle of 70° to the inbound leg of a standard right-hand pattern. (The pattern would be on the left side of the inbound leg for a nonstandard left-hand pattern, and the 70° line would be drawn in the opposite direction.)

The 70° line helps divide the area surrounding the holding fix into three "entry sectors," as indicated by the shading. The type of entry chosen depends upon the entry sector in which you approach the holding fix. I have flown with a few amazing people who can calculate entry sectors quickly in their heads down to the last degree. But this is the exception, not the rule. Most of us can't "do it by the numbers."

Fortunately, three other ways to plan holding entries have evolved over the years:

- Pencil in the holding pattern on the chart (if it isn't already printed). Draw in an approximate 70° line—it's 20° less than the perpendicular line to the inbound course. Then draw the inbound course to the station. This will show which entry sector you occupy and which type of entry to choose. Remember, you don't have to draw the lines perfectly down to the last degree to set up an acceptable entry.

- Use one of the many plastic holding overlays available. Place the overlay over the holding fix on the chart. Align the holding pattern on the overlay with the inbound leg of the holding pattern. You will see clearly which entry sector you occupy. These clear plastic overlays usually have standard (right-hand) patterns on one side and nonstandard (left-hand) patterns on the other side.

- Use the heading indicator. Visualize the inbound leg of the holding pattern, then take a pencil or other straight object and rotate it 70° counterclockwise to this inbound leg. This will reveal the entry sectors very quickly. For a nonstandard (left-hand) pattern, turn the pencil clockwise 70° to indicate the entry sectors.

Finally, *visualization* is the key! Visualize the holding pattern in relation to the fix and where you are. Upon arrival at the fix simply make the shortest possible turn to join the race track of the holding pattern.

IMPORTANCE OF ALTITUDE CONTROL

ATC is more concerned about altitude control than making a perfect pattern entry because other airplanes might be flying the same pattern at different altitudes above and below. You might be in a pattern at 5,000 feet with planes at 3,000, 4,000, 6,000, 7,000, 8,000, and so on. Stacks such as this are a common occurrence around busy airports during landing delays due to weather. (FIG. 11-3)

In a stack, planes are cleared to proceed from the bottom, like dealing from the bottom of a deck of cards. Each time a plane is cleared to leave the stack, all the others are cleared sequentially to descend one level at a time. A vertical separation of 1,000 feet is maintained at all times. Obviously, there must be no confusion about altitude assignments, nor any sloppiness in maintaining assigned altitudes, with several planes all hold-

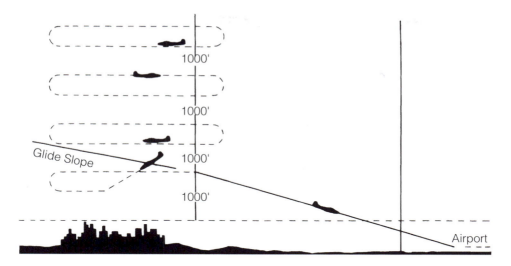

Fig. 11-3. *Stack or shuttle descent in a holding pattern for an instrument approach.*

ing in the same racetrack pattern around the same fix, separated only by altitude. The goal of 2, 2, and 20 becomes a realistic requirement in this situation.

HOLDING PATTERN VARIATIONS

It is not unusual for an instrument flight to be kept in a holding pattern 20 minutes or more. When this happens it is perfectly all right to ask ATC for a pattern with 2-, 3-, or even 5-minute legs. ATC will usually try to grant the request.

Longer legs make it easier to fly the pattern. More attention can be paid to altitude control and to establishing a precise inbound course with the needle centered. Longer legs are also much easier on passengers. Repetition of a straight leg for 1 minute, followed by a 1-minute 180° turn could cause airsickness.

For planning purposes, a pattern with 2-minute legs will take six minutes to complete, a pattern with 3-minute legs will take eight minutes, and one with 5-minute legs will take 12 minutes. If you expect to be in a holding pattern for any length of time, adjust power and relean the mixture to keep fuel consumption to a minimum.

INTERSECTION HOLDS

ATC will frequently issue en route holds at VOR intersections. Some are depicted on en route charts; be mentally prepared to use them. Intersection holds are easily managed. To reduce cockpit confusion, always set up the holding course on the top (No. 1) nav receiver. Set up the intersecting bearing on the bottom (No. 2) nav receiver.

If, on the No.1 nav, you always set up the course *to the station*, not the radial, the needle will always be located in the same direction as the VOR. If the needle is to the right, the station will also be to the right. This is more a matter of reducing cockpit confusion than anything else. Set up the two VORs as described above and you will always have a clear picture of exactly where you are in the pattern. (These are also good procedures even when no holding is involved and you want to keep track of the intersections along an en route leg.)

You have arrived at the intersection when the needle on the No. 2 nav receiver centers. Do the Five Ts checklist.

Start timing the outbound leg when the No. 2 nav needle moves back on the same side as the VOR indicating that you have reached the abeam position. In short, with the No. 2 nav receiver set to the radial (FROM), if the needle and the VOR are on the same side, you have not arrived. If they are on opposite sides, you have passed the intersection.

One-VOR intersection

It is possible to identify VOR intersections and fly holding patterns around them using only one navigation receiver. Sounds difficult at first, I know, but many pilots are quite accomplished with this procedure. In the early days of VOR navigation having even one VOR in the cockpit was considered a luxury and dual VORs were almost unheard of. So one VOR did the work of two.

It's not a good idea to spend any time on one-VOR intersection holds at this point. If one VOR fails in flight, notify ATC immediately. Consider landing as soon as possible because if the other VOR also fails, you would have a job on your hands getting down safely. ATC would have to do a lot of fast shuffling to reroute the traffic in your vicinity to ensure safe separation.

If you would like to try single-receiver intersection holds, have an instructor coach you through a few in the simulator.

Instructor note. It's not a good idea to practice single-receiver intersection holds until the student has thoroughly mastered the art of working close to a VOR station. Here is the procedure for a one-VOR intersection hold.

Track the inbound leg with the appropriate wind correction established and the needle centered. One minute before ETA at the intersection, reset the navigation receiver to the frequency of the station providing the cross-bearing and reset the bearing to that of the second station.

On reaching the intersection, proceed normally. When established on the outbound leg, reset the nav frequency and bearing back to the first station. At 1 minute outbound, turn again and get reestablished on the inbound leg with the needle centered. Fly inbound for 30 seconds, then reset the frequency and bearing to anticipate the cross bearing.

Obviously, a lot depends on the student's ability to get established quickly on the inbound bearing with the needle centered.

DME HOLDING PATTERNS

Another holding pattern variation is based upon DME distances, rather than time. If you have indicated on the flight plan that you have DME aboard (code letter A), you

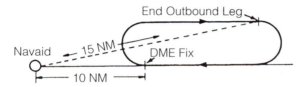

Fig. 11-4. *Holding pattern toward a DME facility.*

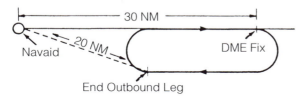

Fig. 11-5. *Holding pattern away from a DME facility.*

can expect to be issued a DME hold. DME holding uses the same entry and racetrack procedures except that distances (in nautical miles) are used in lieu of time. ATC specifies the distance of the fix from the navaid and the length of the outbound leg.

For example, if heading *toward* the VOR/DME (FIG. 11-4) with the fix distance 10 nm and the outbound leg 15 nm, you will enter the racetrack when the DME reads 10. You will end the outbound leg when it reads 15, and commence the turn back to the inbound leg.

Clearance for a DME hold would be something like this:

"Cessna three four five six Xray, hold 10 north of Carmel VOR on the three six zero degree radial, five-mile legs, expect further clearance one five four five, maintain five thousand."

DME holding patterns can also be established with the inbound leg heading *away* from the VOR/DME, as shown in FIG. 11-5. In this example, the DME fix is 30 nm from the station and the end of the outbound leg is 20 nm from the station.

DME holding patterns are certainly a lot easier to manage, and this should be a factor when considering whether or not to invest in DME equipment. But don't throw away that stopwatch yet! You'll need it to time nonprecision approaches.

12
Stalls, unusual attitudes, and partial panel

THIS CHAPTER COVERS IFR STALLS, STEEP TURNS, CRITICAL ATTITUDE recovery, and control of the airplane under partial panel conditions. Naturally, you will work on these situations and conditions under the hood.

Stalls under the hood?

Well, why not? Many people are surprised that my course includes stalls under the hood; however, it is very helpful for an instrument student to become familiar with the instrument indications and the feel of the airplane as its controllability degrades from minimum controllable airspeed into the power-off full stall.

In the real world of IFR you might find yourself in a power-off stall if you become distracted and reduce power too much or you allow the airspeed to drop too low on an approach. Likewise, a power-on stall could develop at the missed approach point with the rapid application of power with gear and flaps down, especially if a considerable amount of nose-up trim has been set. *Fly the airplane first* then trim out the pressure.

PRACTICING STALLS

You have been practicing flight at minimum controllable airspeed since Flight Lesson 2. Stall work will simply be an extension of minimum controllable airspeed. In practicing the power-off stall, fly at minimum controllable airspeed with power set at 1500 RPM or 15 inches of manifold pressure and flaps and gear down if applicable.

Instructor note. Pick a convenient safe VFR altitude for practicing stalls and *always* do clearing turns. The student should do the clearing turns under the hood while setting up for minimum controllable airspeed. This will provide valuable additional practice in handling the airplane on instruments through a wide range of changing control forces. The instructor has sole responsibility for collision avoidance.

When straight and level with flaps and gear down and power at 1500 RPM, reduce power to idle. Cross-check with the turn coordinator and add rudder pressure to keep the ball centered. Allow the airspeed to decrease while holding altitude constant until a full stall occurs.

STALL RECOVERY

As the airplane stalls, effect the stall recovery by applying full power. Reduce back pressure to reduce the angle of attack. Don't push forward on the yoke—you will lose too much altitude if you do. Reduce back pressure to pitch down slightly (as seen on the attitude indicator) then return to straight and level flight promptly without inducing a secondary stall. Use the attitude indicator to maintain straight and level. Use rudder pressure, not ailerons, to hold heading while the airspeed is low. (Smoothness is very important, as always!)

In a stall, one wing will frequently drop. The reaction of many pilots in critical situations and at critical airspeeds is to use the ailerons to raise that low wing. This is incorrect in most airplanes and might make things worse. The proper procedure is to use opposite rudder to add a little speed to the slower descending wing and give it lift.

Instructor note. Because of deficient earlier VFR training, students might require extra practice and instruction in the use of rudder rather than ailerons to raise a low wing during a stall.

As the power becomes effective, start raising the flaps in increments. When climb airspeed (V_y) is attained in a straight and level attitude, gradually pitch up to the first line above the horizon on the attitude indicator. (In a retractable gear airplane, delay raising the landing gear until a positive rate of climb has been established.)

Retracting the gear

Flaps are raised before the gear is retracted for two reasons. First, the flaps create much more drag than the landing gear at slow speeds. A climb can be established sooner when the flaps are raised first.

Second, when you execute the missed approach, you want to establish a positive rate of climb before retracting the landing gear. If you have erroneously allowed the airplane to drift down, you will touch the runway with the wheels, not the first 6 inches of the propeller blades!

This stall maneuver combines the power-off stall that might occur on the final of an instrument approach with the full-power, flaps-and-gear-down situation of a missed approach. The IFR stall maneuver should be practiced on instruments until you can quickly recognize the stalled condition when you begin to lose altitude, then effect a prompt recovery with minimal additional loss of altitude. After you become skillful at this, you should lose no altitude. Losing more than 25–50 feet in the recovery is unsatisfactory. Recognize the stall quickly and execute the recovery procedures promptly, correctly, and automatically.

STEEP TURNS

The next maneuver was required to obtain a private pilot certificate: steep turns at 45° bank. The difference now is steep turns with the hood on, solely by instruments.

Instructor note. The traditional method is to establish a 45° bank and then complete a 360° turn in one direction, followed immediately by a 360° turn in the opposite direction, rolling out on the original heading.

Common errors are altitude control, especially during and immediately following the roll-out, and changing to the opposite direction. Students will usually lose altitude at 60° into the turn when executed to the left and gain altitude in a right turn because of "P" factor. During the change of direction, there is also a tendency to gain altitude due to the excess back pressure required. An excellent training maneuver to overcome this problem if it persists is to have the student do a series of entries to steep (45° bank) turns in opposite directions until the problem is solved.

Steep turns require a faster scan to make sure you absorb all the information that the instruments are showing you when you need it. You can't afford to fixate on any single instrument because everything will be happening quickly. You will use the same control and support instruments for steep 45°-bank turns as standard rate turns, but move your eyes around the panel faster using the same scan pattern—attitude, heading, altitude—but do it faster.

One problem with this maneuver is that the inner ear senses a 90° bank with each change in direction. Beware of nausea; when airsickness occurs, all learning ceases.

Altitude control

Cross-check with the altimeter to maintain altitude. On this maneuver ±100 feet is allowed. However, it is much easier to limit the variation in altitude to ±20 feet than to allow the altitude to vary by 100 feet. Continue the turn for a full 360°. When established in this steep turn, of course, the bank tends to increase. If you are not paying attention or your scan is too slow, expect a rapid loss of altitude and an ever-steepening bank. Correct this by reducing bank to 20° or 30° to recover the lift lost in the steep turn. It's almost impossible to regain the lost altitude unless you decrease the angle of bank.

Trim

Students almost always ask what to do about trim in steep turns. I recommend that the maneuver be performed a few times without adjusting the trim to feel what the control pressures are like and what it takes to cope with the control pressures solely by instruments.

Add a little power—100 RPM or so—in the turn to help keep the nose up. Add a touch or two of nose-up trim. With practice the combination of added power and nose-up trim will result in smoother, easier steep turns.

After completing a 360° turn in one direction, make a smooth transition to a steep turn in the opposite direction. Do not pause during the reversal and do not fly straight and level between one turn and the other.

Keep the attitude indicator dot right on the horizon as you roll and do not allow altitude to vary more than ±20 feet for an easier time managing the control pressures in the reversal. You have been applying so much back pressure (or adding power and nose-up trim) to maintain altitude during the 45° bank that when you roll into this reversal, it feels like you have to push the nose forward to keep from climbing. Also, lift increases when the wings roll through the level position. Fix that dot on the horizon line and visualize rolling around it from one direction to another.

COPING WITH VERTIGO

Steep turns under the hood might induce vertigo. Mild vertigo can make a pilot feel as if the airplane is in a climbing or diving turn when flying straight and level. In extreme vertigo cases, it might feel like a straight-and-level descent when the airplane is actually plummeting down in a tight, ever-steepening turn—the infamous *graveyard spiral* that claimed so many early air mail pilots flying at night or in the clouds before the introduction of adequate flight instruments. The spiral also claims many VFR pilots who continue flying into IFR conditions.

Fortunately, with reliable instruments and better knowledge about human physiology, there is no reason to succumb to vertigo, or *spatial disorientation* as it is technically called. One cardinal rule for coping with any kind of vertigo, dizziness, or confusion between what your eyes see on the instruments and what your other senses are telling you is trust your instruments. Accept the fact that in many circumstances the senses are wrong when instruments are correct.

Here's what happens: When flying VFR, eyes are the main source of information about motion and position, just as eyes are primary sensors on the ground. Pilots control the attitude of an airplane by seeing that the nose is above, below, level with, or banked in relation to the horizon and other outside references. The mind subordinates the input of other senses—hearing, smell, motion—to the messages coming from the eyes.

When flying by reference to instruments, the eyes are seeing symbolic rather than actual references. The input is neither as strong nor as vivid as the real world. Other senses—particularly the motion senses—tend to take over the eyes' role.

Semicircular canals

Semicircular canals located in the inner ear sense motion. The canals are three tiny tubes set roughly at right angles to each other. The tubes are filled with a fluid that is set in motion with movement. The fluid's direction is sensed by tiny hairs and transmitted to the brain.

But the fluid does not keep moving. Unless there is further acceleration or deceleration, the fluid slows down and motion ceases. The hairs return to their normal position. If you roll into a turn and hold it steady for 30–45 seconds, the fluid in the semicircular canals stops moving and the hairs cease sending the turning message.

If your senses are screaming "straight and level" while actually in a turn, you might be tempted to increase the angle of bank. This is how the ever-tightening turns of the graveyard spiral begin.

With more experience controlling the airplane solely by reference to instruments, the sense of sight will override the sense of motion. But vertigo is an insidious problem. It can creep up on even the most experienced instrument pilots if fatigue sets in. Illness, medication, alcohol, sleep loss, and mild hypoxia (lack of oxygen) can increase susceptibility to vertigo, as can extended uncoordinated climbs with the ball off center.

Returning now to the 45° banked-turn maneuver, start from straight and level at some convenient cardinal heading such as 360°. Roll into the turn and establish a 45° bank with the attitude indicator. Place the dot right on the horizon line and visualize the airplane nose rolling on that dot during reversal to the opposite direction.

UNUSUAL ATTITUDES

A steep turn with a bank that increases beyond 45°, coupled with a rapid loss of altitude, opens the realm of unusual attitudes. From the beginning, my students work on unusual attitudes with the attitude indicator covered up. It is advisable not to practice initially with a full panel. (This is the instructor's decision.)

Let's consider what happens when an attitude indicator begins to fail. The first point is that it takes time for the gyros to wind down. Even if there is a sudden failure of the vacuum system powering the gyros of the attitude indicator, the gyroscopes in the instrument will lose momentum slowly.

You might not be aware that a failure has occurred. The attitude indicator doesn't suddenly roll over and die at a dramatic angle; the indicator gradually drifts off. You might continue to use the attitude indicator as the control instrument while it is gradually leading you astray.

The first assumption dealing with an unusual attitude is that the attitude indicator has failed. Don't stop to analyze the failure. Assume that it has occurred, deal with the unusual attitude immediately, and when everything is under control again, try to figure out what went wrong.

RECOVERY PROCEDURES

The first instrument to check in an unusual attitude is the airspeed indicator. Its indication will determine what actions to take. If the airspeed is *increasing* the airplane is in a dive and might run out of altitude during an approach or departure, or exceed the redline airspeed (V_{ne}, never exceed) if flying at cruise altitude.

The recovery procedure for a *diving* unusual attitude is:

1. Reduce power to idle.
2. Level the wings with rudder and aileron. Center the needle of the turn coordinator to control the bank. Keep the ball centered.
3. Raise the nose to stop the descent. Refer to the airspeed indicator to increase pitch and stabilize the airspeed at cruise.
4. When cruise airspeed is attained, apply cruise power and establish straight and level flight on partial panel. (*See* the subsection regarding partial panel procedures in this chapter.)

Let's cope with an unusual attitude in which the airspeed is *decreasing*. The recovery from a *climbing* unusual attitude is:

1. Add full power to increase airspeed and reduce the risk of a stall.
2. Lower the nose.
   ~ Don't run out of airspeed and get into a power-on stall.
   ~ Decrease the angle of attack.
   ~ Use the airspeed indicator to decrease pitch and return to cruise airspeed.
3. Level the wings with rudder and aileron.
   ~ Stabilize the turn coordinator to control the bank.
   ~ Keep the ball centered.
4. When the airspeed reaches cruise, reduce to cruise power and establish straight and level flight by partial panel.

Unusual attitude recovery procedures summary

Resumption of control is initiated by reference to airspeed, altimeter, VSI, and turn coordinator.

Nose low, airspeed increasing:

1. Reduce power
2. Level the wings
3. Raise the nose

Nose high, airspeed decreasing:

1. Increase power
2. Lower the nose
3. Level the wings

Common errors:

1. Fixation. Staring at the least important instrument. (Keep eyes moving.)

2. Improper trimming or continuously holding pressure against the trim. (Learn proper trimming.)

3. Cockpit disorganization. (Plan to take every step in proper sequence.)

4. Attempting recovery by the seat of the pants, which produces misleading sensory inputs. (Believe the instruments.)

Finally, as the famous violinist told the budding musician who inquired about how to get to Carnegie Hall: Practice! Practice! Practice! Practice!

DEGREE OF REALISM

Practice unusual attitudes in realistic circumstances. I have students start a turn and lower their head to try to find an obscure intersection on an en route chart or review a missed approach and holding instructions. Trying to find something complicated while in a turn is certainly a realistic and common scenario in the real world of IFR.

Usually when I say "OK, you have the airplane," the student's head snaps up so quickly that the motion induces a touch of vertigo to add further realism to the scenario. The maneuver need not be violent. A nose-high or nose-low 45° bank is usually sufficient to train the student to make the correct responses.

I would like to simulate failure of the attitude indicator by turning it off and letting it spin down. But because this is not possible I cover the attitude indicator with a round card cut to fit over the face. (The card can easily be cut from a worn out file folder.) Affix a piece of electrician's tape to the top of the disk to hang the disk on the rim of the panel just above the instrument.

With this arrangement I can lift the disk up without removing it to get confirmation about how well the student is doing by comparing the partial panel with the attitude indicator. This confirmation works very well in building a student's confidence when he or she first begins to work without a full panel. On many panels there is even a little thumb-indent below each flight instrument that makes it easy to lift the disk up momentarily whenever the instructor wants the student to make a comparison.

Instructor note. Covering instruments is also a good technique during full panel practice if a student habitually falls into the habit of fixating on one or two instruments. For example, if you observe the student paying more attention to the movements of the altimeter or VSI than to the attitude indicator, cover the altimeter and VSI occasionally to help the student become accustomed to the function of the attitude indicator as the correct control instrument for pitch.

Failures of the heading indicator are also very rarely dramatic. You might be flying along in fine shape and wonder about problems maintaining a VOR course while holding the reference heading perfectly. You reset the heading indicator to agree with the magnetic compass, but the problem doesn't go away.

It is hard to generalize about characteristics of a dying gyro in the heading indicator or in the attitude indicator or in both, which is the net result of having the vacuum pump fail. The best way to get an idea of how these two gyro instruments behave when they fail is to sit in the airplane for a few minutes after engine shutdown at the conclusion of a flight and watch the slow demise of the attitude indicator and heading indicator as the gyros wind down.

PARTIAL PANEL PROCEDURES

Recognition of instrument failure becomes the first step in partial panel work, regardless of whether the failure becomes apparent in an unusual attitude or in erroneous readings that gradually become more serious. Following recognition of the failure, make a positive decision to disregard erroneous instruments and turn to other instruments to supply the missing information (FIGS. 12-1 and 12-2).

Some pilots find it helpful to cover a failed attitude indicator and heading indicator and they carry little rubber suction cup soap dishes on every flight just for that purpose. This is a good idea. In a tight situation you don't want to inadvertently include failed instruments in the scan. If you don't have any soap dish suction cups handy, just tear off pieces of paper and use them to cover the failed instruments.

The next step in the real world of IFR would be to land immediately after a major instrument failure. Don't hesitate to declare an emergency and get vectors from ATC to the nearest VFR airport. If no VFR field is available, head for the nearest IFR airport that has your personal minimums. There is absolutely no point in continuing an IFR flight with major instrument malfunctions. Things will only get worse as fatigue sets in.

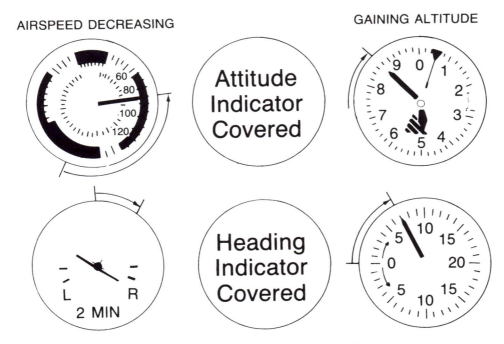

Fig. 12-1. *Nose-high unusual attitude on partial panel: steep right turn.*

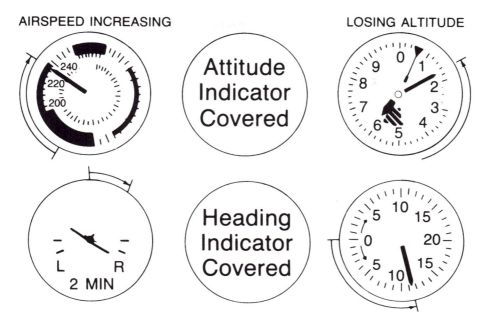

Fig. 12-2. *Nose-low unusual attitude on partial panel: steep right turn.*

PARTIAL PANEL CONTROL INSTRUMENTS

Because you can't always count on going VFR immediately after an instrument malfunction, be prepared to fly on partial panel. Other instruments replace the functions of the attitude indicator and heading indicator when no longer available (FIG. 12-3). As with full panel (FIG. 8-5), a primary instrument provides the most pertinent information about how well you are doing and it does not move when flying precisely.

Use the same scan pattern for partial panel as full: attitude, heading, altitude. The only difference with partial panel is that you scan two instruments for attitude information rather than one.

Note in FIG. 12-3 how the instruments are used in straight and level flight. The control function, formerly performed by the attitude indicator, is now divided between the airspeed indicator and the turn coordinator.

Pitch is controlled with the airspeed indicator. If the airplane is 100 feet below the desired altitude, use back pressure on the yoke to reduce airspeed 5–7 knots. This will pitch the nose up slightly and regain the lost altitude. Conversely, if the airplane is 100 feet above the desired altitude, increase airspeed by a very slight forward pressure on the yoke and pick up approximately 5 knots, indicating a slight nose-down pitch. This will gradually take the airplane back to the desired altitude.

When you reach the desired altitude and the airspeed stabilizes at the desired cruise speed, a minor trim adjustment might be required to help hold that altitude. Keep wings level with the turn coordinator. As with full panel, keep both feet on the rudder pedals and

CONTROL		PRIMARY	SUPPORT
STRAIGHT and LEVEL			
Pitch	Airspeed Indicator	Altimeter	VSI (Rate of Climb)
Bank	Turn Coordinator	Magnetic Compass	ADF
Power			RPM/MP
STANDARD RATE TURN			
Pitch	Airspeed Indicator	Altimeter	VSI
Bank	Turn Coordinator	Sweep Second Hand	ADF
Power		Airspeed Indicator	RPM/MP
CONSTANT AIRSPEED CLIMB			
Pitch	Airspeed Indicator	VSI	Altimeter
Bank	Turn Coordinator	Magnetic Compass	ADF
Power			RPM/MP
CONSTANT AIRSPEED DESCENT			
Pitch	Airspeed Indicator	VSI	Altimeter
Bank	Turn Coordinator	Magnetic Compass	
Power			RPM/MP
CONSTANT RATE DESCENT (ILS)			
Pitch	Airspeed Indicator	Glide Slope Needle	VSI
Bank	Turn Coordinator	Localizer Needle	ADF
Power		Airspeed Indicator	RPM/MP

CONTROL = Main reference instruments
*PRIMARY = Key quality instrument
SUPPORT = Back-up or secondary instrument

*The Primary Instrument is always the instrument that gives the most pertinent information and is not moving when flying precisely.

Fig. 12-3. *Control, primary, and support instruments for partial panel. Simulating loss of vacuum affecting attitude indicator and heading indicator.*

apply pressure if the coordinator leans steadily in one direction or the other. The turn indicator is very sensitive; don't attempt to make a correction every time it moves, or else you will begin chasing the needle and very quickly lose control of the heading. As always, keep the ball centered with rudder pressure.

The poor man's heading indicator

Here is an item I bet you won't find in any other book. Note that the ADF is listed as a support instrument for bank in straight and level and several other flight conditions. Why is this? Because the ADF is an excellent source of bank information. When an ADF is tuned to a station in front of the airplane, the ADF will very quickly indicate drifting left or right, the same way the heading indicator would if it were functioning. I call the ADF the poor man's heading indicator.

The ADF can also help you make standard rate turns. For example, if the ADF is tuned to a station ahead and you want to make a 30° turn to the right, simply turn and begin the rollout approximately five degrees before the ADF needle has moved 30° to the left. There's more to the ADF than meets the eye!

With practice you will automatically look to the correct instruments for control references to replace the attitude indicator. As you make corrections and enter climbs, descents, and turns, be careful to make minimal control movements. The greatest problem I find with instrument students flying partial panel is overcontrolling. Know exactly where to look for control references and minimize control movements and you will rapidly build up skills in the fine art of partial panel flying.

MAGNETIC COMPASS TURNS

Note how the magnetic compass starts to come into play as a primary instrument (FIG. 12-3). The magnetic compass is one of the most familiar and perhaps least understood instruments in the cockpit. It is extremely reliable, even though it bounces around a lot, and it is the only source of heading information that operates completely independently of all electrical, vacuum, and pitot-static systems. Pilots who understand the behavior of the magnetic compass and make a point of practicing with it frequently can achieve amazing flying precision with this instrument.

However, the magnetic compass "lies." You must know when and why. VFR training taught you how to use magnetic compass lag and lead to roll out on headings accurately. Here is a review:

1. When on a northerly heading and you start a turn to the east, the magnetic compass will initially show a turn to the west and will gradually catch up as the turn progresses to give an accurate indication when passing through 090°. Conversely, when on a northerly heading and you start to turn to the west, the compass will initially show a turn to the east, then gradually catch up.

2. When on a southerly heading and you start a turn toward the east, the magnetic compass will initially indicate a turn to the east, but will exaggerate the turn, gradually reducing the error and will be accurate when passing through 090°. When turning to the west, the compass behaves similarly by exaggerating the amount of turn initially, but in the correct direction.

   This lagging and leading behavior of the magnetic compass in 1. and 2. above is *dip error* caused by the earth's magnetic lines of force and their effect on the magnetic compass when it is not precisely level. (A rough formula: dip rror is equal to the closest latitude in degrees.)

3. When flying on an easterly or westerly heading, aircraft acceleration results in a northerly turn indication; deceleration results in a southerly turn indication.

Remember this by the acronym ANDS:

Accelerate

North

Decelerate

South

Magnetic compass procedures and exercises

1. When turning to a heading of south, maintain the turn until the compass passes south the number of degrees of dip error (degrees of latitude) minus the normal rollout lead (one-half the angle of bank).

2. When turning to a heading of north, lead the compass by the amount of dip error (latitude) minus the normal rollout lead.

3. When turning to a heading of east or west, anticipate the rollout by the normal method.

The main point to remember is that on southerly headings, the magnetic compass precedes or leads the actual turn. On northerly headings, the magnetic compass lags behind the turn. Figure 12-4 shows the procedures for making magnetic compass turns in graphic form.

Turns to a heading based upon indications from the magnetic compass are imprecise at best; however, you should practice magnetic compass turns on instruments until you can roll out on a specified heading ±10° using the magnetic compass alone. But don't waste time trying to go beyond this. Concentrate on learning and recognizing the errors affecting the magnetic compass and never believe it unless you are straight and level and in stabilized flight.

This is especially important during the departure and approach phases of an instrument flight when the heading indicator must be accurately set. Do not adjust the heading indicator unless the airplane is in stable and straight and level flight. In a normal approach, the last opportunity for determining heading indicator accuracy might be the outbound (1 minute) leg of the procedure turn.

TIMED TURNS

Partial panel heading changes don't always work so neatly that you can do them with the magnetic compass alone. Suppose that the airplane is on a heading of 035° and you needed to turn to 155°. How would you compute the rollout using the magnetic compass?

In this case you might be better off to use a standard rate turn timed with the sweep-second hand of the clock. Turn 120° to get from the present heading of 035° to the new heading, 155°. At 3° per second for a standard rate turn, you would execute a timed turn for 40 seconds.

Plan the rollout so that wings are level when time is up. If the rollout begins 5 seconds before the 40 seconds is up, the heading change is exactly 120°. Make minor adjustments to the new heading as necessary after the magnetic compass has settled down.

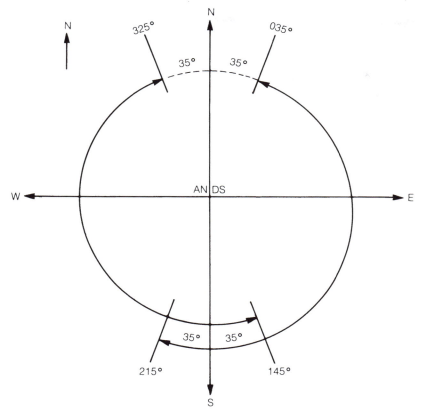

Fig. 12-4. *Magnetic compass leads (to the south) and lags (to the north) assuming a latitude of 30° for the dip error and one-half a bank angle of 10° for leading the roll in, with acceleration/deceleration errors on east and west headings.*

For small changes of heading, use a half-standard rate turn ($1\frac{1}{2}°$ per second). Roll into the half-standard rate turn and count "one one thousand, two one thousand, three one thousand" up to the number of degrees of heading change that you wish to accomplish. This will also bring you close enough to make minor adjustments after the magnetic compass settles down.

Practice steep turns, unusual attitudes, partial panel, and magnetic compass turns from Flight Lesson 6 through the end of the program. The test standards require demonstration of these techniques during the instrument flight test.

13
The NDB unmasked

THE NONDIRECTIONAL BEACON (NDB) IS FOR EVERYONE. I LOVE IT, you should too. Not too long ago many people thought the NDB was on its way out, a relic of the past as far as modern IFR flying was concerned. But this has not proven to be the case. NDBs have traditionally made it possible for small airports to have inexpensive and reliable instrument approaches when ILS and VOR installations are not feasible or too costly. (For example, as chief flight instructor for a school at Lincoln Park, New Jersey, I obtained FAA approval for Lincoln Park's first instrument approach, an NDB approach based upon the compass locator for the Morristown ILS 23 approach.)

In addition to serving small airports, NDBs serve many other crucial functions at larger airports: fixes for holding patterns, procedure turns, missed approaches, and as compass locators for ILS approaches. For example, New York state has 65 airports with instrument approaches. Of these, 26 have NDB approaches. If the plane you fly on your instrument flight test has an ADF, chances are you will be asked to demonstrate an NDB approach.

For some reason NDBs and the airborne ADF receivers have been a mystery for many years. The first aviation direction finding equipment was very complicated, no doubt about it. You had to turn a wheel in the ceiling of the cockpit that rotated a loop antenna attached to the fuselage. In photographs of older planes—as well on the older airplanes in museums—this loop is quite prominent.

That was not an ADF loop, it was a *DF loop*. There was nothing automatic about it. It had a direction finding antenna, but the pilot had to turn it manually, listen to the signal build and fade, then interpret the signal to determine the direction to or from the station.

The introduction of the automatic direction finding (ADF) system was a great advance. Now the indicator of the direction finding instrument automatically points to the station, no matter where the airplane is and the ADF indicator is on the instrument panel not overhead or in back.

ADF ORIENTATION

This leads to the first concept to be stressed in ADF work: The needle always points to the station. This is an obvious point, but many people do not clearly understand its implications. First of all it means that when tuned to an ADF station for homing, tracking, intersections, holding, or an approach, you never have to touch the system. Unlike the VOR, there's no OBS to think about and no need to "twist" anything at station passage or at any other time. This makes ADF much simpler to use.

Second, with ADF you always know where you are in relation to the station. There is no TO-FROM to interpret, no confusion about radials and bearings, no way to set the wrong OBS numbers. ADF orientation is much simpler than VOR orientation. The head of the ADF needle always points to the station. With the azimuth set on 0 (zero)—straight ahead—the ADF needle will always indicate *relative bearing* to the station. (Relative bearing is the number of degrees that the station is from the nose of the airplane.)

To determine magnetic bearing to the station, simply add the *relative bearing* to the *magnetic heading* shown on the *heading indicator*, which equals the *magnetic bearing* (course) to the station. You undoubtedly learned this in your primary training, but let's do a quick review now for some hints to simplify the process.

If you are on a magnetic heading of 030° and the ADF needle is 90° to the right, the magnetic bearing to the station is 120°.

$$030 + 090 = 120$$

Turn to 120° and the ADF needle will point straight ahead.

You don't even need to make a turn to confirm this. Take a medium-length pencil and place it on the needle of the ADF, much in the same manner as a parallel ruler. Move it onto the heading indicator and the pencil will point to the magnetic bearing to the station, eliminating the arithmetic. This is one of the shortcuts used in flight to simplify a visualization of "where we are now."

An inexpensive feature on many ADF indicators is a third, even simpler method of determining magnetic bearing to the station: the rotating azimuth ring. Simply rotate this ring manually to line up the magnetic heading with the mark at the top of the ADF indicator and the ADF needle will automatically point to the magnetic bearing to the station.

Let's say you are on a heading of 300° and the needle is pointing to the right wing of the airplane at 090°. Apply the formula and add 300° plus 090° and come up with 390°.

If the answer is more than 360°, all you have to do is subtract 360° from the total for the magnetic bearing to the station.

$$300 + 090 = 390 - 360 = 030$$

Instructor note. Check the ADF visualizing devices and teaching aids available through aviation supply companies. These simple and inexpensive plastic or cardboard devices are invaluable for demonstrating ADF problems on the ground. Students can practice with them at home. This will save time and money compared to doing the same exercises in the air. And it will make your job easier.

Have your student diagram several examples using two circles, one for the heading indicator and the other for the ADF. This is an excellent practice drill for the student and for you to determine if the student really understands the ADF and how to use it. This will develop their skill at visualization.

Be careful, as with a VOR, to tune the station correctly and verify its Morse code identifier. Turn up the audio sufficiently to hear the ADF signal faintly in the background—but not so loud as to interfere with communications—and keep the volume at that level.

When there is a disruption in VOR signals, warning flags appear on the face of the nav instrument; this does not happen on the ADF indicator. Continuously monitor the identifier to detect any signal disruptions. Keep the volume low enough to hear the signal in the background. This is the only way to be sure that the ground station has not gone off the air for some reason or that the ADF unit in the airplane has not malfunctioned.

Check the heading indicator against the magnetic compass at least every 15 minutes and reset it as necessary throughout all phases of IFR and VFR flight. This is critical in ADF work because it is impossible to determine the magnetic bearing if the heading indicator has drifted off. Prior to a magnetic bearing determination, verify accuracy of the heading indicator.

ADF TIME/DISTANCE CHECKS

An exercise to develop NDB orientation awareness is the ADF time/distance check. Start in a simulator, which often reinforces the ability to orient around an NDB so well that a little additional practice in flight is needed.

Instructor note. For airborne ADF practice of the time/distance check and other ADF maneuvers, use commercial broadcast stations that lie well outside airways, approach and departure corridors, and traffic patterns. Very few are shown on sectional charts. Look them up in a publication called *Flight Guide* sold by many FBOs and aviation supply houses. It comes in Eastern, Central, and Western U.S. editions. For further information contact:

Tel: 1-800-FLY-FLY1
Web: www.flightguide.com

Pick stations with the highest output for the best results. The needle might wobble with a commercial station, but this usually clears up closer to the station and it is not a problem with powerful stations.

Tune the station and determine the relative bearing. Turn the number of degrees necessary to place the ADF needle at either 090° or 270° relative—the right wing or left wing.

Note the time and fly a constant heading until the bearing changes 10°. Note the number of seconds it takes for the bearing to change 10° then divide by 10 for the time to the station in minutes.

$$\frac{\text{Time in seconds}}{10} = \text{Minutes to station}$$

You can determine distance to the station with this formula:

$$\frac{\text{TAS} \times \text{minutes flown}}{\text{Degrees of bearing change}} = \text{Distance to station}$$

INTERCEPTING A BEARING

Let's take an example from the real world of IFR. Suppose you have executed a missed approach and want to return to an NDB via a specified bearing to be in a good position to make another approach.

First, turn to the desired bearing.

Second, note the number of degrees of needle deflection to the left or right of the 0° position on the face of the azimuth card (ADF indicator) and double this amount to determine the *intercept angle*.

Third, turn toward the head of the needle the number of degrees determined for the *intercept angle*. As you turn toward the needle this predetermined number of degrees, the needle will pass through the 0° position and on *to the other side* of the 0° position on the face of the ADF indicator.

Wait a minute, the needle always points to the station. Why does it appear to move? The answer is that the needle doesn't move, the airplane moves and the ADF indicator face is attached to the airplane. In a turn, the needle continues to point to the station, but the airplane is moving under the needle.

A good, simple way to visualize this is to place a book or other object on the floor to simulate an NDB station. Stand a few feet away from the object, and point toward it so your arm simulates the ADF needle. Your nose becomes the zero point on the indicator face, matching the nose of the airplane.

Now turn your body to a new "heading" while continuing to point toward the "station." Your arm will behave the same way as the ADF needle, apparently moving away from your nose. But you will quickly see that it is really your body that is turning while your arm continues to point steadily at the "station."

Back to intercepting the bearing.

The fourth step is to maintain the new *intercept heading* until the needle is deflected on the opposite side of the nose the same number of degrees as the *intercept angle*.

Then turn to the desired heading, which is the magnetic course inbound. Hold this heading until you notice a drift of the needle, which indicates *wind drift*. The procedure outlined above may be repeated as often as necessary. As you become more proficient through practice, you will be able to determine a wind correction angle and make corrections as you proceed to keep on the magnetic course. (This is not as difficult as it seems at first. Have faith and the mystery gradually unfolds. This is also where thorough flight instructors are worth their weight in gold.)

Let's say you want to head inbound to the station on a 360° bearing (FIG. 13-1). Turn first to a heading of 360° and note the 15° deflection of the needle to the left (A).

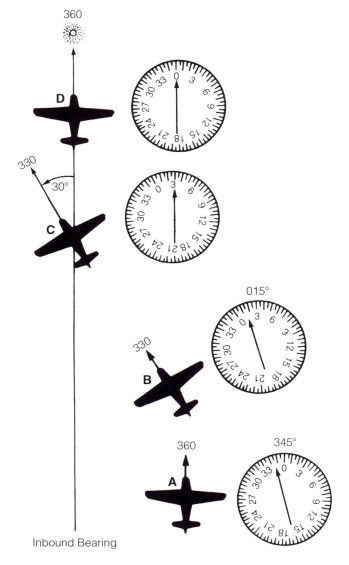

Fig. 13-1. *Intercepting an ADF bearing.*

Next, double the 15° to the left for an *intercept angle* of 30°. Now turn 30° toward the left (B). Note that the needle has now passed to the other side of the nose and the airplane is now heading 330°. Maintain this intercept heading until the ADF needle has deflected 30° to the right as at position (C).

Finally, turn inbound on a heading of 360°, the same as the inbound bearing to the station (D). (Lead the turn by 5° to avoid overshooting.)

With no wind, all you would have to do is continue the inbound heading to the station, but there is always wind. So let's proceed to examine the techniques used to correct for the effects of wind.

HOMING IS UNACCEPTABLE

Heading inbound you can "home" in to the NDB by placing the ADF needle on the nose and keeping it there with heading adjustments. With the wind constantly pushing from one side, you will have to constantly change the heading as you proceed toward the station to keep the needle on the nose.

Figure 13-2 is an illustration of the homing method of reaching an NDB. The airplane starts heading inbound on the 360° bearing at A. As the wind blows from the left, the heading has to be adjusted to maintain the needle on the nose (B and C). At D, the airplane has been blown so far off the inbound bearing that it is flying a heading of 315° instead of the desired 360° to keep the needle on the nose.

Homing is unacceptable for IFR navigation because the airplane strays too far from the intended course. The wide, looping course shown in FIG. 13-2 might lead into the side of a hill, a radio tower, or other obstruction at the minimum altitudes of an NDB approach. Figure 13-2 is not an exaggeration. Many poorly prepared instrument students do this on the instrument flight test. (This is a certain failure on a flight test!)

TRACKING AND BRACKETING

When you have intercepted the desired bearing, hold that heading and see what affect the wind has. Let's continue with the previous example—interception and tracking of the 360° bearing to the station—and see what happens (FIG. 13-3). As the airplane proceeds toward the station, the wind from the left blows it off the bearing. At B the airplane has drifted off the bearing by 15° relative. To get back on course, double the drift noted at point B and turn toward the needle that amount, in this case 30° (C). You are reintercepting the bearing.

The ADF needle will then swing over to a relative bearing of 015°. Hold the intercept heading (330°) until the relative bearing reads 030°. That puts the airplane back on the desired bearing to the station (D). Now reduce the corrections by half—15° in this case—to compensate for the wind (E). (You should also lead the turn back by approximately 5°.)

This method of correcting for wind drift is called *bracketing* and you might have to "bracket" several times to establish a reference heading that will remain on the desired bearing, especially if a long distance from the station. The initial wind correction might be too large or too small to stay on the bearing. If so, adjust the correction. In the example

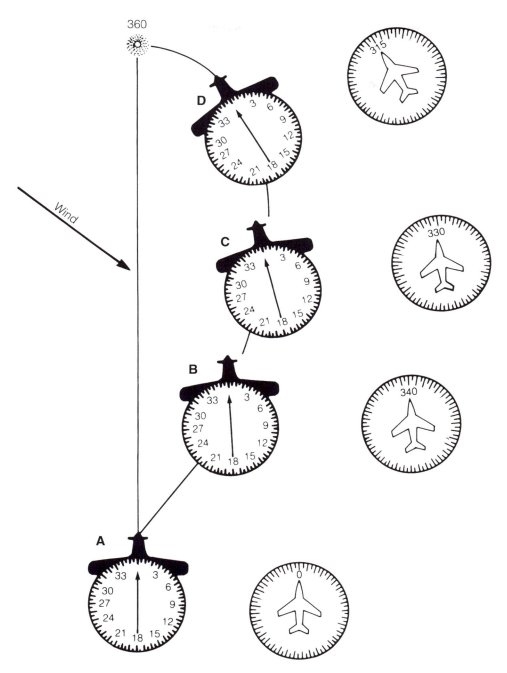

Fig. 13-2. *The problem with ADF homing is that it takes you off course to a degree that is unacceptable for IFR flying—especially on NBD approaches.*

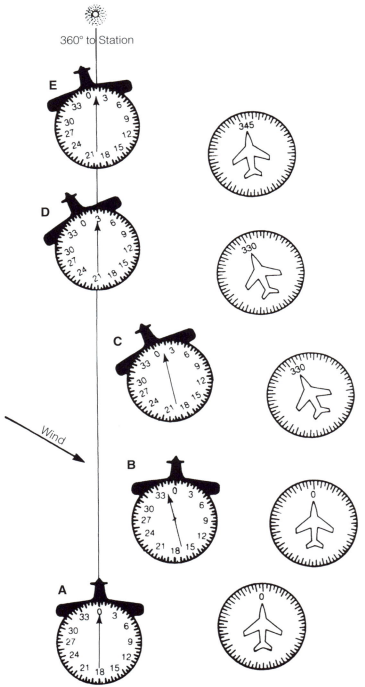

Fig. 13-3. *ADF tracking corrections utilizing the bracketing procedure.*

above, for instance, if a 15° correction proves to be too much, reintercept the bearing and try a 10° correction.

Chasing the needle is a common mistake in ADF intercepts, tracking, and bracketing, just as in VOR navigation. It is so tempting to follow that moving needle! Resist the temptation. Hold the heading steady until the needle reaches the relative bearing you want, then make the turn.

Near the station the ADF needle will become "nervous" and start oscillating, becoming more sensitive. Don't chase the needle, just fly the reference heading. Passing over the station the needle will commence a definite swing to the right or left. Note the time and start timing the outbound leg.

Wait until the needle has definitely swung around to verify station passage—at least 5–10 seconds past the ADF station. At this point you should turn to the outbound magnetic course to determine which side of the bearing you are now situated and how much you will have to correct. It can be fun!

OUTBOUND BEARINGS FROM THE NDB

The procedures for intercepting, tracking, and bracketing outbound from the station are almost identical to procedures for the inbound magnetic bearing.

Turn to the outbound bearing and determine how many degrees you are off the bearing. Then double the error and turn toward the desired bearing by this amount. If off the bearing by 10°, turn toward the bearing 20°.

When the angle of the ADF needle off the tail and the intercept angle are the same, the airplane is on the desired bearing. Turn toward the outbound bearing and bracket outbound to determine the wind correction necessary to hold that bearing.

Remember that the needle always points to the station. Never put the needle on the tail by changing the heading. This will cause you to miss the bearing completely and as you will see during an NDB approach if you lose that outbound bearing you will miss the airport.

Try to visualize where you are at all times. If in doubt, turn to the outbound bearing and check whether you are to the left or to the right of course, and by how much. Then double this amount and reintercept the outbound bearing. Practice this at altitude until you become thoroughly proficient at tracking outbound before commencing NDB approaches.

Better yet, if you are having a problem tracking NDB bearings outbound, try the visualization exercise again. Place an object on the floor of a large room or parking ramp to represent the NDB and walk through the entire procedure of tracking a bearing inbound, then station passage, then tracking the same bearing outbound, using your right hand to always point to the NDB and your nose as the nose of the airplane. Believe me, this works!

I am very serious about practicing this visualization exercise until you understand clearly what the ADF needle is indicating, especially outbound from the NDB. This is the critical final approach leg on an NDB approach. One of the most frequent causes for failures on flight tests is confusion on the final leg of the approach to the airport. Time and again a candidate will track inbound to the NDB accurately, then turn the wrong way after passing the station.

In so many elements of IFR flight, visualization is the key to making the correct moves. The visualization exercise doesn't cost a cent, no matter how often you practice it.

PRACTICE PATTERNS

A good exercise for sharpening your skill at tracking and bracketing inbound and outbound is the simple pattern depicted in FIG. 13-4. It is a good idea to use a commercial broadcast station for practicing this and other patterns to avoid straying into busy airspace.

Pick a *cardinal heading* such as 270°, and intercept the bearing inbound to the station as shown. (A cardinal heading is one of the four directional points of a compass: north, east, south, west.) After station passage, track outbound correcting for the wind for 3–4 minutes.

Then reverse course with a 90-270. Make a standard rate turn in one direction for 90°, then reverse smoothly into a standard rate 270° turn in the other direction. The 90-270 is a quick and efficient 180° change in heading. And if you have maintained a steady bearing outbound, the 90-270 will place you close enough to the inbound bearing so that only small corrections will be required.

Track inbound to the station, then outbound on the other side for the same amount of time. Do another 90-270, and repeat the process until you can maintain steady bearings with corrections. A standard procedure turn may be substituted for the 90-270 course reversal.

Other good ADF exercises are Pattern A and Pattern B set up on an NDB or commercial broadcast station. Start each pattern over the station and orient the pattern on cardinal headings, at least in the beginning. Plan for each straight leg to return over the station.

These two patterns will provide plenty of practice in intercepting bearings and tracking inbound and outbound. And because the patterns contain all elements of an instrument approach they are good introductions to NDB approaches, which are discussed in Chapter 14. Patterns A and B are also good exercises to practice in a simulator.

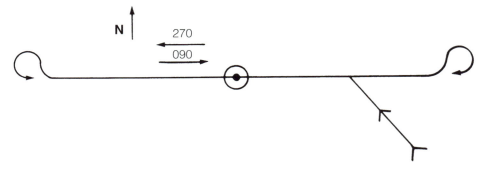

Fig. 13-4. *Pattern for inbound-outbound ADF tracking practice, with 90-270 course reversals.*

ADF HOLDING PATTERNS

Nothing could be simpler or easier than an NDB holding pattern. (Procedures for entering a VOR holding pattern apply to entering an ADF holding pattern.)

Track inbound on the desired bearing, making wind corrections as needed to maintain the bearing. At station passage, begin a standard rate turn to the right (left in a nonstandard holding pattern). Roll out on the reciprocal of the inbound bearing and double the wind correction to account for the effect of the wind on the outbound leg and in the two turns.

Mark station passage when the ADF needle reaches the 90° position. Adjust the timing of the outbound leg (nonprecision side) to produce a 1-minute inbound leg. Now here's a good cross-check that many instructors overlook. After flying outbound 1 minute, the needle should point 30° off the tail: 30° off to the right in a standard holding pattern, 30° on to the left in a nonstandard pattern.

Completing the outbound leg, turn again, rolling out on the inbound heading. Note the bearing error and correct for it.

Once again, the quickest way to visualize ADF holding patterns is to run through the floor exercise described earlier in this chapter. Review VOR holding for methods of entering holding. The same procedures apply to entering ADF holding patterns.

Instructor note. Students will get the picture on ADF holding patterns much quicker if you start them with direct entries, then progress to teardrop and parallel entries. Note also that holding patterns are excellent exercises if a student has difficulty with ADF procedures close to the station. You might extend the legs of the pattern to 2–3 minutes to provide more practice in bracketing and tracking. If you are in an ATC environment, get permission to extend the legs.

There is nothing like intensive ADF work to rivet a student's attention on the instruments. This is all to the good. But it also means that the instructor or safety pilot must make a greater effort to watch out for other aircraft, particularly if the NDB you're using is part of an instrument approach. If this is the case, practice at a higher altitude above the approach. And, as suggested before, invite another pilot along to act as an additional set of eyes.

14
Approaches I: Approach basics and NDB approaches

ONE OF THE MOST SATISFYING THINGS YOU WILL EXPERIENCE AS AN instrument pilot is breaking out at minimums on an IFR approach with the runway straight ahead and a comfortable landing assured. The average VFR pilot—and most passengers—considers this nothing short of miraculous!

But it isn't, really. By the time you begin to polish the fine points of IFR approaches, you will have learned to use the ADF and VOR equipment with great precision. And, if your instructor has been following the syllabus, you will have made several unhooded instrument approaches—enough to see the "big picture" of an instrument approach.

This chapter covers the basics of approaches. I will show how to analyze an approach while planning for an IFR flight and move step by step through representative examples of ADF approaches.

Understand, however, that every approach is different. The examples will illustrate the usual sequence of events in nonprecision approaches plus the dialogue with approach control that accompanies this representative sequence. You will be able to apply the approach procedures, techniques, and communications to all nonprecision approaches. But always remember that approaches in the real world of IFR will differ from these representative examples. Adjust your procedures and communications accordingly.

NONPRECISION APPROACHES

A nonprecision approach is defined in the Pilot/Controller Glossary as "a standard instrument approach procedure in which no electronic glideslope is provided." ADF, VOR, DME, and several less common types of approaches fall into the nonprecision category because they do not provide electronic glideslopes. The only electronic guidance they provide is for the approach course.

PRECISION APPROACHES

A precision approach, on the other hand, is defined in the Pilot/Controller Glossary as "a standard instrument approach procedure in which an electronic glideslope/glidepath is provided." An ILS provides an electronic glideslope and is thus a precision approach. Precision approach radar (PAR) depicts on the radarscope an electronic glidepath along which the airplane is guided by the final approach controller. In the military, this is known as GCA, or "ground controlled approach." I will discuss precision approaches in Chapter 16, however, nonprecision and precision approaches have many elements in common.

Common elements

First and foremost, all instrument approaches have an altitude below which you cannot legally descend unless the airplane is in a position to make a safe landing. This altitude is called *minimum descent altitude* (MDA) for nonprecision approaches; *decision height* (DH) for precision approaches. Other common elements include an *initial approach fix* (IAF), a *final approach fix* (FAF), a final approach course, a missed approach procedure, and very often one or more intermediate fixes between the IAF and the FAF.

The term *segment* is used frequently. Here is what the different segments mean:

Initial approach segment. The segment between the IAF and an intermediate fix, *or* between the IAF and the point where the airplane is established on an intermediate course or the final approach course.

Intermediate approach segment. The segment between the IAF and the FAF.

Final approach segment. The segment between the FAF and MAP (missed approach point).

Missed approach segment. The segment between the MAP, or arrival at the DH, and the missed approach holding fix.

ALTITUDE MINIMUMS

Let's explore the question of how altitude minimums are derived. The most important consideration, for obvious reasons, is safe obstacle clearance. This is spelled out in the FAA's *United States Standard for Terminal Instrument Procedures* (TERPS), which is the "bible" on instrument approach tolerances. Along the centerline of the approach course, minimum obstacle clearance is provided for nonprecision approaches as follows:

- NDB located on airport, 350 feet
- NDB off airport with FAF, 300
- VOR located on airport, 300 feet
- VOR off airport, 250 feet
- DME arc as final approach course, 500 feet
- Localizer, 250 feet
- ASR radar (no glidepath), 250 feet
- DF steer approach, 500 feet

Obstacle clearance is provided for a "primary" area on either side of the final approach course centerline. The width of the primary area varies with the type of approach and distance from the field. But it is never less than 1 mile on either side of the final approach course centerline of an NDB, VOR, or other nonprecision approach.

These are comfortable obstacle clearances for the precise pilot, but there is not much room for error or sloppy procedures. Coming in from the FAF on an NDB approach, for example, you will clear obstacles a mile on either side of the inbound course by only 350 feet.

One of the most frequent reasons for failing the instrument flight test is going below minimums on an approach. I don't mean just momentarily dipping below a minimum because of turbulence, then correcting right away. What always surprises me are the candidates for an instrument rating who consistently fly 25, 50, or even 100 feet below minimums without taking corrective action, or have not determined the correct minimums to begin with.

Minimums are so basic, yet many pilots seem to have problems with them. Why is this so? I believe it is because pilots do not always use a systematic procedure to analyze the minimums.

ADJUSTMENTS TO MDA

Let's make a step-by-step analysis of a conventional nonprecision approach. If you follow these steps every time you plan a flight, you will develop the good habits that will enable you to quickly size up an unfamiliar approach that might be assigned by ATC at the last minute, perhaps due to a runway change. The example for this exercise is the NDB RWY 26 approach at Pittsfield, Massachusetts, an uncontrolled field in the Berkshire hills of western New England (FIG. 14-1). Let's analyze the Pittsfield approach using a systematic, six-step method. As you will see, this is going to take some detective work.

Fine print

(1) *Read the fine print first*. Don't leave the fine print until last because you might miss something very important. Consider items A and B on the Pittsfield RWY 26 approach chart:

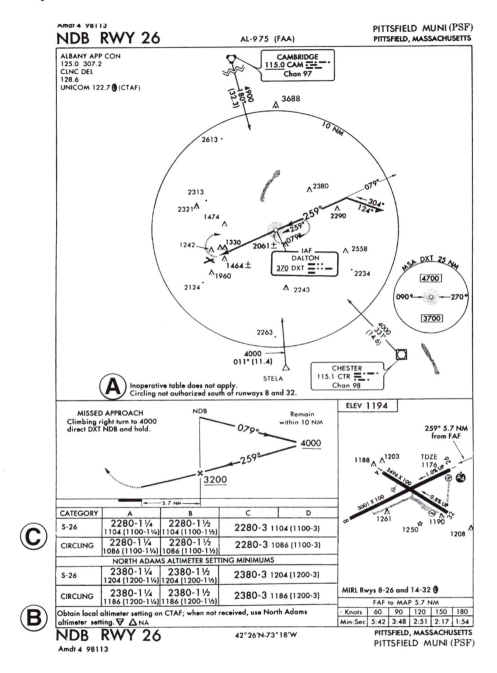

Fig. 14-1. *A typical NDB approach is NDB RWY 26 at Pittsfield, Massachusetts.*

A. "Inoperative table does not apply. Circling not authorized south of runways 8 and 32."

B. "Obtain local altimeter setting on CTAF; when not received, use North Adams altimeter setting."

In item A, "inoperative table does not apply" means that the airport has no approach components whose outage would require higher minimums. I'll have more to say about the inoperative components table later in this chapter. But a statement like this should alert you to the fact that the airport has only minimal lighting. Check the lighting information in the airport box at the lower right of the chart and note what lighting there is. Would this be sufficient in terms of your "personal minimums" for an actual IFR approach down to minimums? How about at night?

The second sentence in item A is very important: "Circling not authorized south of runways 8 and 32." This means that obstacle clearance is not provided in this sector at circling minimums. Stray into this area and you might hit something!

Let's turn now to item B, which presents more of a challenge.

You must have an accurate altimeter setting for every instrument approach. More and more uncontrolled airports, including Pittsfield, have an automatic weather reporting system, either "ASOS" or "AWOS." ASOS stands for *automated surface observing system*. AWOS is the acronym for *automated weather observing system*. These systems provide highly accurate altimeter settings along with other weather information of concern to incoming pilots. Use ASOS and AWOS information as you would ATIS. Detailed information on these automated systems is found in AIM.

As noted in item B, altimeter settings for Pittsfield are provided through the *common traffic advisory frequency* (CTAF). But when this is not available, you must get an altimeter setting from North Adams and use higher minimums.

Here is where the detective work comes in. North Adams has no instrument approach, so there is no handy approach chart giving the North Adams frequency for weather information. You must look up North Adams under Massachusetts in the AF/D. There you will find the frequency on which the ASOS information is broadcast. It is 134.775.

Finally, you should look up and learn the meaning of the triangular symbols that follow the fine print on altimeter settings.

The information contained in the fine print for the NDB RWY 26 approach to Pittsfield must be ferreted out on the ground when you plan the flight. Can you imagine what it would be like trying to look up all these things in flight?

And the fine print can be very important. Here is a "gem" from the instrument approach to Indian Mountain Air Force Station, Alaska, as reported by Barry Schiff in *AOPA Pilot*:

"CAUTION: Rwy located on slope of 3,425' mountain...successful go around improbable."

Take a few minutes to browse through the approach charts for your area and highlight the fine print at the airports you are likely to use. There might be some surprises; mark them and they won't surprise you during an approach.

Height of obstacles

(2) *Check the height of obstacles in the vicinity of the airport.* These obstacles determine the MDA. Note how many obstacles rise above 2,000 feet MSL in the vicinity of Pittsfield. Check the airport diagram at lower right to see how many rise above the touchdown zone elevation (TZDE) of 1,176 feet MSL for runway 26. They do not appear to be a problem at Pittsfield.

Aircraft approach category

(3) *Pick the published minimum for your aircraft category and type of approach,* either straight-in or circling (C). Aircraft approach categories are explained in the front section of every set of NOS approach charts. The explanation is clear and simple:

"Speeds are based on 1.3 times the stall speed in the landing configuration of maximum gross landing weight. An aircraft shall fit in only one category. If it is necessary to maneuver at speeds in excess of the upper limit of a speed range for a category, the minimums for the next higher category should be used. For example, an aircraft which falls in Category A but is circling to land at a speed in excess of 91 knots, should use the approach Category B minimums when circling to land. See following category limits:"

Maneuvering Table

Approach Category	A	B	C	D	E
Speed (Knots)	0–90	91–120	121–140	141–165	Abv 165

Most general aviation propeller-driven airplanes fall in either Categories A or B. Categories A and B are the same at Pittsfied (FIG. 14-1), but this is not always the case.

Straight-in vs. circling. MDAs for each category are further classified by the type of approach, either straight-in ("S-26" at Pittsfield) or circling. Straight-in approaches are allowed when the angle of convergence between the final approach course and the extended runway centerline does not exceed 30°. If the angle is greater than 30°, you must use circling minimums. Note that the pilot does not make the decision as to whether an approach is straight-in or not. Yes, you may break off a straight-in approach and circle to land on another runway (using the higher minimums), but the designation of an approach as straight-in or circling is based upon the layout of the airport, the angle between the final approach course and the landing runway, the location of the electronic facilities, and the design of the instrument approach.

With these points in mind, you can establish the basic minimums for each category airplane for Categories A and B for the NDB approach at Pittsfield:

The numbers mean:

- 2280-1 $1/4$ are MDAs and the minimum visibilities for both straight-in and circling approaches (using the local altimeter setting).

- 1104 and 1086 are the heights above the airport (HAA) at the MDA.
- (1104-1 $\frac{1}{4}$) and (1104-1 $\frac{1}{2}$) are military minimum ceilings and visibilities and are not applicable to civilian aircraft.

Inoperative components

(4) *Check inoperative component changes in minimums.* If any component of an approach listed on this table (FIG. 14-2) is out of service, the minimums might have to be increased. The table is published on the inside front cover of every set of NOS approach charts. Definitions and descriptions of MM, ALSF, MALSR, etc., are in the front section of the NOS sets on page L1 entitled "Approach Lighting System—United States."

Nonprecision approach visibility minimums increase $\frac{1}{4}$ and $\frac{1}{2}$ mile when certain approach lights and runway lights are inoperative. Check the lighting legend in the front section of the NOS approach chart sets (FIG. 14-3) against the airport diagram to see if the airport has any lighting systems affected by the inoperative components table.

This looks a little intimidating at first, but an instructor can help you sort things out. If you make a habit of checking the destination against the inoperative components table and the lighting legends every time you file IFR, you will soon be able to handle this problem quickly and easily. You will also broaden your understanding of the roles played by the various components, and what the wide variety of approach and runway lights look like.

The inoperative components check for Pittsfield reveals no lights affected by the "Inoperative Components Table." The minimums remain at 2280-1 $\frac{1}{4}$ for both straight-in and circling approaches regardless of lighting.

Approach adjustments

(5) *Make adjustments required by the fine print.* As noted earlier, here is where Pittsfield throws a zinger at the unwary pilot. MDAs must be increased 100 feet at Pittsfield if a local altimeter setting is not available and the North Adams setting is used as a substitute.

Pittsfield should provide the altimeter setting on the CTAF, 122.7. If the Pittsfield altimeter setting is not available for any reason, the fine print will apply; obtain the North Adams setting on 134.775 and use the North Adams limits.

It is a good rule of thumb in your flight planning to automatically add the difference required by alternate altimeter settings (100 feet in this case). If it turns out that you can get a local altimeter setting, it will be a simple matter to glance at the approach chart and drop down to the lower MDA. Better to add the difference in the quiet of the planning room than fumble around for the correct NMA during the approach!

Altimeter error

(6) *Add the altimeter error.* For reasons discussed in Chapter 6, always *add* the altimeter error, regardless of whether it is plus or minus. For purposes of illustration, you find an error of 30 feet when you check the altimeter. Add the altimeter error of 30' to the 100'

INOP COMPONENTS
97198

INOPERATIVE COMPONENTS OR VISUAL AIDS TABLE

Landing minimums published on instrument approach procedure charts are based upon full operation of all components and visual aids associated with the particular instrument approach chart being used. Higher minimums are required with inoperative components or visual aids as indicated below. If more than one component is inoperative, each minimum is raised to the highest minimum required by any single component that is inoperative. ILS glide slope inoperative minimums are published on instrument approach charts as localizer minimums. This table may be amended by notes on the approach chart. Such notes apply only to the particular approach category(ies) as stated. See legend page for description of components indicated below.

(1) ILS, MLS, and PAR

Inoperative Component or Aid	Approach Category	Increase Visibility
ALSF 1 & 2, MALSR, & SSALR	ABCD	¼ mile

(2) ILS with visibility minimum of 1,800 RVR.

ALSF 1 & 2, MALSR, & SSALR	ABCD	To 4000 RVR
TDZI RCLS	ABCD	To 2400 RVR
RVR	ABCD	To ½ mile

(3) VOR, VOR/DME, VORTAC, VOR (TAC), VOR/DME (TAC), LOC, LOC/DME, LDA, LDA/DME, SDF, SDF/DME, GPS, RNAV, and ASR

Inoperative Visual Aid	Approach Category	Increase Visibility
ALSF 1 & 2, MALSR, & SSALR	ABCD	½ mile
SSALS, MALS, & ODALS	ABC	¼ mile

(4) NDB

ALSF 1 & 2, MALSR	C	½ mile
& SSALR	ABD	¼ mile
MALS, SSALS, ODALS	ABC	¼ mile

CORRECTIONS, COMMENTS AND/OR PROCUREMENT

FOR CHARTING ERRORS CONTACT:
National Ocean Service/NOAA N/ACC1, SSMC-4, Sta. #2335 1305 East-West Highway Silver Spring, MD 20910-3281 Telephone Toll-Free (800) 626-3677 Internet/E-Mail: Aerochart@NOAA.GOV

FOR CHANGES, ADDITIONS, OR RECOMMENDATIONS ON PROCEDURAL ASPECTS:
Contact Federal Aviation Administration, ATA 110 800 Independence Avenue, S.W. Washington, D.C. 20591 Telephone Toll-Free (800) 457-6656

TO PURCHASE CHARTS CONTACT:
National Ocean Service NOAA, N/ACC3 Distribution Division Riverdale, MD 20737 Telephone (800) 638-8972

Requests for the creation or revisions to Airport Diagrams should be in accordance with FAA Order 7910.4B.

INOP COMPONENTS
97198

Fig. 14-2. *An inoperative components table is found in the front section of every set of NOS instrument approach procedures.*

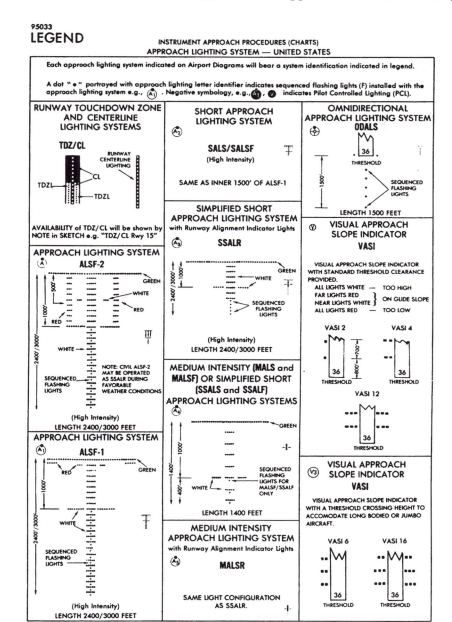

LEGEND

Fig. 14-3. *Approach lighting system codes and descriptions found in the front section of every set of NOS instrument approach procedures charts.*

adjustment if no local altimeter is available. This would yield adjusted MDAs of 2410-1 $^1/_4$ for both straight-in and circling MDAs at Pittsfield.

To summarize the step-by-step method of analyzing minimums:

1. *Read the fine print*

2. *Check the height of obstacles*

3. *Pick the correct minimums for airplane category and type of approach*

4. *Check adjustments for inoperative components table*

5. *Make adjustments required by fine print*

6. *Add the altimeter error*

OPERATION BELOW MDA

I urge students to fully analyze the MDAs for the approaches they expect to make because a pilot *cannot descend below an MDA at any time during a nonprecision approach unless certain very specific requirements are met* as prescribed in FAR 91.175 (c). The regulation can be summarized:

No pilot may operate an aircraft below the authorized MDA (or continue an approach below the DH) unless:

(1) The aircraft is continuously in a position from which a descent to a landing on the intended runway can be made at a normal rate of descent using normal maneuvers.

(2) The flight visibility is not less than that prescribed for the approach being used.

(3) At least one of the following visual references for the intended runway is distinctly visible and identifiable to the pilot:
   (i) The approach light system, including the red terminating bars or the red side row bars.
   (ii) The landing threshold.
   (iii) The threshold markings.
   (iv) The threshold lights.
   (v) The runway end identifier lights (REIL).
   (vi) The visual approach slope indicator (VASI).
   (vii) The touchdown zone or touchdown zone markings.
   (viii) The touchdown zone lights (TDZL).
   (ix) The runway or runway markings.
   (x) The runway lights.

VISIBILITY MINIMUMS REQUIRED FOR LANDING

Even if you can see the runway (or one of the other visual references listed above) as you approach the field at MDA, you may not legally make a landing if the visibility is less than that prescribed for the instrument procedure being used. This is the regulation, and it is stated in FAR 91.175 (d).

Visibility, not ceiling, determines whether or not you can land. MDA establishes the altitude below which you cannot descend unless you have one of the prescribed references in sight. Visibility tells you whether or not you can legally land when you have one of those prescribed references in sight.

Visibility is expressed in miles and fractions of a mile or in feet of *runway visual range* (RVR). (*See* Appendix C, Glossary for complete definitions for visibility.) Visibility is the prevailing horizontal visibility near the surface as reported by an accredited observer. ATC tower controllers are qualified to report visibility.

RVR is measured by a transmissometer located alongside a runway. If a runway has a transmissometer, the visibility minimums listed on the approach chart will be expressed as a two-digit figure representing feet of RVR. If there is no transmissometer, the visibility will simply be expressed in miles, as is the case with the Pittsfield NDB approach.

If the transmissometer is out of service, the published RVR minimums must be converted to miles and fractions of a mile according to the table in the front section of the NOS approach chart sets (FIG. 14-4).

Missing RVR also increases the minimum visibility for some precision ILS approaches, as noted on the inoperative components table (FIG. 14-2).

If making an approach at an uncontrolled field, the pilot must decide if the visibility meets the requirements. Check the length of the landing runway, which is a good reference for estimating visibility: a statute mile long (5,280 feet); or half a mile (2,640 feet); or a mile and a half (7,920 feet). If you can see to the end of a mile-long runway when the descent begins, you may legally land when the visibility minimum is one mile. But if you can only see partway down that runway, a landing might be illegal, and you should execute a missed approach. (This is a subject of much controversy and misunderstanding in the aviation community.)

If you are approaching a controlled field, the tower will inform you of the visibility. When it drops below the prescribed visibility minimums, the runway involved—or the entire airport—might be closed to landing traffic.

MISSED APPROACH PLANNING

The *missed approach point* (MAP) is no time to fumble for the approach chart and try to figure out what to do next. All attention must be riveted on controlling the airplane

RVR/Meteorological Visibility Comparable Values

The following table shall be used for converting RVR to meteorological visibility when RVR is not reported for the runway of intended operation. Adjustment of landing minima may be required — see Inoperative Components Table.

RVR (feet)	Visibility (statute miles)	RVR (feet)	Visibility (statute miles)
1600	¼	4000	¾
2000	⅜	4500	⅞
2400	½	5000	1
3200	⅝	6000	1¼

Fig. 14-4. *Table for converting RVR to miles and fractions found in the front section of every set of NOS instrument approach procedures.*

during the first few moments of a missed approach—add full power, stop the descent and initiate a climb, raise the flaps and gear, and maintain a steady course.

Diverting attention to the fine print of the missed approach procedure at this time could start a chain of events leading to a collision with an obstacle or the ground. While planning the flight, always assume that you will be required to make a missed approach and plan accordingly. Review the procedure again en route, before making the approach.

So plan ahead. I find that students have little trouble coping with a missed approach if they break it into five phases:

1. *Transition* to a stabilized climb. You add full power, stop the descent, raise the flaps in increments, get the gear up, and initiate a normal climb straight ahead. (Or start a turn if directed to in level flight at minimum controllable airspeed. You have been practicing minimum controllable airspeed under the hood; now all that practice becomes very valuable.)

2. *Climb.* Do you climb straight ahead or make a climbing left turn or a climbing right turn? The missed approach for Pittsfield prescribes a climbing right turn (FIG. 14-1). What is the level-off altitude? At Pittsfield it is 4,000 feet.

3. *En route* to the holding fix. What is the holding fix? Is it a facility you already have tuned in, as at Pittsfield? Or is it a VOR fix that might require resetting frequencies and OBS numbers? Do you proceed direct? Do you have to intercept a bearing or radial to get to the fix?

4. *Holding.* What type of pattern entry will you use? What outbound heading do you turn to when you reach the fix? Write it in big numbers on the approach chart.

5. *Departure* from holding. Plan for two alternatives: returning for another instrument approach or diverting to the filed alternate. When approach control asks "What are your intentions?" have your mind made up and respond promptly what you intend to do, including an abbreviated flight plan with route and altitude to the alternate if that's what you decide to do. You cannot depart the missed approach holding pattern until cleared by ATC.

Once again, visualization is the key to success in working out the moves made on a missed approach. Visualization is also the key to success in making the basic approach. A good instrument approach, which always includes the missed approach procedure, begins the night before, along with your planning for the departure and en route phases. Mentally fly the approach step-by-step, or even better, walk through it by placing objects on the floor to simulate the airport and the approach and missed approach fixes.

As the final step in approach planning, run through a MARTHA check:

MA (Missed approach procedures)

R (Radios—nav and com frequencies and OBS settings)

T (Times from FAF to MAP)

H (Heading of final approach course)

A (Altitude of MDAs, adjusted as discussed above)

This abbreviated approach checklist will also come in handy in the air near the destination while preparing for the approach.

With a little practice, you will find that planning an approach takes far less time than reading about it!

NDB APPROACHES

I always introduce students to NDB approaches before VOR or ILS approaches. This might come as a surprise, but it really shouldn't. The two lessons in the syllabus that precede approaches are devoted to ADF procedures; therefore, ADF is still fresh in the student's mind. And because VOR is the backbone of the federal airway system, most students start instrument training with far more VOR experience than ADF. So, I pay extra attention to ADF as we move through the course. By introducing NDB approaches before the others, I can make sure the student is skillful, confident, and comfortable with them. If NDB approaches are introduced later in the course, they might not get the attention they require.

Let's return to the NDB approach at Pittsfield (FIG. 14-1) and talk through the procedure one step at a time. Because this is the first approach discussed in detail in this book, I will also introduce material on approach control, communications, and flight procedures that apply not only to NDB, but also to approaches in general.

RADAR VECTORS

In the real world of IFR, you will be handed off from the ATC center controller to the appropriate approach controller at a comfortable distance from your destination. You will frequently be cleared to a lower altitude just before or just after the handoff to approach control. Leaving an assigned altitude is one of the occasions for a *required* report whenever this occurs during an IFR flight. The readback to ATC will be like this:

"Cessna five six Xray contact approach control, descend to five, report leaving seven."

Remember to use the full call sign on initial contact with approach control. Approach control will give you an expect further clearance or expect approach clearance for use in case of lost communications. The time they give will also help you plan the approach. If you don't get a "further" time, request it.

Approach control will issue vectors to intercept the final approach course (259° at Pittsfield) 1–5 miles outside the final approach fix (DALTON NDB), where you will be "cleared for the approach." This will give you time to establish yourself on the final approach course before reaching the FAF, to slow to approach speed, and to prepare for the final descent and landing. (Sometimes, as at Pittsfield, the FAF and IAF are the same.)

Treat radar vectors as commands. They are issued as required to provide safe separation for incoming traffic; therefore, do not deviate from the headings and altitudes issued by approach control.

Sometimes it becomes necessary for ATC to vector you *across* the final approach course for spacing or other reasons. This is not unusual at busy airports with a mix of slow traffic and high-speed traffic. It is much easier to move you out of the way of a rapidly closing jet than to have the jet break off the approach. After the jet has passed, you will be vectored back to the final approach course with minimum disruption.

You will normally be informed when it becomes necessary to vector you across the final approach course. If you see that interception of the final approach course is imminent and you have no further instructions, question the controller. Simply give your call number and "final approach course interception imminent, request further clearance." You will be cleared either to complete the approach or to continue on present heading for separation from incoming traffic. Do not turn inbound on the final approach course unless you have received an approach clearance.

THE FULL APPROACH PROCEDURE

In the beginning of your intensive work on NDB approaches, skip the radar vectors and request the "full approach procedure" to become completely skilled in all the elements of the approach. Make this request for the full procedure on initial contact with approach control after the handoff from center. You will probably be cleared direct to the FAF and receive an expect further clearance time or expect approach clearance time.

If a direct course to the FAF is within 10° of the final approach course, go ahead and intercept the final approach course and proceed directly to the FAF. Approach control will expect you to do this and will clear you for the approach before reaching the FAF.

PROCEDURE TURNS

If you can't line up with the final approach course and then proceed directly to the FAF, you will need to execute a *course reversal*. There are two ways of doing this—in a *procedure turn* or in a holding pattern.

At Pittsfield, the course reversal must be made in a procedure turn, as indicated by the arrowhead to the northeast extending out from the 079° bearing from Dalton NDB. Fly outbound on the 079° radial for one minute and make a 45° turn to the right as shown on the chart to a heading of 124°. Reset the OBS to the inbound course to the FAF, 259°. This is the beginning of the procedure turn, an easy, reliable method of course reversal that will return you to the inbound course with a minimum of corrections.

Fly outbound on the 124° heading for one minute, adjusting for the wind, then make a 180° turn to a heading of 304°. Intercept the inbound course using the bracketing procedure described in Chapter 10. Hold the 304° intercept heading until the needle is about three-quarters of the way from full-scale deflection toward the center, then begin a turn to the inbound course, 259°. Correct for the wind and establish a reference heading that will hold the inbound course to the station.

It should be noted that there is no "right" way to make a procedure turn. Nowhere is it written that you must use the 45° procedure published on the approach charts. All that is required is that somehow you must get turned around and headed back on the inbound course within the mileage limit published on the chart, usually 10 nm.

But it makes good sense to use the 45° procedure published on the approach charts. The 45° headings are printed on the chart; so there is no guesswork about headings. And the 45° method will enable you to intercept and get established on the inbound course quickly and easily.

Note that the fine print in the profile section of the Pittsfield NDB 26 approach chart says "Remain within 10 nm." This means that you must complete the procedure turn within 10 nautical miles of the NDB.

How far is 10 nm? Work it out on the circular slide rule while in the planning room. At a ground speed of 90 knots it takes 6 minutes 40 seconds to cover 10 nm; at 100 knots ground speed it takes 6 minutes exactly. You should complete the procedure before the times expire, depending on your ground speed. If you stray beyond the 10-nm radius, obstacle clearance is not guaranteed.

The normal procedure is to fly outbound for 1 minute, depending on the wind, then begin the procedure turn. Descend to the procedure turn altitude (4,000 feet at Pittsfield) while heading outbound and during the procedure turn.

Do not descend below 4,000 feet until you intercept the inbound course of 259°. Again, this is for obstacle clearance reasons. After intercepting the inbound course, you may to descend to 3,200 feet en route to the FAF. On reaching the FAF, do a Five T check and continue descent to the MDA.

Sometimes a holding pattern is mandated for a course reversal instead of a procedure turn. The race-track pattern shown on the approach chart will be printed with a much darker and heavier line than other holding patterns as shown in FIG. 14-5, the NDB 34 approach at our planning destination, Binghamton, NY. Note that this is listed as an "NDB or GPS RWY 34" approach. This is an example of a GPS "overlay" approach that can be flown with a Global Positioning Satellite (GPS) receiver substituting for an ADF receiver. I'll have more to say about GPS approaches in the next chapter.

If holding patterns are depicted with dotted lines (as is the case at Pittsfield), they are not available for course reversals. Instead, you must use the procedure turn shown on the chart. Both methods of course reversal—the holding pattern and the procedure turn—are widely used in NDB and VOR approaches. For obstacle clearance reasons, you do not have the option of substituting one type of course reversal for another.

It is very important to the success of an NDB approach to get lined up on the final approach course as soon as possible with an accurate wind correction. If you are not lined up properly at the FAF, the chances of making a successful approach are very slim indeed. A well-executed course reversal is the secret to success in quickly establishing good lineup.

APPROACH SPEEDS

Slow to approach speed and lower approach flaps, usually one increment, during the course reversal and in steady flight during a full approach or as you head toward the FAF with radar vectors. One hundred knots is a comfortable approach speed for most light airplanes. This will usually result in a 90-knot ground speed in typical winds. The exact speed doesn't make

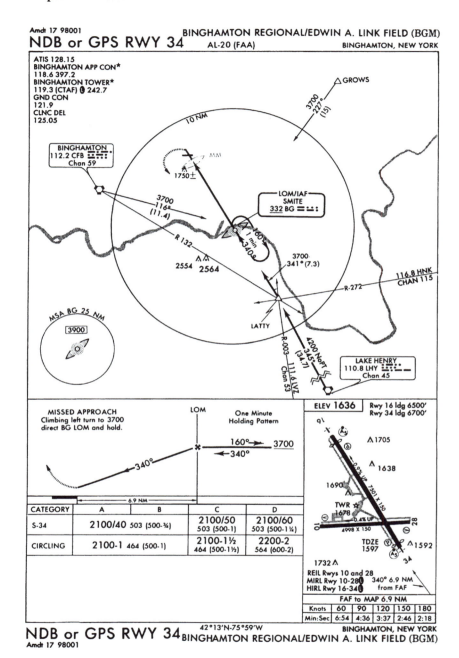

Fig. 14-5. *NDB approach with course reversal in holding pattern.*

too much difference as long as it is a comfortable speed that you can hold constant throughout the approach, including the descent to MDA after leaving the FAF.

If you are aware of faster traffic behind you, maintain cruise airspeed and keep the flaps up. Plan to make a high-speed final approach at 110 knots, or even 120 knots if safe. Don't worry about coming in too fast and using up too much runway! If the runway is long enough for the jets behind you, that runway will be plenty long enough for you!

On the other hand, it is preferable to use that last one-half mile or the middle marker to transition to your most comfortable airspeed and configuration. This way you will come over the runway threshold in a normal manner. Consistency makes better landings.

Complete the approach checklist as you head toward the FAF, and run through the MARTHA check again. All approach charts for the destination should be on the clipboard with the probable approach chart on top. It is very helpful to clip approach charts to the yoke for quick reference throughout the approach and missed approach. Some airplanes come equipped with a yoke chart clip; you may also purchase a clip at aviation supply companies and many FBOs.

APPROACH COMMUNICATIONS

If you have to hold in a depicted holding pattern, you must make another *required* report while entering the hold, as follows:

"Approach control, Cessna five six Xray, (name of fix), entering hold, level at three."

As the expect further clearance time approaches, you can anticipate that approach control will either clear you for the approach, issue a revision of the time, or—at a controlled airport—hand you off to tower. If the latter is the case, you will be given the tower frequency.

At an uncontrolled airport, ATC will ask how you plan to terminate the approach. You have three options:

- Land
- Make a low pass and cancel IFR
- Execute a missed approach

With the first option—a landing—approach control will tell you "report landing or landing assured." You will remain on the approach control frequency until advised "frequency change approved." You must then switch to the CTAF frequency and report your position on the CTAF to alert other traffic about your position and that you are inbound on a specific instrument approach. CTAF is also the frequency to get the weather at the airport and learn the runway in use. When you report landing or landing assured to ATC, the IFR flight plan will be closed by ATC. If the landing is at a remote airport without communications to ATC or a flight service station, a telephone call to an FSS might be required to close the IFR flight plan.

With the second option—low approach and cancel IFR—the IFR flight plan will be canceled when you announce "cancel IFR" to ATC.

With the third option—missed approach—you return to the ATC frequency at the MAP. Executing a missed approach is the occasion for another *required* report. But you don't have to make this report as soon as you add power for the missed approach. Wait until the climb is stabilized and everything is under control, then report. Always remember: aviate, navigate, communicate!

Expect an abbreviated clearance from approach control for returning to the missed approach holding fix. It is not a good idea to try to copy a clearance while you still have your hands full controlling and cleaning up the airplane in the transition phase of a missed approach. Wait until you are in a stable climb, then contact approach control.

FLYING THE NDB APPROACH

So far we have discussed elements of the instrument approach that also apply to all non-precision and precision approaches to one degree or another: radar vectors, approach speeds, course reversals, and terminating the approach. Now let's back up and discuss how you will actually fly the full procedure in the example, the NDB 26 at Pittsfield.

Proceed to the IAF, Dalton NDB, following the clearance from approach control. On the way to Dalton, slow to approach speed and review the MARTHA check. On reaching Dalton, run through the "Five Ts" as you always do at a fix or when making a change in course or altitude:

- Write down the *time* of arrival at Dalton on the approach chart. You will also need to start timing the outbound leg.

- *Turn* to the appropriate heading outbound for the procedure turn.

- *Twist* is not necessary for this NDB approach. Instead, use this item as a reminder to adjust the volume on the identifier to hear it faintly in the background. Monitor the NDB identifier continuously throughout the approach to detect a failure of either the transmitter or receiver.

- Reduce *throttle* for 100 knots if you have not done so already.

- *Talk*: Report as requested by ATC, for example, "Cessna five six Xray, Dalton procedure turn outbound."

WHEN TO DESCEND

When approach control has cleared you for the approach, you may begin a descent to the altitude prescribed on the approach chart—4,000 feet at Pittsfield—as soon as you depart Dalton outbound on the procedure turn. If you have not been cleared for the approach, you must remain at your assigned altitude—5,000 feet in this example—until approach control clears you to a lower altitude, or says "cleared for the approach." This reason for remaining at the assigned altitude is obvious because there might be other airplanes in a holding pattern below.

Make a normal, stabilized, constant-airspeed descent. Slow the airplane to the approach speed you have selected, say 100 knots (if you haven't already done so). When stabilized, reduce power 100 RPM (or 1" of manifold pressure) for each 100 feet per minute you want to descend. A reduction of 500 RPM—from 2300 RPM to 1800 RPM, for example—will produce a rate of descent of 500 feet per minute. A reduction of 5" of manifold pressure will also produce a 500-foot-per-minute descent.

It's always a good idea to start a descent as soon as you are cleared to do so. The sooner you get down to the desired altitude, the more time you have to stabilize altitude, airspeed, and heading. This becomes very important as you descend to the MDA on the final approach course. If you have a large amount of altitude to lose, descend at 1,000 feet per minute until 1,000 feet above the desired altitude, then reduce the rate of descent to 500 feet per minute.

Make the procedure turn and descend in the turn to 3,200 feet if you have been cleared for the approach. As you head inbound toward Dalton, intercept the inbound course (259°) and begin bracketing to establish a reference heading that will correct for the wind and maintain an inbound course.

Upon reaching the FAF, follow through with five important steps:

- Start timing the final approach segment

- Adjust heading as necessary

- Make sure the volume is correctly set to faintly hear the identifier throughout the approach

- Reduce power 500 RPM (or 5" of manifold pressure) to begin a 500-foot-per-minute descent to the MDA

- At a controlled airport, contact tower if you haven't already done so. At an uncontrolled airport, report on CTAF passing the FAF. Always report position and intentions on CTAF to alert local traffic. Make frequent additional reports on final as needed.

TIMING THE APPROACH

I recommend using a stopwatch to time the final approach segment from FAF to MAP, or a digital timer on the instrument panel that you can start as you pass the FAF. A stopwatch (or digital timer) is also handy for timing the legs of a procedure turn or a holding pattern.

You can also use the sweep-second hand of the conventional clock to time the approach. But many students find it confusing trying to keep track of how many minutes have passed on a long final approach segment. On some long finals the time from FAF to MAP might be more than 5 minutes.

As you concentrate on maintaining the MDA and the final course, it is easy to forget how many times the sweep-second hand of the conventional clock has gone around. (Is it three? No, that was last time. Must be four. But it's taking so long! Maybe I've already gone five minutes!)

Eliminate the confusion altogether. Buy a timer at the beginning of your IFR training and use it on every approach. You will soon find that accurate timing ceases to be a problem. (Occasionally use the panel timepiece—conventional or digital—to maintain proficiency in case a handheld stopwatch fails.)

The times from FAF to MAP at different speeds are located in the lower right corner of the NOS approach charts. These times are based upon no-wind conditions, so you must adjust them for the estimated ground speed. If you are good with numbers you can interpolate and quickly determine the time to match the ground speed. If you make this calculation part of the MARTHA check when approaching the airport, it will save a lot of fumbling at the FAF. None of these calculations will be accurate if you cannot fly a constant airspeed during the descent, then level off at the MDA.

Tip: When the ceiling and visibility are well above minimums, say 600 overcast and 2 miles, use the next faster speed when timing the FAF to MAP segment.

FINAL APPROACH COURSE

Another problem that I see frequently when giving flight tests for the instrument rating is pilots getting so disoriented on the final approach course after passing the FAF that they cannot find the field. The basic problem here is poor training or lack of practice in tracking and bracketing the NDB. If you understand and practice the NDB procedures discussed in Chapter 13, you should have little difficulty in this phase of the approach.

As you head toward the FAF you should have enough time to bracket the inbound course and determine a reference heading that will correct for wind and maintain that course. At station passage, don't chase the needle. Maintain the reference heading outbound from the FAF until the needle settles down and you have completed the "Five Ts" checklist. After that you may make minor adjustments in the reference heading if necessary.

If you have an accurate reference heading when passing the FAF, all you have to do is maintain that reference heading and you will see the landing runway when the time has expired, ceiling and visibility permitting.

There are times, however, when even the best pilots are unable to establish an accurate reference heading as they fly toward the FAF. The wind might be changing rapidly or approach control might turn you inbound so close to the FAF that there is not enough time to get established on course.

Do the best you can heading inbound in this situation. Then at station passage turn immediately to the inbound course, wait 10 seconds for the needle to settle down, note the number of degrees the needle is off to the left or right, and then reintercept and bracket the outbound course using the procedures described in Chapter 13. That will get you back on course before you get to the MAP.

MISSED APPROACHES

There might be times when even this won't work. If you cannot establish yourself on the final approach course for any reason—or if you have lost track of the timing—you must execute an early missed approach. The procedure for an early missed approach is different from a missed approach at the MAP. In an early missed approach, add full power, clean up the airplane, establish a normal climb, report to ATC, and transition to the published missed approach procedure. Above all, *do not make a turn until you have reached the MAP*. You are not guaranteed the full obstacle clearance associated with that approach if you depart from the final approach course.

A decision to make a normal missed approach at MAP is based upon several variables.

Very often missed approaches are required because you didn't get down to MDA and were unable to establish the visual references spelled out by FAR 91.175 (c) for a further descent. A missed approach is also required if the visibility is below that required for the approach. Or perhaps you weren't lined up properly and just caught a glimpse of a corner of the airport as you flew by.

You might also have to make a late missed approach if you have the required visual references and the visibility needed to land but find, after beginning the descent for landing, that you cannot land for some reason, perhaps another airplane on the runway.

Remember the five phases of a missed approach:

1. *Transition* to a stabilized climb
2. *Climb* straight ahead or in a climbing left or climbing right turn, as prescribed in the published missed approach procedure
3. *En route* to the holding fix
4. *Holding* at the designated holding fix
5. *Departure* from the holding fix for another approach or to an alternate

You should carefully study the missed approach procedure the night before the flight, along with your research on the rest of the approach. Note the missed approach instructions on the approach charts. How will you proceed after pullup? You will have three choices: climb straight ahead, make a climbing left turn, or a climbing right turn, as is the case with Pittsfield. Where will you hold and how will you enter the holding pattern? At Pittsfield there is a holding pattern at Dalton NDB for the missed approach, but not for a procedure turn. You must think these points through in your preflight planning. The cockpit is not the place for original research on missed approaches!

There is no limit, other than the amount of fuel on board, to the number of approach attempts. If you missed the approach for reasons other than weather—poor lineup, for example—go back and try again. But if you reached MDA and the weather was below minimums, or if the weather was obviously deteriorating, the smart move would have been to proceed to the alternate.

CIRCLING APPROACHES

You are on the final approach at MDA and, as you run out of time, hopefully you run out of clouds and the airport is in sight. You do a landing check passing through 500 feet above the airport, lower full flaps, and land.

But somewhere prior to completing the landing check and lowering full flaps, you might find that you have to land on a different runway. If landing at a controlled airport, the tower will make the decision and issue a clearance: "Circle and land Runway (as assigned)." At an uncontrolled field, however, the pilot must make the decision. Several variables affect this decision.

In some cases there is no choice. Many approaches are not sufficiently aligned with the runway to permit a true straight-in approach. When a procedure does not meet the criteria for straight-in approaches, it is designated A, B, C, and so on (NDB-A, NDB-B, NDB-C, etc.), and no straight-in minimums are published. An example of this is the NDB-A approach at Perkasie/Pennridge, Pennsylvania (FIG. 14-6). The final approach course is 181° and the only runway at Perkasie is 8-26. So a circling approach is the only alternative available.

Conditions at the time of reaching the MAP might dictate a circling approach. If the crosswind is too great for the straight-in runway, for example, you should choose a landing runway that is closer into the wind, if one is available. If there isn't a better runway, execute a missed approach.

Sometimes you learn that airplanes in the landing pattern ahead of you are using a different runway. Learn the active runway from unicom, or from other airplanes in the pattern as they report their positions on the CTAF. If a different runway is in use when you arrive, you will have to make a circling approach and fit into the traffic pattern. If there is traffic in the landing pattern, it better be VFR; so you should break out in VFR conditions well above the circling MDA. Nothing says you can't fly a circling approach higher than the circling MDA as long as you are clear of clouds.

You may get right down to the straight-in MDA before deciding to make a circling approach. Perhaps another airplane taxis out and dawdles on the runway just as you are about to land. Add power and go around, just as you would under VFR conditions. But you are still IFR and must circle around again for another attempt at landing. In this case you must climb back up to the published circling minimums, or traffic pattern altitude, in order to continue.

If that puts you back in the clouds, you will have to execute a missed approach. You must keep the runway of intended landing in sight at all times during a circling approach or execute a missed approach.

Circling approach patterns

The recommended circling approach patterns are shown in FIG. 14-7. Pattern A may be used when the final approach course intersects the runway centerline at less than a 90° angle and you see the runway clearly enough to establish a base leg.

If you see the runway too late to fly pattern A, circle as shown on B and make either a left downwind or a right downwind. Fly pattern C if it is desirable to land opposite the direction of the final approach course and the runway is seen in time for a turn to the downwind leg. If the runway is sighted too late for a turn to the downwind as shown in C, fly pattern D.

So far this all sounds very reasonable; however, *the circling minimums might be as much as 500 feet lower than a VFR traffic pattern* for the same runway flown at 1,000 feet above field elevation. Some people call the circling maneuver legal scud running. It takes some very careful maneuvering to make a safe approach and landing from a low altitude.

Furthermore, the circling minimums guarantee an obstacle clearance of *only 300 feet* within the *circling approach area*. This is a very small area and you must remain within it. TERPS describes how the circling approach area is constructed (FIG. 14-8). The circling approach area is based on the same aircraft approach categories A, B, C, D, and E

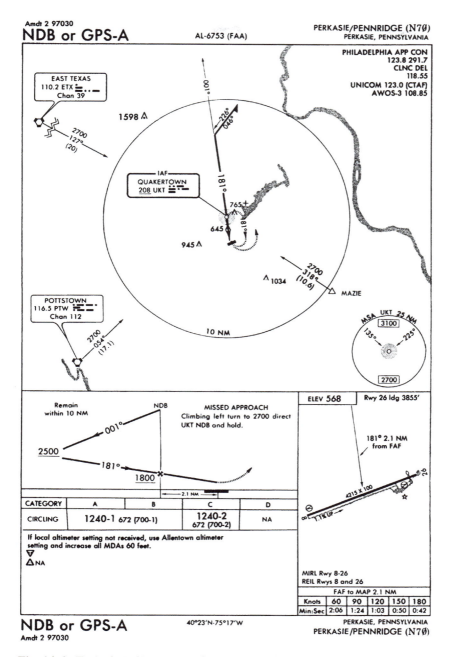

Fig. 14-6. *Typical circling approach procedure when runway is not aligned with final approach course.*

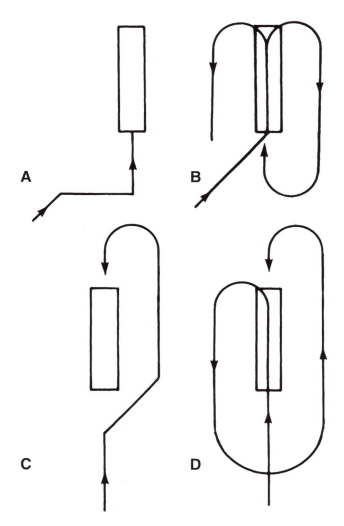

Fig. 14-7. *Circling approach patterns. Use A when final approach course intersects runway centerline at less than 90°; use B if you see runway too late to fly pattern A; use C to land in opposite direction from final approach course; use D if you see the runway too late to use pattern C.*

that appear in the minimums sections of instrument approach charts. For category A, which most of us use for instrument training, the circling area has a radius of only 1.3 nautical miles from the end of each runway.

Figure 14-8 shows how arcs drawn from these radii outline the area in which obstacle clearance is provided. *Outside this area, there is no obstacle clearance protection.*

Keep the circling approach pattern within the safe area. Use the runway length to help visualize 1.3 nm.

You are required to perform a circling approach on the instrument flight test. And circling approaches are often the only kind allowed at many small airports. Master the skills necessary to carry out this maneuver at circling MDA but realize that circling approaches are imprecise and might be dangerous if not performed properly.

Instructor note. Practice, as always, is the best way to build confidence in circling approaches. In VFR conditions, make it a routine to terminate one instrument approach on every flight with a circling approach at the VFR pattern altitude so that the student will learn to sequence with other traffic.

In actual IFR, take advantage of every opportunity to have the student make circling approaches at circling MDA at an uncontrolled airport. The tower at a busy airport probably won't let you do this for practice, but will insist that you land.

Be sure to brief the student about your intentions. Students won't get much out of the practice if they are totally confused about what is going on.

NDB ON AIRPORT

Some NDBs are located right on the airport, as seen in FIG. 14-9, the NDB RWY 22 approach at Easton, Maryland, a busy field on Maryland's popular Eastern Shore.

An NDB approach with the NDB on the airport is a very simple approach. Proceed to the NDB, which is the initial approach fix (IAF) and the MAP. (There is often no FAF when the NDB is located on the field.) Turn outbound on the indicated course, in this case 048°, the reciprocal of the inbound course. A procedure turn is indicated rather than a course reversal in a holding pattern.

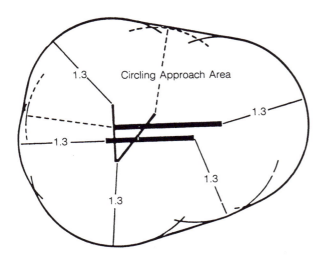

Fig. 14-8. *Circling approach area within which obstacle clearance is provided.*

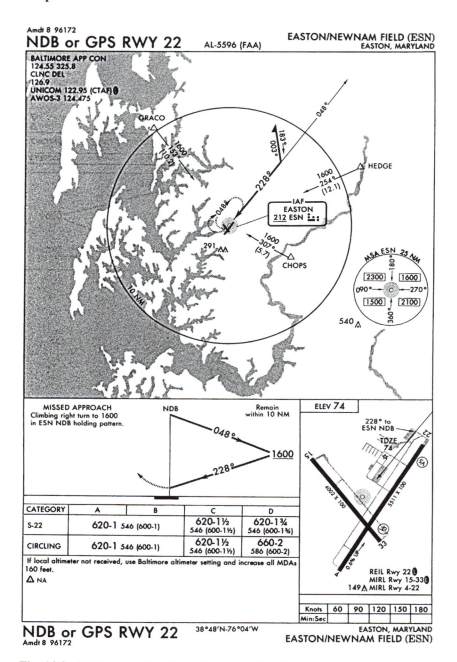

Fig. 14-9. *NDB approach with station located on the airport.*

Do not descend below 1,600 feet until you intercept the inbound course of 228°. Again, this is for obstacle clearance reasons. After intercepting the inbound course, you are free to descend to 620 feet, which is the MDA for straight-in and circling approaches at Easton.

Note that no times are given from FAF to MAP because there is no FAF and the MAP is the NDB itself. Therefore, there is no need to time the inbound leg. Just stay on the inbound course of 228° at the MDA of 620 feet (as adjusted in your planning) until station passage occurs. If you don't have the visual references you need to descend below MDA when the needle reaches the 90° position, you must execute the missed approach.

Planning steps, communications, MARTHA and Five T checks, approach speeds, and missed approach phases will be the same for an on-airport NDB approach and an NDB approach with an FAF some distance from the field.

15
Approaches II:
VOR, DME, and GPS

ONE WAY TO GAIN A GOOD OVERVIEW OF APPROACHES IS TO EXAMINE them as short cross-country flights. A VOR approach is nothing more than a miniature cross-country. You proceed from the last en route or feeder fix to the IAF, to the FAF, and then to the MAP flying along predetermined courses, making turns and changing altitudes as required.

It might take 2–5 minutes to get from the en route or feeder fix to the IAF and then another 5–10 minutes to arrive at the FAF. Remember that you will be tracking inbound and outbound by VOR during the approach as if tracking VORs inbound and outbound on a cross-country.

The planning, communications, MARTHA and Five T checks, approach speeds, and missed approach phases remain the same for VOR approaches as for NDB approaches.

FLYING THE VOR APPROACH

Let's work our way step-by-step through the VOR-A approach (FIG. 15-1) at Poughkeepsie/ Dutchess County, New York, which is also a GPS "overlay" approach. This approach has one item of special interest: It contains a "dogleg." It also has a procedure turn rather than a course reversal in a holding pattern.

Once again, request the full procedure in order to get the maximum training benefit out of this exercise. Proceed to the IAF, Kingston VOR, do the Five Ts and turn outbound and get established on the 037° radial, the reciprocal of the inbound course, to commence the procedure turn.

Fly outbound on the 037° radial for one minute and make a 45° turn to the left as shown on the chart. Reset the OBS to the inbound course, 217°.

Fly outbound on the 352° heading for one minute, adjusting for the wind, then make a 180° turn. This will establish the 172° course to intercept the inbound course to FAF, 217°. Hold the 172° intercept heading until the needle is about three quarters of the way from full-scale deflection toward the center, then begin a turn to the inbound course. Correct for the wind and establish a reference heading that will hold the inbound course.

As noted in the discussion of NDB approaches, a holding pattern might be prescribed in lieu of a procedure turn. If an approach has the note "NoPT," no procedure turn is permitted and *you cannot execute it* without clearance from ATC. A few VOR approaches state flat out, "Procedure Turn NA"—not authorized. Don't even think about it! There is probably a big mountain or a tall radio mast or a power line precisely where you would normally expect to make a procedure turn.

You may commence a descent to the procedure turn minimum altitude as soon as you pass the IAF. The procedure turn minimum altitude is 2,800 feet on the Poughkeepsie VOR-A approach (FIG. 15-1). Do not go below the procedure turn minimum altitude until established on the inbound course. "Established" means a "live" needle, not necessarily centered. When established you may descend to the FAF minimum altitude, 1,600 feet in this case.

On reaching the FAF, begin a descent to the MDA for the approach. Get down to the MDA as quickly as you comfortably can to give yourself the maximum opportunity to see the airport and pick out the landing runway. Make a constant airspeed descent so you do not throw your timing off.

NORMALLY AFTER COMPLETING a procedure turn you can expect to fly a straight-line course to the FAF and then on to the MAP. But not on the approach at Poughkeepsie. On reaching the FAF, make a right turn to 242° and proceed toward the MAP on this new course. That's why this is designated an "A" approach; that dog leg does not meet the criteria for a straight-in approach, even though you might end up lined up for a landing on Runway 24, if you fly a perfect approach!

You will be very busy at the FAF as you run through the Five Ts: time, turn, twist, throttle, talk. You must start *timing* the final approach leg; *turn* to intercept 242°; *twist* the OBS to 242°; *throttle* back 500 rpm (or 5" of manifold pressure) to begin a 500 fpm descent at the approach speed; and then *talk* to tower. The report will be "Cessna five six Xray, Kingston inbound." Aviate, navigate, communicate!

The key to coping with complications like this "dogleg" is to spot them while planning the flight, then talk yourself through the approach until you understand clearly the

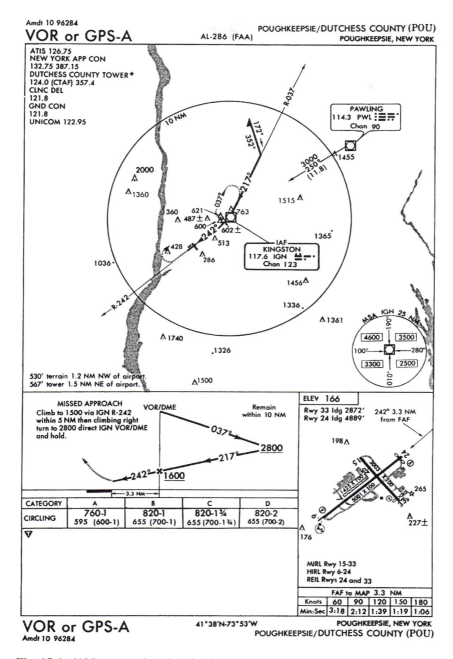

Fig. 15-1. *VOR approach with a dog leg.*

course, turns, descents, and reports. Again, think of it as a miniature cross-country and be sure to include the missed approach procedure as part of the cross-country. Fortunately you won't encounter too many dogleg VOR or NDB approaches, but be prepared to handle them.

A more common variation on the VOR approach is the VOR located on the field. This is the case at Bridgeport, Connecticut (FIG. 15-2). This airport is located on a point of land jutting out into Long Island Sound. The only place approach facilities could be located is on the field or they would be underwater. When a VOR is located on the field, the MAP is reached when the TO-FROM indicator flips to FROM.

The Bridgeport VOR RWY 24 approach also has another frequently seen feature. Many of the fixes are VOR intersections, including the IAF at MILUM and the missed approach holding fix at STANE. Tune Carmel VOR (116.6) on your No. 2 VOR receiver as you approach the area. Carmel VOR will provide the cross-bearings for the IAF, the course reversal in a holding pattern, and the MAP holding pattern.

As you talk through an approach such as this in the planning room, consider writing out a separate sequence of OBS settings for all these fixes on the flight log. Be sure to include other intersections shown on the approach chart (such as BAYYS on the Bridgeport VOR 24 approach chart) in case approach control specifies them in a clearance. (Preparation is 90 percent of the law in the legal world, as is success in aviation.)

DME AND DME ARC APPROACHES

Proficiency in the use of distance measuring equipment (DME) is not a requirement of the *Instrument Rating Practical Test Standards*. But ATC expects you to be competent and able to use any equipment in the airplane. If you file equipment code A (DME and transponder with altitude encoding capability), ATC will issue clearances with DME points. Be prepared to make DME approaches, some of which might surprise you if you haven't practiced them.

DME indications sometimes appear on VOR approach charts as an aid to making a conventional VOR approach; however, if DME is not included in the name of the approach procedure—if the name of the approach is simply VOR RWY 28, for example—then DME distances are just aids, and DME is not required for the approach. On the other hand, if DME is included in the name of the approach procedure (FIG. 15-3) you must have DME to execute the approach, unless ATC agrees to call out the DME fixes.

Note that the two IAFs for the VOR/DME RWY 15 approach at Johnstown/Cambria County, Pennsylvania, are located where two Johnstown VOR radials intersect the 10 mile DME arc. Then you fly the 10 DME arc around to intercept the 326° radial and turn inbound. The FAF is HINKS intersection, DME 4 on the inbound 146° course.

Flying a DME arc is not as difficult as it looks on the approach chart. You won't have to do the impossible and fly a smooth, continuous, perfect arc. Instead, fly a series of short, straight tangents to the arc, as you would on a time/distance check. These short tangents will keep you close to the 10 miles specified.

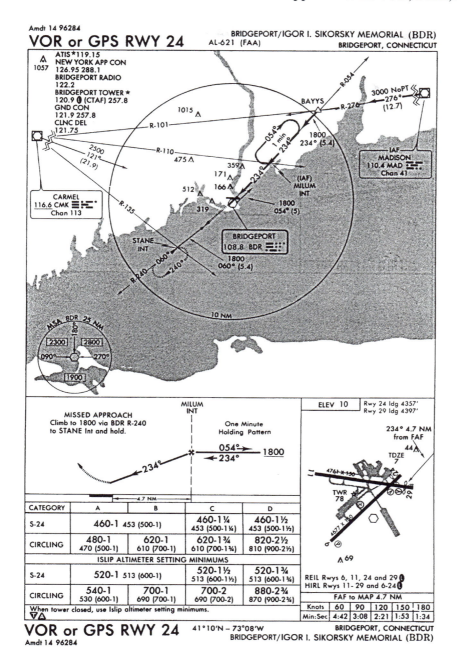

Fig. 15-2. *VOR approach with fixes at VOR intersections.*

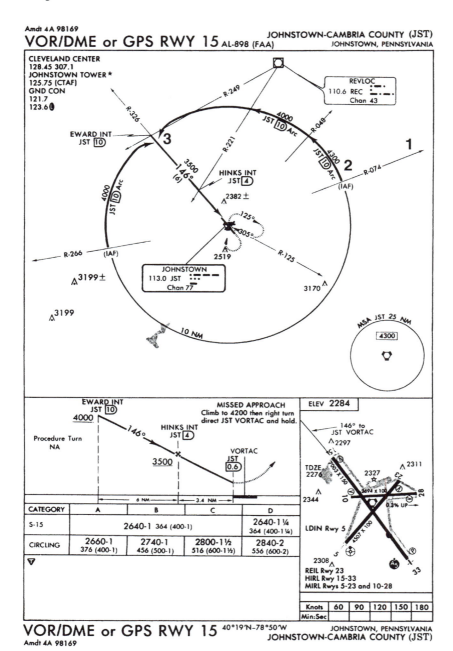

Fig. 15-3. *Procedure for flying VOR/DME RWY 15 at Johnstown, Pennsylvania. Intercept radial 074 at (1), turn onto arc at (2). turn to inbound course at radial 326 (3).*

FLYING THE DME ARC

Let's work our way through the Johnstown VOR/DME RWY 15 approach, arriving from the northeast (FIG. 15-3). First, intercept the 074° radial and turn inbound (1) on the reciprocal bearing to the station, 254°. As the DME mileage clicks off, anticipate a turn at 10.5 DME. This will enable you to lead the turn onto the 10-mile arc by half a mile.

At 10.5 DME, turn right 80° (2) to a heading of 334° (FIG. 15-3). When you complete the turn you will be on a tangent to the arc at a distance of 10 miles or very close to it. Rotate the OBS 10° *opposite* to the direction of the first 90° turn (left to a setting of 064°). As you continue on the 344° heading, the DME mileage will begin to increase.

Remember that when you set radials on the OBS, the CDI needle will start out on the *same side as the station* and move to the opposite side as you pass the radial. In this case, the CDI needle will move from left to right as you approach and pass the 064° radial.

When the needle centers, turn left 10° to a new heading of 324°. Reset the OBS 10° to read 054°. In a no-wind condition, the DME distance will decrease to 10 miles after the turn, then begin to increase again as you fly the tangent. Continue with these 10° heading and OBS changes as you track around the arc toward the 326° radial.

Naturally, the wind will tend to blow you toward the station or away from it, depending on its direction. If you find the DME distance increasing, you are being blown away from the station. Make the next 10° heading change sooner or make the turn more than 10°. This will bring the airplane back inside the curve.

If the DME distance decreases, you are being blown toward the station. Reset the OBS for 20°. Make the next heading change 10° as usual. You will fly a longer tangent before the needle centers again. This will correct for the wind blowing toward the station.

Lead the turn onto the inbound course by 5°. In the Johnstown example, establish a 146° course inbound to the station. After passing R334, set the OBS to 151° to lead the inbound course (3). Do the Five Ts and start the turn inbound when the needle centers. After completing the turn, reset the OBS to 146° and track this course inbound.

NOTE HOW DME distances are used to fix the FAF at HINKS intersection and the MAP at .6 DME. Normally, when the VOR is located on the field, the MAP occurs at station passage. But in this case, you must execute a missed approach before reaching the VOR to avoid obstacles.

With the Johnstown VOR/DME RWY 15 approach, the DME arc is used to position the airplane on a conventional VOR course, and for the FAF and MAP. In some cases, believe it or not, the DME arc *is the final approach course.*

The VOR DME RWY 14 at Baltimore/Martin State, Maryland, (FIG. 15-4) is a fairly simple approach to fly, despite the way it appears on the chart. Intercept the Baltimore 331° radial, fly inbound to the 14.7 DME arc, turn left 90°, then begin making 10° tangents all the way around until the airport lies ahead. The various radials provide the descent points and the MAP.

It should be clear at this point that there is nothing conventional about a DME approach. Everyone of them is different, sometimes radically so. But if you have DME

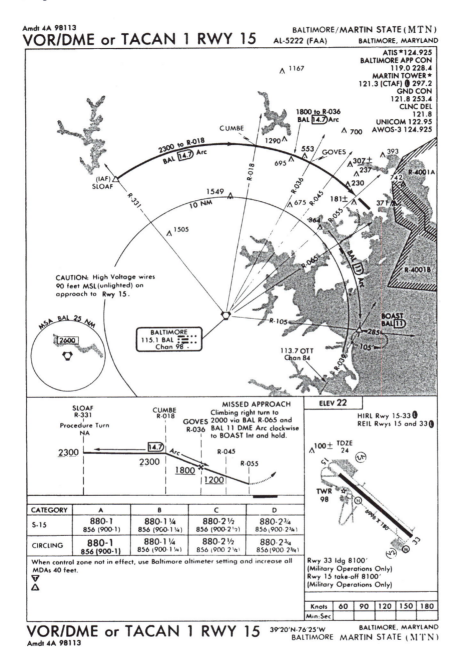

Fig. 15-4. *VOR/DME at Baltimore/Martin State, Maryland, with DME arc as final approach course.*

aboard and take the time to practice DME approaches and get acquainted with them, you will gain access to a large number of approaches that you might otherwise not be able to use. Believe it or not, DME arcs are easier to fly than to explain in written text, and they are great fun.

Instructor's note. If there are no DME arc approaches available for practice in your area, or if they are too far away, select a nearby VOR en route facility with DME and set up DME arcs around that. Remain VFR at all times and work in a quadrant that will keep you away from all airports and instrument approaches associated with the VOR. And be sure to operate at an altitude that guarantees obstacle and terrain clearance.

In fact, with a little research you should be able to practice a published DME approach (such as those in FIGS. 15-3 AND 15-4) on a nearby VOR DME. Superimpose the published DME approach on the nearby VOR and see if it will conflict with any other airports and approaches. If there are no conflicts, find the Minimum Obstruction Clearance Altitude (MOCA) for the area as shown in large numerals on the IFR En Route Low Altitude chart. Use the MOCA as the field elevation and add it to the MDA and other altitudes on the approach chart.

GPS APPROACHES

The day will come eventually when the Global Positioning Satellite (GPS) system will replace VOR en route navigation as well as NDB, VOR, DME, ILS and all other types of approaches except radar. We are in the midst of a revolution in air navigation that will make instrument flying simpler and much safer. Think of it! Instead of learning separate techniques for all of the above, we will only need to learn GPS. And thanks to the wonders of powerful small computers, GPS offers the promise of being much easier to learn and use than anything we have now.

But a few words of caution are in order at this point. We have a long way to go before GPS becomes the standard system for air navigation. Despite the hype surrounding GPS in the last couple of years, the Federal government has not developed a clear policy for the implementation of GPS as the air navigation system of the future. Nor do we have realistic goals on the way toward the achievement of an all-GPS air navigation system. The cockpit equipment is still very costly, and using GPS in a single-pilot, single-engine situation can be extremely work intensive.

GPS uses timed signals from 24 U.S. military NAVSTAR satellites to provide precise position information through sophisticated, high-tech receiver/processors. (A good discussion of GPS basics may be found in AIM, Chapter 1.) GPS provides two levels of service: "Standard Positioning Service" and "Precise Positioning Service." The standard service is accurate to 100 meters (328.1 feet) or less, which is acceptable for en route navigation and nonprecision approaches. Standard service is available to all users.

The precise service is accurate to 16 meters (52.49 feet), but its use is restricted to military and other national security applications. Even if the precise service was made available for civilian use, the signals would have to be corrected—"augmented"—to meet the course and glideslope requirements for ILS precision approaches.

GPS signal corrections eventually will be provided by "differential GPS" (DGPS). DGPS works through precisely located monitoring stations on the ground that compare the predicted GPS signals for that precise location with the satellite signals actually coming in. The differences between predicted signals and actual signals are processed by the ground stations and converted to differential corrections.

The current plan is to provide corrected GPS signals to airborne receivers through the "Wide Area Augmentation System" (WAAS). "WAAS will consist of 24 monitoring stations that will sample signals from GPS satellites passing overhead," notes the AOPA Air Safety Foundation in its recent "Safety Advisor" booklet, *GPS Technology*. "The data will be sent to three control stations, which will rapidly analyze the information and uplink corrective signals to three geostationary satellites covering the United States. The satellites will broadcast corrected GPS signals data to airborne WAAS-capable receivers."

So we're looking at a new system that—in addition to our present array of 24 GPS satellites—will require 24 additional monitoring stations and 3 control stations on the ground, plus 3 geostationary WAAS satellites, *plus* all new avionics for every aircraft that uses the national airspace system. For the greater precision required of ILS Category II and III approaches, a supplemental "Local Area Augmentation System" (LAAS) will be installed at selected high density airports.

AOPA and the Air Transport Association (ATA), which represents the nation's airlines, are supporting the FAA's plans to implement both WAAS and LAAS. And much of the work has already been done. There are now hundreds of GPS nonprecision approaches available throughout the country, with many more on the way.

GPS APPROACH BASICS

There are two types of GPS approaches in use these days. The most common type is the "overlay" approach which is identical to an existing NDB or VOR approach except that GPS is the means of navigation. See the following approaches, which were illustrated previously:

- Fig. 14-5: NDB or GPS RYW 34, Binghamton, NY, page 182
- Fig. 14-6: NDB or GPS-A, Perkasie, PA, page 189
- Fig. 14-9, NDB or GPS RNW 22, Easton MD, page 192
- Fig. 15-1: VOR or GPS-A, Poughkeepsie, NY, page 197
- Fig. 15-2: VOR or GPS RYW 24, Bridgeport, CT, page 199
- Fig. 15-3: VOR/DME or GPS RWY 15, Johnstown, PA, page 200

Fixes, courses, frequencies, minimum altitudes, course reversals, and missed approach procedures are the same for GPS as for the underlying NDB or VOR approach.

The second type of GPS nonprecision approach is the "stand alone" type which may be encountered at airports that have no underlying NDB or VOR approaches, or where there are differences that apply to the GPS approach and not to the others. See FIGS. 15-5a

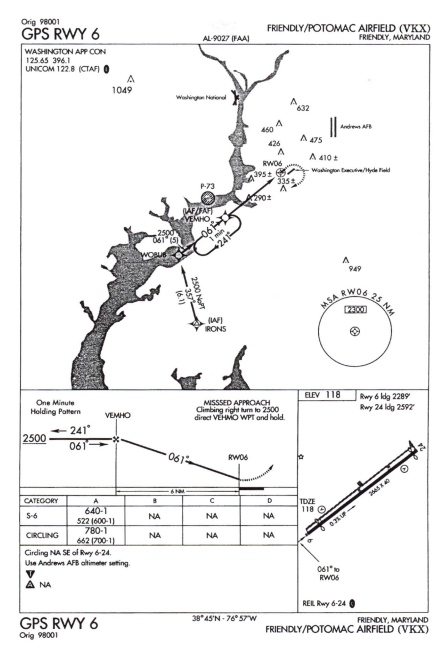

Fig. 15-5a. *GPS stand alone approach to Friendly/Potomac Airfield, Maryland.*

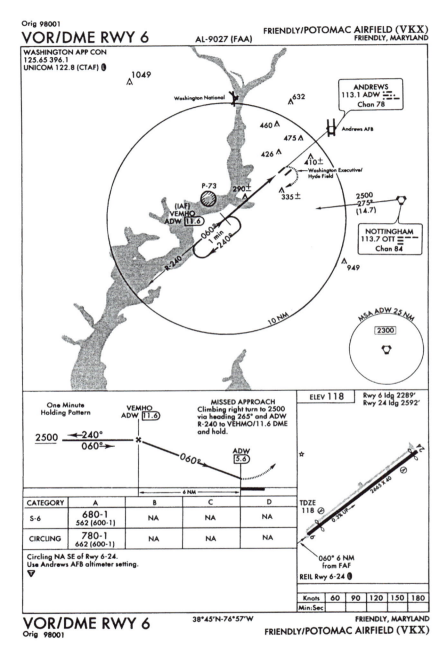

Fig. 15-5b. *Conventional VOR approach to Friendly/Potomac Airfield, Maryland.*

and 15-5b, the two approaches to RWY 6 at Potomac Airfield, Friendly, MD. Look closely and you will see that:

- The MDAs are different, with the GPS MDAs being slightly lower.
- The courses are slightly different.
- There are additional waypoints for the GPS approach at IRONS and WOBUB.
- The missed approach instructions are slightly different.

GPS APPROACH PLANNING

Flying the GPS approach is pretty simple—you just intercept and track the courses indicated on your OBS indicator or Horizontal Situation Indicator (HSI). You descend as shown on the approach chart at the various waypoints until you reach MDA and land or execute a missed approach. Sound familiar?

The big difference with GPS is that you are dealing with a computer for course and waypoint information, not fixed signals from the ground. The GPS receiver/processor gets its signals from a universal source in the sky then reinterprets these signals according to your instructions. The computer in your GPS system needs to know where you want to go and what you want to do in order to lead you in the right direction. The choices offered by GPS are almost unlimited; so you must enter your instructions very carefully. Or else you might find yourself being taken to some place you don't want to go!

In addition to the approach planning outlined in Chapter 14, we must add another layer of planning for GPS. It's easiest, I believe, to think of a set of scenarios, such as the following:

Vectored approaches. How will I set up my GPS computer for vectors to the Final Approach Fix?

Full approaches. What steps do I take to instruct my GPS computer to handle course reversals in a procedure turn? In a holding pattern? (Some systems require that you put GPS tracking on "hold" while executing these maneuvers.)

Changes of clearance. Suppose Approach Control clears me to a different waypoint—or a different runway—than I was planning on. What do I have to do to reset my computer for the new clearance?

Missed approaches. Two things here: How do I instruct my computer to return for another pass? Or what do I need to do when I must proceed to my alternate?

These moves cannot be researched in the cockpit. Each of them must be rehearsed beforehand for every flight, and the key instructions written on your planning log. Talk yourself through each of these scenarios and simulate the "knobology" needed to enter the correction instructions into your GPS computer. Think about setting up a dummy GPS panel to help you make the right moves with the knobs and buttons. Or go to your plane, turn the GPS on, and rehearse the inputs with the real thing while you are on the ground. Some systems have a built-in simulator mode.

Experienced instrument instructors say that it takes 15 to 20 hours of GPS instruction before you are ready to use GPS confidently on an IFR flight, and sometimes more.

Each GPS manufacturer has configured its equipment slightly differently; so you must learn how your particular equipment does the job, in addition to mastering the basic inputs common to all.

In addition to the scenarios above, there are other details that must be considered every time you plan a GPS approach:

Is your database up to date? Revised GPS digital approach databases are issued every 56 days by NOS, the same as your paper NOS or Jeppesen Instrument Approach Procedures. You must have a current database in your GPS receiver/processor. At this point, the major GPS manufacturers have different types of cards for updating chart information. Contact Jeppesen at 1-800-621-5377 for a free catalog listing the different types of data cards currently available and their subscription prices.

How do I tell if my equipment is operating properly? All receiver/processors are required to provide "receiver autonomous integrity monitoring" (RAIM). RAIM checks to see if there is a sufficient number of satellites available for positioning and that their information has not been corrupted. RAIM provides several levels of warning, with the time factor becoming more and more critical in the approach phase. Study your equipment and learn when and how RAIM warnings appear and the actions you should take when RAIM information appears.

This leads us to a final point:

Monitor the underlying NDB or VOR approach while you conduct the GPS approach. Strictly speaking, this is no longer mandatory for GPS overlay or stand-alone approaches. But you must have "alternate means of navigation" aboard your aircraft, such as NDB or VOR. And you must be prepared to use it if you get a RAIM warning, or RAIM capability is lost. Furthermore, if your flight plan requires an alternate airport, this alternate must have an approved approach other than GPS, and you must be prepared to execute this approach in the event of a RAIM problem.

If your GPS has a moving map display—and that is really the way to go these days!-it is easy to become complacent and let GPS do all the work. But the sharp instrument pilot will always cross-check every phase of the flight, especially an approach, with VOR and NDB and be prepared to switch to them instantly if a GPS problem arises. VOR and NDB alternatives should always be a part of your preflight planning, and you can count on your instrument check-ride designated examiner marking you down if you don't do this.

The future for GPS is very bright, and when coupled to such features as moving map displays, HSIs or Flight Directors, and three-axis autopilots, the future promises to eliminate many of the uncertainties, frustrations, and anxieties of instrument flying. And the future might be nearer than you think! These elements are all available now and though expensive, they are seeing increasing acceptance by general aviation. The revolution is here—but the best is yet to come!

TIPS ON FLYING APPROACHES

The successful outcome of the approach is usually assured by thorough preflight planning, by carefully studying the approach that will probably be used, and by having all approach charts for that airport readily available.

Know instantly where to look for all significant items on the approach chart.

Be prepared for the next step of the approach. Think ahead about the segment you are about to fly.

Don't try to comprehend or digest the entire approach chart all at once.

Always be prepared for a possible missed approach. "Gotta' landitis" prevents some pilots from growing older!

Keep the approach technique simple.

Slow to approach or holding speed before commencing the approach or during course reversal and lower approach flaps.

Fly "by the numbers" at predetermined airspeeds and power settings to attain a trimmed configuration.

Determine the wind correction before reaching the FAF and fly the reference heading ±2°–5° to maintain the desired track.

Perform a prelanding check prior to reaching the FAF. Lower the landing gear at the FAF or make a power reduction in a fixed-gear airplane. Then note the time over the FAF, check heading and turn as necessary, change OBS if required, and report to the controlling facility.

Know where you are at all times! Continuous situational awareness at all times is the key to confident, safe flying.

16
Approaches III:
ILS, localizer, and radar

THE ILS APPROACH IS THE MOST PRECISE APPROACH AVAILABLE TO the general aviation instrument pilot. It is also the easiest to master. It must be easiest because very few instrument students seem to have problems with it! This might seem like a puzzle at first because the ILS approach is fairly complex and requires an extra degree of skill for heading and altitude control. What happens, I think, is that instrument students become enamored with the ILS and practice it more than any other approach. As is the case with everything else in instrument flying, the more you practice something, the better you become at it.

The ILS is a precision approach because it incorporates an electronic glideslope. An ILS approach will bring you in exactly on the runway centerline if you fly the approach properly and it will take you down to within 200 feet of the runway when you break out of an overcast at minimums. To do this, ILS provides very precise indications that you must respond to very precisely.

Chapter Sixteen

NEEDLE SENSITIVITY

By the time you have reached the point in instrument training where you concentrate on ILS approaches, you probably will have practiced several without the hood. You know those needles are sensitive. The vertical needle is approximately four times more sensitive when set for the localizer of an ILS than for a VOR. And the horizontal glideslope needle is about four times more sensitive than the localizer indicator needle.

At the outer marker, a displacement of one dot equals approximately 300 feet on the localizer and 50 feet on the glideslope (FIG. 16-1). At the middle marker one dot equals 100 feet on the localizer and about eight feet on the glideslope. Only eight feet!

Now more than ever you can begin to understand the importance of the standard of 2, 2, and 20—±2 knots, ±2° and 20 feet. If you have been working toward these goals throughout your instrument training, you should have no difficulty coping with the sensitivity of the ILS needles.

FLYING THE ILS

On an ILS final approach segment the basic instrument techniques must be very sharp. Overcontrolling will peg the needles and cause a missed approach. To center the localizer needle, plan the turn onto the final approach course to roll out of the turn just as the needle centers. Quickly establish a reference heading that will correct for the wind, then use rudder pressure alone to make minor heading adjustments to the reference heading. Any bank at all will displace you from the localizer centerline so fast that the needle will probably peg. Keep that localizer needle centered all the time. Avoid the temptation to make heading adjustments with bank.

When you begin the descent, set up a reference descent rate that will maintain the glideslope. The next question is: What is the best descent rate for the approach? How do you determine what rate of descent will keep you on the glideslope? If you can find out, you will know what sort of power adjustment is necessary to set up that rate of descent.

A good method is to take the best estimate of the ground speed, divide by 2, and multiply by 10 for the rate of descent.

$$(80 \text{ knots} \div 2) \times 10 = 400 \text{ fpm}$$
$$(90 \text{ knots} \div 2) \times 10 = 450 \text{ fpm}$$
$$(100 \text{ knots} \div 2) \times 10 = 500 \text{ fpm}$$
$$(120 \text{ knots} \div 2) \times 10 = 600 \text{ fpm}$$

When you get the ATIS information for the landing runway, use this rule of thumb to estimate what the ground speed will be for the approach speed. If there is no ATIS and you can't estimate the ground speed accurately, use the indicated airspeed less 10 knots as the next best thing.

Then, as you begin descent on the ILS, reduce power 100 rpm (or 1" manifold pressure) for each 100 feet rate of descent sought. If you estimate that your ground

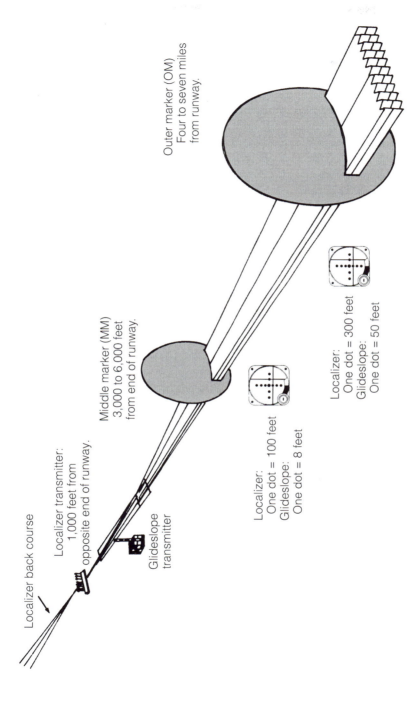

Localizer back course

Localizer transmitter:
1,000 feet from
opposite end of runway.

Glideslope
transmitter

Middle marker (MM)
3,000 to 6,000 feet
from end of runway.

Outer marker (OM)
Four to seven miles
from runway.

Localizer:
One dot = 100 feet
Glideslope:
One dot = 8 feet

Localizer:
One dot = 300 feet
Glideslope:
One dot = 50 feet

Fig. 16-1. *Configuration of a standard ILS approach.*

speed will be 80 knots, reduce power 400 rpm (or 4" manifold pressure) to set up a 400 fpm descent.

It's interesting to note that if you are flying a Cessna 172 at an airspeed of 90 knots and you have to maintain a 600-fpm rate of descent to stay on the glideslope, that means the ground speed is 120 knots. You have a strong tailwind and if you have a short runway you might have to circle to land or you could run off the end.

Once you set up a reference descent rate that more or less maintains the glideslope, leave the power alone. (Throttle jockeying is a form of overcontrolling.) Don't worry about the airspeed. Use elevator pressure alone to make minor pitch adjustments. "Pitch to the glideslope—power to the airspeed." Just as easy as flying precise altitude on a cross-country flight.

The glideslope needle becomes the "altimeter" for pitch; if you go above glideslope use forward pressure to decrease pitch slightly and return to the glideslope; if you descend below glideslope use back pressure to establish level flight and reintercept the glideslope. If you go below both the glideslope and the MDA, *execute an automatic missed approach immediately*. Obstacle clearance is not provided below MDA unless you are in a position to make a normal descent to a landing.

Remember how sensitive the glideslope needle is. You don't need to make a large correction to move 8 feet in the vicinity of the middle marker.

If you are flying a retractable, intercepting the glideslope is even simpler: as you intercept, lower the gear. That will automatically produce the proper descent rate to stay on the glideslope, with minor adjustments. It doesn't matter if the plane is a Mooney, Arrow, Aztec, Baron, Seneca, Aerostar, or an Aero Commander, drop the gear and that will set up a good rate of descent to stay on the glideslope.

ANALYZING AN ILS APPROACH

Let's turn now to the ILS RWY 6 approach at Allentown-Bethlehem-Easton, Pennsylvania, (FIG. 16-2) and analyze it using the step-by-step process applied earlier to ADF and VOR approaches.

1. **Read the fine print.** In the lower left of the profile you will find the glideslope angle (GS 3.00) and the threshold crossing height (TCH 56). This information is provided for all ILS approaches. The threshold crossing height is the altitude in feet above ground level where the glideslope crosses the threshold.

2. **Check the height of obstacles.** Two tall obstacles are within 10 nm of the airport (1,610 feet and 1,598 feet) and several rise above 500 feet in the vicinity of the final approach course.

3. **Pick the correct minimums for airplane category and type of approach.** Now the advantages of the ILS's greater precision become obvious. The altitude minimum for the straight-in approach to Runway 6 is only 594 feet. The visibility minimum is 24. This is a transmissometer-measured visibility of 2,400 feet (less than one-third the length of the landing runway).

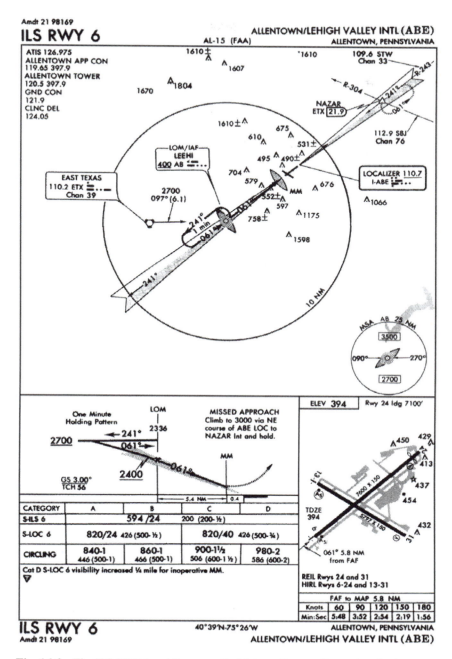

Fig. 16-2. *The ILS RWY 6 at Allentown-Bethlehem-Easton, Pennsylvania, is a typical precision approach.*

DECISION HEIGHT

The 594-foot altitude minimum for the straight-in approach to Runway 6 is a *decision height* (DH), not a minimum decent altitude. Decision height is the height at which a decision must be made during an ILS or other precision approach to continue the approach and land or to execute a missed approach. DH does not allow the maneuvering that is possible with MDA. You cannot level off at DH and continue in the hope of seeing the field and landing. At DH, you *must* decide to land or make a missed approach. These are the only options available at DH.

Even though DH is the minimum altitude on an ILS approach, you must also determine the MDA. If you drop below both DH and MDA on an ILS approach, a missed approach is mandatory. Obstacle clearance is not provided below MDA unless you are in a position to make a normal descent to a landing.

To continue the approach after reaching DH, you must comply with the criteria listed in FAR 91.175 (c) regarding "operation below DH or MDA." Summarized:

- The aircraft must continuously be in a position from which a descent to a landing can be made
- The visibility is not less than that prescribed for the approach in use
- One or more of the nine defined visual references must be distinctly visible and identifiable

Note that there is an additional approach possible with an ILS: a localizer approach shown as S-LOC 6 at Allentown. A localizer approach utilizes the high precision localizer beam for course guidance, but has no glideslope information. If the glideslope transmitter at the airport goes off the air or if the glideslope receiver in the airplane fails, you may continue the approach on the localizer alone. But the approach becomes nonprecision without the glideslope.

The minimum altitude for a localizer approach is a minimum descent altitude (MDA), not a DH. The circling minimums listed below the localizer minimums apply only to the localizer approach. You cannot circle to land out of a full ILS approach with the glideslope, unless the descent is stopped at the circling minimums shown on the approach chart for that specific approach.

Check adjustments for inoperative components table. The inoperable components table is carried in the front section of each set of NOS Instrument Approach Procedures (FIG. 14-2). The visibility minimum increases with the outage of various approach and runway lights. The approach light code for Allentown Runway 6 is shown on the airport diagram (FIG. 16-2) at the approach end of the runway. Use the approach lights table in the front pages of the NOS Instrument Procedures booklet (FIG. 14-3), to identify the type of lighting for the ILS landing runway. For Allentown Runway 6, the system is MALSR (A_5). If it goes out, the visibility minimum increases $1/4$ mile for Category A airplanes on the ILS approach. That certainly makes sense. If there is a failure of the lights, you will need more visibility to find the runway, especially at night.

Be sure to examine higher minimums that might be required for inoperative components in the localizer approach. You will find the localizer increases lumped together in Section 3 of the table with many other nonprecision approaches.

Make adjustments required by fine print. None in this case.

Add the altimeter error. If you detect an altimeter error prior to flight, make no adjustments. Just *add* the error to the minimums on the approach.

Now run through the MARTHA check and make sure you understand all these elements as they apply to the intended approach, in this case the ILS RWY 6 approach at Allentown.

MA-Missed approach. "Climb to 3,000 via NE course of ABE LOC to NAZAR Int. and hold." This is different. The procedure calls for you to track outbound on the back course of the localizer. When tracking a localizer back course outbound, the normal tracking procedure is used—turn toward the needle the same as the ILS front course (normal sensing continues when outbound on an ILS back course).

R-Radios. The No. 1 nav will be set on the localizer frequency, 110.7. When you tune the localizer frequency, the glideslope is automatically received. Note that the localizer frequency is underlined. This indicates no voice transmission capability. The identifier is I-ABE. All localizer identifiers have the prefix I to eliminate any confusion between localizers and VORs.

The No. 2 nav will be set for the radial that establishes the holding fix on the missed approach. The station is SBJ (Solberg VOR) on 112.9. You will dial the Solberg 304 radial with the OBS.

ADF will be set to the ILS Runway 6 compass locator, LEEHI, identifier AB.

MARKER BEACONS

Marker beacons send up a very narrow VHF beam to fix an airplane's position on the ILS final approach course. Beacons are tuned automatically whenever the receiver is operating. The outer marker transmits a continuous series of two audible dashes and a light flashes blue when you pass over the marker.

The middle marker transmits a continuous series of audible alternating dots and dashes and an amber light flashes. (Students find it easy to remember the code if they think of it as saying "You're HERE, you're HERE, you're HERE.") Some ILS approaches—mainly at the larger and busier airports—also have an inner marker. The inner marker transmits a continuous series of dots and flashes white.

Back to the MARTHA check.

T-Time. Pick the time from FAF to MAP based upon the best estimate of ground speed. All ILS approaches should be timed. If the glideslope goes out you can continue with a localizer approach without resetting anything. Use the published MDA instead of DH and the MAP will be determined by timing the final approach segment from the FAF to the MAP.

H-Heading. The final approach course heading in this case is 061°.

A-Altitude. DH for the straight-in ILS 6 approach is 594 feet. MDA for the straight-in localizer 6 approach is 820 feet. MDA for circling approach out of the localizer approach is 840 feet.

ILS TIPS

Request the full procedure where available to get the most out of ILS training. Large, busy airports will probably turn you down because of the heavy flow of traffic. Search out an uncontrolled airport with an ILS where you can practice as many full approach procedures as you wish.

On VFR cross-countries (and at your home airport if it has an ILS) contact approach control and request a "practice" ILS. Remain VFR and fly the approach unhooded when you don't have a safety pilot. (Don't forget collision avoidance—somebody must be looking!) Practice approaches will help you see the "big picture" of how ILS proceeds from step to step at different airports. In spite of common basic elements, all approaches—including ILS—are slightly different.

Practice holding on the localizer course (as shown on the Allentown ILS RWY 6 procedure, for example) without using the compass locator. It takes a little extra practice to set up a holding pattern on that very sensitive localizer needle. Turn outbound when the marker beacon starts to fade.

Always be prepared to switch from the full ILS to the localizer approach at any time, should the glideslope fail.

Instructor note. Some airplanes have circuit breakers that you can pull to simulate a glideslope failure. If you can't do this, simulate the failure by turning the receiver off. But be fair to the student. Turn it off early so the student can retune a second receiver to the localizer frequency.

Be prepared to switch from an ILS or localizer approach to an NDB approach if there is a compass locator at the outer marker. Place the NDB approach chart beneath the ILS chart on the clipboard or yoke chart clip so you can look at it quickly if necessary.

Not all ILS approaches have compass locators or NDB approaches to the same runway. But if there is an NDB approach collocated with the ILS, it is excellent backup in case of transmitter or receiver failure.

When you have tuned and identified the codes of the localizer and any VOR you might need, turn the volume down or the audio off. Failure in these two systems will cause warning flags to appear. On the other hand, adjust the volume on the ADF—after identifying the NDB—to hear the ID faintly in the background. The only way to recognize an ADF or NDB failure is listening to the identifier. As long as you can hear the ID, all is well. (Unless you have inadvertently switched the ADF to REC instead of ADF.)

Remember that a power reduction of 100 RPM (or 1" of manifold pressure) produces a descent of 100 fpm minute; a power reduction of 500 rpm (or 5" of manifold pressure) produces a 500 fpm descent at constant airspeed.

Avoid overcontrolling on the final approach course by using rudder pressure only—no banking—to keep the localizer needle centered. Use gentle elevator pressure to keep the glideslope needle centered. (Heading changes should be limited to 2°, or at most 5°, at any one time. Because the rule of thumb is "never bank more than one-half the degree of heading change," there is no way you can see a 1° bank angle. So, why bother?)

"Pitch to the altitude" and "power to the airspeed" on the glideslope.

ILS/LOC identifier signals are usually not clearly audible until you are at least within 40° of the final approach course. When you are abeam the transmitter site, all you hear is a lot of scratch, which might cause you to miss important communications.

BACK COURSE APPROACHES

ILS localizer antennas are located on the runway centerline about 1,000 feet beyond the far end of the approach runway. The localizer signal radiates in two directions:

- The "front course" is used for the ILS approach
- The "back course" (FIG. 16-2) provides a nonprecision approach path to the opposite end of the runway

The back course cannot be used for instrument approaches unless a specific approach procedure has been approved for that back course. Allentown has a back course approach—LOC BC RWY 24 (FIG. 16-3)—based upon the ILS RWY 6 localizer. Note the words **BACK COURSE** printed on the chart in large bold type. The reason for this warning is that back course approaches resemble conventional localizer front course approaches on approach charts. But they cannot be flown like front course approaches because the needle moves in the opposite direction when heading inbound and no altitude (glideslope) information is available.

The reasons why the localizer needle moves in the opposite direction on a back course approach are fairly complicated questions to fully answer and have to do with the way the localizer signal radiates from the antenna. It is more important to understand that when you make a back course approach, the needle moves *opposite* the way it does on a front course.

Always turn away from the needle to make a heading correction during a back course approach; turn to the left when the needle moves to the right; turn to the right when the needle moves to the left. This is the opposite of VOR or ILS corrections where you always turn toward the needle.

Many students find it simpler to imagine that "they are the needle" and turn toward the bull's eye at the center of the instrument for correction.

A simple way to remember this is that when you are traveling in the same direction as the course for the normal ILS, you make corrections in the same direction as the needle. When you are traveling in the *opposite direction* from the normal ILS, as you do on a back course approach, you make *opposite corrections.*

One or two practice sessions with back course approaches will make all this clear. Some other points to consider when working with back course approaches are:

- Although a back course does not have glideslope sensing, the glideslope needle might come alive periodically. These are false indications. Ignore them.

- The back course needle will be more sensitive than the front course because the localizer antenna array is usually located at the far end of the ILS front course runway (FIG. 16-2); thus, you will be operating just that much closer to the transmitting antenna during the back course approach.

- Some back course approaches have a marker beacon at the FAF to indicate where the approach descent begins. These back course marker beacons might be coded differently than beacons on the front course ILS. Back course markers transmit a continuous audible series of two dots and the white light flashes.

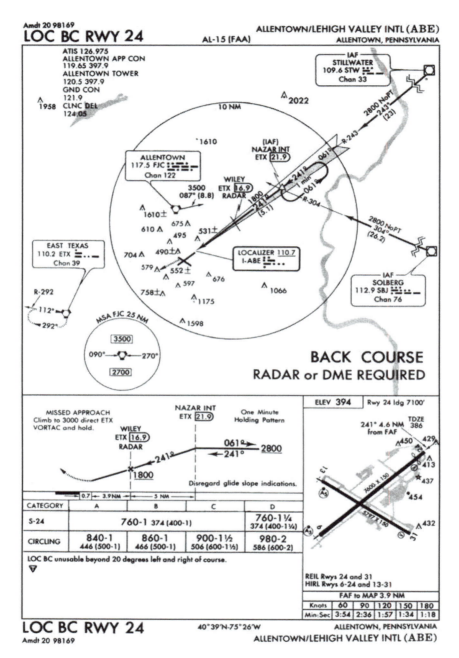

Fig. 16-3. *LOC BC RWY 24 is the back course approach at Allentown.*

LOCALIZER, LDA, AND SDF APPROACHES

As we have seen, localizer approaches can be made on an ILS system whenever the glideslope is out. Many airports have localizers with no glideslope transmitting equipment. This enables an approach with very precise centerline guidance into an airport with terrain and obstacles that rule out the glideslope required for the full ILS.

Figure 16-4 shows the localizer approach at Pittsfield, Massachusetts. The approach has three features we haven't encountered.

- The fine print below the profile says "Inoperative table does not apply." The minimums are already high because of surrounding mountains; therefore, inoperative component comments are unnecessary.

- The localizer (I-EIF, 108.3) also has DME. See the profile and note how DME is used as a cross-check at the IAF, FAF, marker, and MAP. DME is not required for the approach; however, it would certainly be nice to have DME with high obstacles all around. Note that the lower minimums apply for a straight-in approach to Runway 6 if DME is available.

- The third feature of interest is a *fan marker* rather than a marker beacon between the FAF and the MAP. This one is coded "R," as shown on the approach chart (dit-dah-dit) and activates the white light on the marker beacon panel. Fan markers are similar to other marker beacons but more powerful—100 watts output, whereas ILS beacons have an output of 3 watts or less.

LDA

A *localizer-type directional aid* (LDA) approach is uncommon but you still need to know about it. An LDA is a conventional localizer that is not aligned with the runway, for instance the LDA RWY 2 approach at Hartford-Brainard, Connecticut (FIG. 16-5). The runway heading is 020°, but the localizer approach course is 002°, too great a divergence to qualify for approval as a localizer approach.

Straight-in approaches are allowed with an LDA when the divergence between the localizer course and the runway does not exceed 30°, as is the case at Hartford/Brainard. If the divergence is greater than 30°, only circling approaches may be made.

SDF

A *simplified directional facility* (SDF) (FIG. 16-6) transmits a course similar to a localizer but it is not as precise as a localizer. A localizer beam varies between 3° and 6° to produce a width of 700 feet at the landing threshold. The SDF transmitter is fixed at either 6° or 12°. Think of the SDF sensitivity somewhere between a VOR radial and an ILS localizer. The SDF might also be offset from the runway centerline.

You do not need to demonstrate back course, localizer, LDA, or SDF approaches during the instrument flight test. But if any are in your area, especially back courses, you should fly them whenever you have the opportunity. As a rated instrument pilot, you will

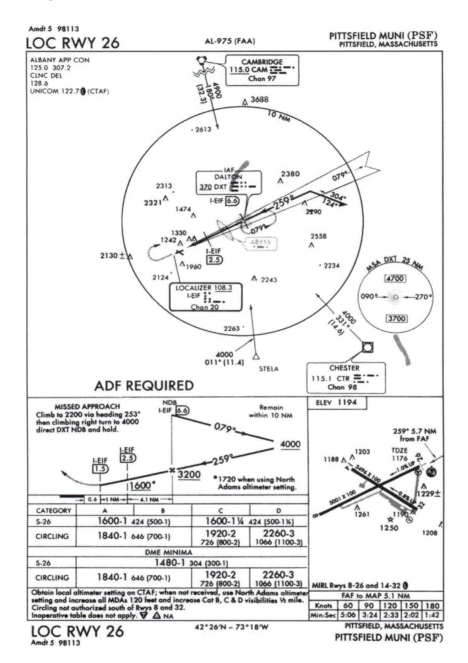

Fig. 16-4. *A localizer approach at Pittsfield, Massachusetts, with two unusual features: DME with the localizer and a fan marker on the final approach course.*

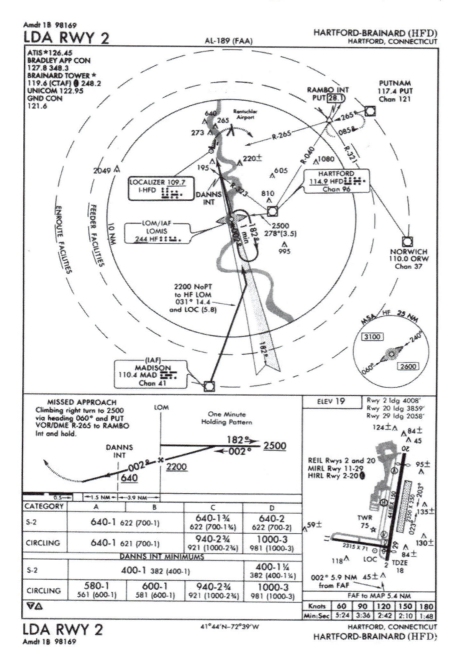

Fig. 16-5. *LDA RWY 2: A localizer directional aid approach at Hartford-Brainard, Connecticut.*

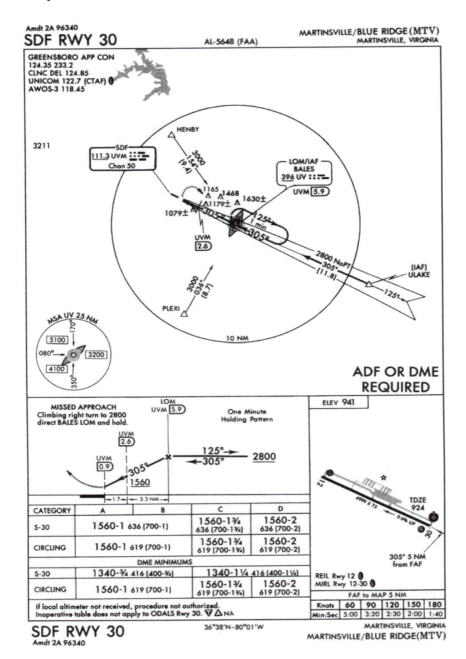

Fig. 16-6. *SDF RWY 30: A simplified directional facility approach at Martinsville/ Blue Ridge, Virginia.*

be expected to execute these approaches whenever they are assigned by ATC. So practice now and avoid surprises and embarrassment later.

RADAR ASSISTS

Expect radar assists on almost every approach. ATC monitors your en route progress with radar, and then hands you off to radar approach control (RAPCON), which then vectors you to the final approach course for the procedure in use. Approach control will often turn you directly onto the final approach course. (Procedure turns are prohibited on radar approaches.)

This doesn't mean that you shouldn't master the full procedures. On the contrary, you must know the full procedure for every approach you fly in order to visualize what approach control specifies. Be prepared to go to the full procedure if you lose radio communications.

This raises an interesting question. Suppose you are receiving radar vectors to intercept the final approach course of an ILS. But you have not been cleared for the approach itself and you lose radio communications at that point. You don't hear anything on either receiver or any voice frequency you might have tuned in. What do you do?

Carry out the lost communications procedure as specified by FAR 91.185 (Summarized):

- **VFR.** "Continue the flight under VFR and land as soon as practicable." In other words, break off the instrument approach and enter the normal VFR traffic pattern.
- **IFR.** Continue with the "route assigned in the last ATC clearance received." In this case the "route" would be "radar vectors to the final approach course." Turn to intercept the final approach course and complete your approach and land or make a missed approach and depart for the filed alternate.

Instructor note. Check the full text of FAR 91.185 and teach it correctly. Remember, if the student hasn't learned, the instructor hasn't taught! Don't assume lost communications just because approach control hasn't talked to you lately. They might have their hands full with an emergency or some other serious situation. Remind ATC: "Cessna five six Xray, final approach course imminent, request further clearance," or whatever covers the situation, and they should respond.

Radar monitoring and radar vectors are not radar approaches. To get a radar approach you must request it; a radar approach might be offered to airplanes in distress or to expedite traffic.

ASR APPROACHES

The most common type of radar approach is the *airport surveillance radar* (ASR) approach, or *surveillance approach*. Look in the front of any NOS instrument approach procedure booklet to find a section that lists radar approaches available in the area covered by the booklet, along with their minimums FIG. 16-7). Note the DH/MDA column. ASR approaches don't get very low.

RADAR INSTRUMENT APPROACH MINIMUMS

ERIE, PA 　　　　Amdt. 7A, OCT 29, 1997　　　　　　ELEV **733**
ERIE INTL
RADAR- 121.0

	RWY GS/TCH/RPI	CAT	DH/ MDA-VIS	HAT/ HAA	CEIL-VIS	CAT	DH/ MDA-VIS	HAT/ HAA	CEIL-VIS
ASR	24	ABC	**1180**-¾	448	(500-¾)	D	**1180**-1	448	(500-1)
	6	ABCD	**1240**/50	508	(600-1)				
CIRCLING		AB	**1340**-1	608	(700-1)	C	**1340**-1¾	608	(700-1¾)
		D	**1360**-2	628	(700-2)				

Inoperative table does not apply to S-6; Categories C and D visibility increased to 1½ miles for inoperative SSALR.
For inoperative MALSR, increase S-24 categories A,B visibility to 1.
When control tower closed, procedure not authorized.
▽
⚠

PHILADELPHIA, PA　　　Amdt. 17, AUG 2, 1984　　　　　ELEV **22**
PHILADELPHIA INTL
RADAR- 128.4 343.6

	RWY GS/TCH/RPI	CAT	DH/ MDA-VIS	HAT/ HAA	CEIL-VIS	CAT	DH/ MDA-VIS	HAT/ HAA	CEIL-VIS
ASR	17	AB	**480**-½	469	(500-½)	C	**480**-¾	469	(500-¾)
		D	**480**-1	469	(500-1)				
	9R	AB	**520**/24	499	(500-½)	C	**520**/40	499	(500-¾)
		D	**520**/50	499	(500-1)				
	9L	AB	**480**-1	466	(500-1)	C	**480**-1¼	466	(500-1¼)
		D	**480**-1½	466	(500-1½)				
	35	AB	**540**-1	529	(600-1)	C	**540**-1½	529	(600-1½)
		D	**540**-1¾	529	(600-1¾)				
	27R	AB	**580**/24	569	(600-½)	C	**580**/50	569	(600-1)
		D	**580**/60	569	(600-1¼)				
	27L	AB	**580**/24	569	(600-½)	C	**580**/50	569	(600-1)
		D	**580**/60	569	(600-1¼)				
CIRCLING		A	**580**-1	559	(600-1)	B	**600**-1	579	(600-1)
		C	**600**-1½	579	(600-1½)	D	**640**-2	619	(700-2)

▽

NE-2

RADAR INSTRUMENT APPROACH MINIMUMS

RADAR MINS

Fig. 16-7. *Radar instrument approach minimums are listed in the front section of every NOS instrument approach procedures set.*

Once approach control assigns an ASR approach, they will tell you *exactly* what to do. If you don't have the ATIS, they will provide it; they will provide lost communications and missed approach procedures; the approach controller will provide radar vectors and runway alignment.

Approach control will issue advance notice of where descent will begin and if the pilot requests it, they also provide recommended altitudes on final approach. You will hear this kind of phraseology:

"Prepare to descend in _____ miles."

"Published minimum descent altitude _____ feet."

"_____ miles from runway. Descend to your minimum descent altitude."

Approach control will keep talking to you on final and inform you if you are deviating from the final approach course:

"Heading _____, on course" (or) *"well left"* (or) *"right of course."*

As you get closer the controller will inform you of your distance from the MAP (and the recommended altitude, if requested):

"_____ miles from missed approach point."

"Recommended altitude is _____ feet."

Surveillance approach guidance can be discontinued when the pilot reports the runway in sight. Approach will say:

"_____ miles from runway" (or) *over missed approach point, take over visually. If unable to proceed visually, execute a missed approach."*

NO-GYRO APPROACHES

A pilot flying with a partial panel can be given a "no-gyro approach." In this procedure, all turns are started and stopped by approach control. The pilot is expected to make standard rate turns until turning onto final approach, when all turns are half standard rate. Turns should be started immediately upon receiving instructions. The instructions couldn't be easier to follow, consisting of such directions as "turn right," "stop turn," "turning final, make all turns one-half standard rate."

Radar controllers need to practice surveillance and no-gyro approaches, so make a point of requesting this service frequently enough to keep everyone proficient.

Instructor note. ASR approaches might be available for airports other than those listed in the front section of the NOS Instrument Approach Procedures booklet. Telephone the RAPCON serving your area—or go visit the facility—and discuss your training needs with a supervisor. Facility personnel can inform you which airports are best for ASR practice and which ones can't handle practice ASR approaches because of heavy traffic.

PAR APPROACHES

It's too bad that *precision approach radar* (PAR) approaches aren't widely available. They are easy to learn, easy to use, extremely accurate, and no needles have to be centered. Ask any current or former military pilot about PAR—known in the military as *ground controlled approach* (GCA)—and you will hear high praise for this precision approach.

With PAR, an electronic runway centerline and an electronic glide path are transmitted from equipment located alongside the runway in use. The precision radar tracks the incoming plane and shows the plane on a scope in relation to the electronic centerline and glide path. The radar is so sensitive that it can detect and display deviations of a few feet. Experienced controllers monitor the displays and tell incoming pilots what action to take to return to the centerline or to get back on the glide path.

A PAR approach begins like an ASR approach. The airplane is vectored onto the final approach course at a specified altitude. As it nears the electronic glide path, the final controller talks the pilot down. The sequence of instructions from the final controller runs something like this:

"Approaching glide path." (Ten to 30 seconds before final descent.)

"Begin descent." (On reaching the point where final descent is to start.)

"Heading _____. On glide path, on course." (To hold the airplane on course and on glide path.)

"Slightly above glide path, slightly left of course."

"Well above glide path, well left of course."

"Above glide path and coming down."

"Left of course and correcting."

"On course. On glide path."

"Three miles from touchdown."

"At decision height."

"Over approach lights."

"Over landing threshold. Contact tower after landing."

I don't know of any PAR approaches routinely available for civilian pilots to practice, nor did research for this book find one. So there is apparently no opportunity to practice this approach.

However, there are still several military airports that have GCA approaches. Their controllers are so skillful at talking down an airplane that you won't need much practice if you ever have to use a PAR in an emergency. Just call the nearest military airport on 121.5, tell them your emergency, request a "GCA," and do what they tell you. They'll get you down safely in an expeditious manner.

VISUAL AND CONTACT APPROACHES

Be prepared to execute two more types of instrument approaches. The first is the *visual* approach. As you arrive on an IFR flight plan, approach control might clear you for a visual approach to the airport or to follow another airplane. ATC cannot issue a visual clearance unless the approach and landing can be accomplished in VFR conditions. Approach control uses the visual approach to expedite incoming IFR traffic when the airport is VFR.

Some visual approaches are so common around big busy airports that approach charts have been developed. Figure 16-8 is the RIVER VISUAL RWY 18 approach to Washington National Airport, Washington, D.C.

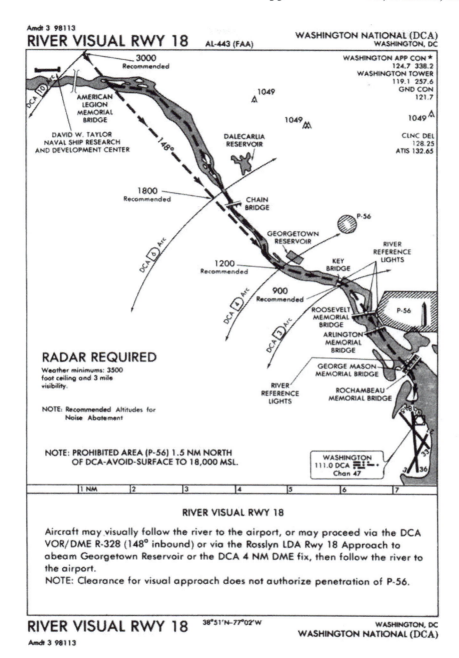

Fig. 16-8. *RIVER VISUAL RWY 18: A published visual approach to Washington National Airport, as depicted on an NOS instrument approach procedures chart.*

A *contact* approach might also be available in good weather conditions. The visual and contact approaches have a notable difference:

- Approach control assigns visual approaches
- A contact approach must be specifically requested by the pilot

Approach control cannot initiate a contact approach. You may request a contact approach if flying an instrument approach and the airplane breaks out clear of clouds with at least 1 mile flight visibility and you can expect to continue to the destination in these conditions.

"One mile and clear of clouds" rings a bell, doesn't it? Right, these are the minimums for special VFR and they are *extremely marginal minimums*. Legal scud running, some would call it.

Never request a contact approach at a strange airport. Contact approaches should only be used at familiar airports. And they should only be used when there is sufficient ceiling and visibility to depart from the instrument approach and enter a comfortable landing pattern.

INSTRUMENT TAKEOFFS

I like to introduce instrument takeoffs about midway in a training course when the student is able to fly the airplane by instruments in a very competent manner. This is usually the point where the student also begins concentrating on precision approaches, so instrument takeoffs are included in this chapter.

FAR 91.175 (f) covers takeoff minimums under IFR. If you read that regulation carefully, you will see that it does not prescribe any IFR takeoff minimums for aircraft operating under Part 91. That's you and me. We may legally take off in any kind of weather. But if we do so when the ceiling and visibility are very low, we might be violating FAR 91.13, operating an aircraft in a "careless or reckless manner."

While it might be technically legal to take off when the ceiling and visibility are below IFR minimums, I think it is very poor judgment to do so. You should always be able to return immediately to the departure airport for an instrument approach if a problem develops. Furthermore, when you expect to climb into actual IFR soon after takeoff, you need a ceiling of approximately 200 feet to get established before entering the clouds.

However, practicing an instrument takeoff with a hood on is an exciting and interesting exercise. (It's always a revelation to students that they can do this, and their confidence grows considerably after they have tried a few.) The procedure is simple. When cleared for takeoff, taxi out and line up on the centerline as usual. Hold the brakes and add full power. Release the brakes when you have three-quarters to full power and anticipate the tendency of the airplane to turn left during the roll by applying right rudder pressure.

Use rudder pressure to keep the airplane rolling straight down the centerline by reference to the heading indicator. Don't try to force the plane into the air. Let the airspeed build up 5 knots or so beyond normal lift-off airspeed, then apply back pressure. Pitch up to the first mark above the horizon on the altitude indicator and hold that attitude until a steady, positive rate of climb shows on the VSI and you have reached 500 feet above the airport. Then adjust the attitude for a normal climb.

You have to make corrections promptly to counter the tendency of the airplane to drift left during the takeoff roll. Fortunately, the heading indicator is sensitive enough so that you can detect very slight changes of heading. React quickly and positively to these slight movements of the heading indicator with rudder pressure.

With a little experience you will find that you can glance out the left window and see whether you are drifting left or right. This comes after you have done two or three instrument takeoffs and are confident enough to take your eyes off the heading indicator for a few seconds then look back to it promptly without being distracted.

17
Putting it all together: The long IFR cross-country

FLYING THE LONG CROSS-COUNTRY TRIP IS THE CULMINATION OF all your efforts so far, as well as a pregraduation introduction to the real world of IFR. It meets the requirements of FAR 61.65 (d) (iii), which calls for "at least one cross-country flight...that is performed under IFR and consists of—

(A) A distance of at least 250 nautical miles along airways or ATC-directed routing;

(B) An instrument approach at each airport; and

(C) Three different kinds of approaches with the use of navigation systems."

When you appear for the instrument flight test, one of the first things the examiner will do is check your logbook to make sure you have accomplished this long IFR cross-country as required.

If this long cross-country is attempted just to meet the bare bones requirements of the FAR, you're missing the true benefit and purpose of the flight. You can meet the qualifications by covering the distance and stumbling through the three different approaches, but you'll shortchange yourself.

This is the point at which instrument students demonstrate to their instructors and themselves that they can do all the planning, fly from one place to another using the ATC system effectively, and make the different approaches down to minimums. Then all the instructor has to do after this lesson is review and polish up those elements that were found a bit lacking on the long cross-country. The student should then be ready for the flight test.

THE VALUE OF ACTUAL IFR

It doesn't make too much difference whether you have an actual IFR day or whether you have to simulate IFR with a hood. The ideal long cross-country has at least one leg of fairly heavy actual IFR. If the entire flight has to be completed in actual IFR conditions, great. It certainly builds confidence in the student to experience the real conditions of a cross-country flight.

It is almost criminal to find that a number of instrument-rated pilots have never seen the inside of a cloud. They file IFR and everything goes just fine until they see a bank of clouds ahead of them. Some will cancel IFR and duck under the clouds, which is just the opposite of what they were trained to do.

Students really haven't earned their instrument ratings unless they have experienced some actual IFR conditions, not merely flown through a few clouds. Actual IFR builds confidence and shows that all those hours under the hood were not make-believe or theory.

Once the initial shock wears off, it's much easier to fly IFR in actual conditions than to simulate it with a hood. The hood is restrictive and uncomfortable, and if you can legitimately get rid of it, the flight will seem more normal. Also under actual IFR the skies are certainly less crowded.

The long IFR cross-country required by the FAR is a basic, straightforward flight. If you have followed the syllabus carefully, you will have planned, filed, and departed on many IFR cross-countries already.

UNCONTROLLED AIRPORTS

File three separate flights at the outset, one for each of the three different types of approaches required. Ideally, one approach should be to an uncontrolled airport. One of the most common errors I see on instrument flight tests is a candidate making an approach at an uncontrolled airport without making traffic advisories on unicom/CTAF. The candidate invades the territory of other pilots and comes barging into their airspace unannounced, then leaves without so much as a hello, thank you, or good-bye. It certainly gets the locals perturbed if they find themselves sharing the final with some pilot who has a hood on! Or worse yet, who sets up a collision course with the traffic using the opposite runway.

One approach to an uncontrolled field will bring out any deficiencies in uncontrolled airport procedures. Most examiners make it a point to evaluate the way an instrument candidate conducts IFR approaches at an uncontrolled airport.

If you haven't had much practice picking up a new clearance in the air, the long IFR cross-country is a good opportunity to do so. An instrument rating should not be attached to a person's pilot certificate without some acquaintance with air filing and picking up a clearance in the air.

Plan to execute a missed approach, which will keep the IFR flight plan open. When ATC requests intentions before commencing the approach, tell them you have an IFR flight plan on file to your next destination. Be sure to have that flight plan handy so you can read it to ATC in case it has gone astray in the computer.

Controllers are usually very considerate of pilots in the air when reading a clearance. They'll read it piecemeal and try not to issue more than two or three parts of a clearance at one time.

VOID TIME CLEARANCES

The long IFR cross-country also provides a good opportunity to practice void time clearances over the telephone when departing from an uncontrolled airport. Your flight plan has already been filed, so you will telephone flight service or some part of the ATC system and ask for the clearance. They might ask you to call back at a certain time for the IFR clearance, release, and void time. Void time means exactly that—the clearance is void after that time. If you see that you can't take off before the void time, you will have to call flight service again and request a more practical release time.

Void time clearances take a bit of forethought. You have to complete the preflight and get the charts and radios all set for departure before calling for the clearance and release time. Position the airplane for a prompt departure.

PARTIAL PANEL

I also like to do some partial panel work during a leg that is long enough so the student is well caught up on the cockpit workload. One thing I like to see on partial panel is the student tuning the ADF to an NDB or commercial broadcast station up ahead that will serve as a backup heading indicator, giving a constant relative bearing in combination with the magnetic compass.

As you plan the flight, ask what you would do if given a simulated vacuum system failure at various points along the way, then note what NDBs or commercial broadcast stations can function as a backup for the heading indicator at these different points. The examiner will certainly be impressed if you demonstrate this on the instrument flight test.

The fine points of IFR cross-country flying are well covered by Background Briefing 16-17 that precedes this flight lesson; however, I would like to discuss two other aspects of cross-country flying in greater detail: fuel management and lost communications.

FUEL MANAGEMENT

Year after year, running out of fuel continues to be a major cause of general aviation accidents. There is no excuse for this. Everyone knows, or should know, how much fuel is in the tanks at the beginning of the flight, the gallons or pounds per hour that will be consumed at various power settings, and how long it will take to complete the flight. Fuel requirements must be computed for every cross-country flight. The FARs require this. Fuel is time in the tanks. When will the fuel be exhausted?

What is the problem? There are really two problems.

First, many pilots rely too heavily on fuel gauges. The fuel gauges in general aviation airplanes are not reliable, especially below one-quarter of a tank when the fuel is sloshing and surging around.

Second, you must break the habit of thinking in terms of "full tank," " half tank," "quarter of a tank," etc. Instead, think of hours and minutes of fuel remaining.

To do this, it is essential that you keep track of the time when tanks are switched on and off. Keep an accurate log of takeoff time and arrival time at all fixes and checkpoints. This isn't very hard because you have been logging these items on all cross-country flights, VFR as well as IFR, since student pilot days.

LOGGING THE FLIGHT

Figure 17-1 shows the filled-out log of the flight to Binghamton, New York, that was previously planned. It contains all the information that was entered during the flight.

Note the block for entering takeoff time (A), and the blocks for logging the times when tanks are normally switched (B). There was no need to switch tanks because we flew an airplane that feeds from both tanks.

The two columns under (C) provide blocks in which to log estimated and actual times en route and arrival for each leg and fix. No actual times were logged for the flight to the alternate because the weather at Binghamton was VFR by the arrival time. Notice the landing ATIS "Whiskey" (D); above that is ATIS information "Bravo" for the departure airport, Westchester County.

While looking at the log, note how the clearance was copied (E), and how frequency and altitude changes were recorded (F) and (G).

Also, in block (A) the time of arrival was logged and the total time en route was computed and entered. To compute fuel remaining, subtract total time en route (1:10) from the hours and minutes in the tank on departure: $5:30 - 1:10 = 4:20$. Think hours and minutes—in this case 4 hours and 20 minutes fuel remaining—not full and half-full tanks.

OBTAINING WEATHER INFORMATION IN FLIGHT

It would be nice if the weather on a cross-country flight always turned out to be the same as forecast when you worked out your flight plan. But this is rarely the case. In many parts of the country and in many seasons of the year, weather can change rapidly, and sometimes violently.

As your flight progresses, you should systematically check out the weather up ahead to make sure you can get through to your destination and make a safe approach and landing, and that your alternate is still available if you can't land where you want to.

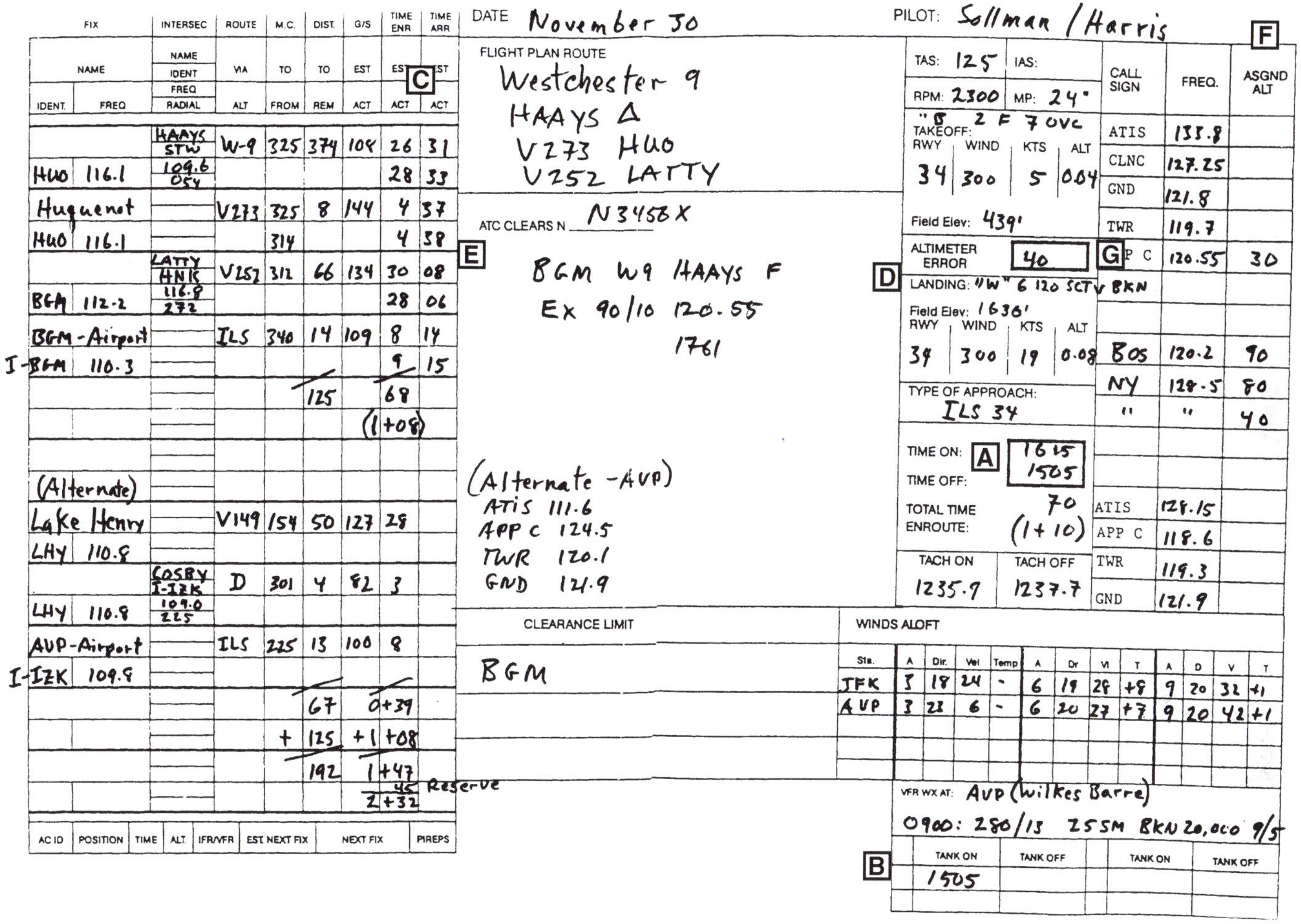

Fig. 17-1. The completed flight log to Binghamton showing how items were entered during the flight.

The best way to do this is to contact "flight watch" on 122.0 MHz. Use the name of the ATC facility serving the area—Boston, New York, Cleveland, etc.—and be sure to get permission from ATC that you are leaving that ATC frequency for a few minutes. ATC might want you to delay your switch to flight watch in order to amend a clearance or hand you off to another facility.

Flight watch is the call sign for the FAA's *en route flight advisory service* (EFAS). When you talk to flight watch you will speak directly to a qualified weather briefer who will answer your specific questions without having to read out all the details. On the flight to Binghamton, for example, you could call New York flight watch half an hour or so before landing and get the latest terminal weather for Binghamton and Wilkes-Barre, the alternate. Be sure to advise ATC that you are back on their frequency after you are finished with flight watch.

Selected VORs also broadcast hazardous in-flight weather advisory service (HIWAS) information. (See Fig. 17-2.) HIWAS is continuous. It disseminates severe weather forecast alerts, SIGMETS, convective SIGMETS, center weather advisories (CWAs), AIRMETS, and urgent PIREPS. Look up the VORs in your area in A/FD and see which provide HIWAS, then listen in a couple of times to find out what is available.

LOST RADIO CONTACT

Complete loss of communications while airborne on an IFR flight is extremely rare. What is more likely to happen is a loss of radio contact. Many times lost radio contact is self-induced. The squelch might be tuned down too low or a wrong frequency might have been inadvertently tuned. Double-check all the navcom settings and the audio panel settings. Also try transmitting on the other radio. It is possible for one to fail and the other to work just fine.

You might be flying through a quiet zone along the airway, or flying low enough so that the VHF line-of-sight transmissions are blocked by mountains or other obstructions.

Here is a typical scenario: an ARTCC to ARTCC handoff. Let's say Boston ARTCC calls you with a transmission: "Cessna five six Xray, contact New York Center, one two eight point five."

You acknowledge the handoff by reading back the clearance: "New York Center one two eight point five, five six Xray," then you tune the new frequency and report, using the full N number because this is an initial contact, and you report your altitude: "New York Center, Cessna three four five six Xray at niner thousand."

No response. Suddenly all is quiet.

You try two or three more times and still get no response. What do you do? Make sure you haven't made some mistake such as switching from speaker to headphones.

Return to the frequency you just left and report Cessna "five six Xray, unable New York Center one two eight point five." Boston should answer and might tell you to remain with them, or to attempt to contact New York Center on another frequency, or to stay with Boston and try New York again in a few minutes or a few miles.

Now suppose you had returned to the Boston Center frequency and couldn't raise them again. Nothing doing on either New York ahead or Boston behind you. Now what?

JOHNSTOWN
FULTON CO   (NY27)   2 E   UTC−5(−4DT)   N42°59.89' W74°19.77'   **NEW YORK**
  881   B   S2   **FUEL** 100LL, JET A   OX 1   TPA−1660 (800)   Not insp.   **L−25B, 26F**
  RWY 10−28: H4000X75 (ASPH)   MIRL   0.4% up E   **IAP**
    RWY 10: REIL. PAPI(P2L). Tree.   **RWY 28:** REIL. PAPI(P2L). Tree.
  AIRPORT REMARKS: Attended Mon−Fri 1300−2200Z‡, Sat−Sun 1300−2300Z‡. ACTIVATE MIRL Rwy 10−28—CTAF.
  COMMUNICATIONS: CTAF/UNICOM 122.7
    BURLINGTON FSS (BTV) TF 1−800−WX−BRIEF. NOTAM FILE BTV.
    ALBANY APP/DEP CON 118.05
  RADIO AIDS TO NAVIGATION: NOTAM FILE ALB.
    ALBANY (L) VORTAC 115.3   ALB   Chan 100   N42°44.84' W73°48.19'   316° 27.7 NM to fld. 275/13W.
    JOHNSTOWN NDB (MHW) 523   JJH   N42°59.96' W74°19.95'   at fld. NOTAM FILE BTV.

JOSEPH Y RESNICK   (See ELLENVILLE)

KAMP   (See DURHAMVILLE)

KATHI   N43°06.55' W78°50.30'   NOTAM FILE IAG.
  NDB (LOM) 329   IA   279° 4.7 NM to Niagara Falls Intl. Unmonitored when Niagara Falls Intl twr clsd.

KEENE
MARCY FLD   (NY29)   3 N   UTC−5(−4DT)   N44°13.25' W73°47.48'   **MONTREAL**
  985   Not insp.
  RWY N−S: 2190X95 (TURF)
    RWY N: Road.   **RWY S:** Trees.
  AIRPORT REMARKS: Unattended.
  COMMUNICATIONS: CTAF 122.9
    BURLINGTON FSS (BTV) TF 1−800−WX−BRIEF. NOTAM FILE BTV.

KENNEDY   N40°37.97' W73°46.28'   NOTAM FILE JFK.   **NEW YORK**
  (H) VOR/DME 115.9   JFK   Chan 106   at John F. Kennedy Intl. 11/12W.   **H−3J, 6I, L−24H, 28G**
  RCO 122.1R 115.9T (NEW YORK FSS)

KINGSTON   N41°39.93' W73°49.34'   NOTAM FILE ISP.   **NEW YORK**
  (L) VOR/DME 117.6   IGN   Chan 123   242° 3.6 NM to Dutchess Co. 580/12W.   **HIWAS.**   **H−3J, 6I, L−25B, 28H**
    VOR unusable 045°−050° byd 35 NM blo 4300'   070°−140° byd 30 NM blo 3400'
  RCO 122.1R 117.6T (NEW YORK FSS)

KINGSTON−ULSTER   (20N)   4 N   UTC−5(−4DT)   N41°59.12' W73°57.85'   **NEW YORK**
  149   S4   **FUEL** 100LL   TPA—1200(1051)   **L−25B, 28H**
  RWY 15−33: H3100X60 (ASPH)   S−10E   MIRL   0.5° up NW   **IAP**
    RWY 15: PAPI(P2L). Thld dsplcd 240'. Trees.   **RWY 33:** REIL. PAPI(P2L).Thld dsplcd 315'. Pole.
  AIRPORT REMARKS: Attended 1300Z‡−dusk except Thanksgiving, Christmas and New Years. PPR for ngt ops call
    914−336−8400. PAEW adjacent twy/rwy. Acft should not taxi off paved area except near ramp. Rwy 33 dsplcd
    thld has faint non-standard markings. Noise abatement procedures in effect call 914−336−8400. Avoid
    overflight of housing development adjacent to N end of rwy. ACTIVATE MIRL Rwy 15−33, PAPI Rwy 15 and Rwy
    33, REIL Rwy 33 and twy lgts—123.3.
  COMMUNICATIONS: CTAF/UNICOM 122.8
    NEW YORK FSS (ISP) TF 1−800−WX−BRIEF. NOTAM FILE ISP.
  Ⓡ **NEW YORK APP/DEP CON** 132.75
  RADIO AIDS TO NAVIGATION: NOTAM FILE ISP.
    KINGSTON (L) VOR/DME 117.6   IGN   Chan 123   N41°39.93' W73°49.34'   354° 20.2 NM to fld. 580/12W.
    HIWAS.

KIRKI   N43°06.73' W76°00.16'   NOTAM FILE SYR.
  NDB (LOM) 242   SY   281°4.6 NM to Syracuse Hancock Intl.

Fig. 17-2. *HIWAS at Kingston VOR indicated by arrow.*

1. Check for other center frequencies in the area (FIG. 17-3). In this case, there is another New York Center frequency available: 134.65.

2. Attempt to raise another aircraft to relay the message. If you hear a specific N-number on the frequency, call that aircraft. If nothing is ever heard, transmit in the blind: "Any aircraft, Cessna three four five six Xray, requesting a communications relay." If someone answers, have the pilot try to reach the new center—New York in this case—and relay their instructions to you.

3. Try contacting the nearest approach control or even a nearby tower and ask them to relay your message to the center.

4. Call flight service on 122.1, and give them a nearby VOR identification and frequency with voice capability that you can hear clearly. Be patient. It might take them a couple of minutes to complete another transmission before they can get to you.

5. Come up on the emergency frequency 121.5, say you're having communications difficulties, and if anyone replies, contact them with your next transmission. Many pilots are intimidated by the emergency frequency. They are reluctant to use it because they think they will have to write a lengthy letter to the FAA in Washington justifying the action. This is usually not the case. You don't have to report to anyone just because you used the emergency frequency. The emergency frequency is intended for lost radio contact as well as for more serious situations. When all else fails, try to reestablish communications on this frequency.

6. When you switch to 121.5 as the last resort, set the transponder to squawk 7600—the code for radio communications lost—until the conclusion of the flight or until ATC communications have been restored. (The squawk code for a flight emergency is still 7700. *See* a current edition of AIM for any additional amendments.)

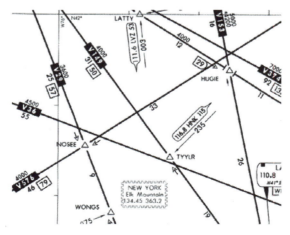

Fig. 17-3. *Box containing sector frequencies for an air route traffic control center (ARTCC).*

7. Try to repair the radio. You might have a hot microphone or a mike button stuck in the keyed position. You can find out very quickly by unplugging the mike from its jack. If you start hearing other transmissions again, the mike is the problem. A muting relay engages when you depress the microphone key. It cuts out incoming signals when you transmit because you can't talk and listen at the same time without getting a howling, screeching feedback that will blast your eardrums out of action. So if the mike button is stuck, the muting relay will be engaged and you won't hear anything.

Because a stuck mike transmits continuously it also blocks everything on that frequency within miles. When you unplug the stuck mike, you will begin hearing normal transmissions again, plus a lot of unkind remarks about the pilot with the stuck mike. So help everyone out when you have a radio problem and check for a stuck mike. If it sounds like you are still transmitting after you take your thumb off the mike button, check for a stuck mike first before trying any of the other alternatives.

Also try giving the transceivers a firm push back into the racks. Sometimes they come loose and connections at the back become disengaged.

TWO-WAY RADIO COMMUNICATIONS FAILURE

If you have tried every method of restoring radio communications and nothing works, then you must assume a two-way radio communications failure. FAR 91.185 (Appendix B) spells out very clearly the procedure you must follow when this happens.

IMPORTANCE OF LOGGING TIMES

Note how important it is in FAR 91.185 to know the estimated time of arrival (ETA) at your destination if you have not received an "expect further clearance time." ATC will expect you to show up for an approach at an estimated time of arrival (ETA) based upon your filed or amended estimated time en route (ETE).

ATC knows when you took off from the departure airport—do you?—and they know the estimated time en route from your filed or amended flight plan. They will add the time en route to the departure time and come up with an ETA. ATC will reserve all approaches at the destination for this ETA plus 30 minutes. Until that time expires, you "own" that airport.

So it becomes extremely important to know what time you took off because without that information, you'll never know when to commence the approach. ("You can't tell when you're going to get someplace if you don't know when you left someplace.") This seems so basic, yet many pilots forget to log their takeoff time. When I notice this on an instrument flight test, you can bet that at some point I will ask the applicant to tell me what the destination ETA will be in the event of lost communications or to detail the steps to take in case of communications failure.

EMERGENCY ALTITUDES

A second point that always raises questions is the altitude for completing the flight in the event of a two-way radio communications failure. ATC expects the flight to continue at

(1) the last assigned altitude, *or* (2) the minimum altitude for IFR operations, *or* (3) the altitude that ATC has advised you to expect in a further clearance. You will fly at the highest altitude of the three choices for any given leg.

You will have to make some choices. These choices hinge on how the "minimum altitude for IFR operations" is determined, and whether or not it is higher than the altitude assigned or advised to expect.

Along an airway, the minimum for IFR operations is the MEA or minimum en route altitude. If the en route low-altitude chart shows that you are approaching a route segment with an MEA that is *higher* than that assigned by ATC or what ATC advised to expect, you would begin a climb to reach the new altitude so that you reach it prior to the point or fix where the new altitude begins. If the MEA drops down below the last assigned altitude, you would descend to the last altitude assigned by ATC or one that you were advised to expect, whichever is higher.

But suppose you were cleared direct. What is the minimum altitude off airways where there are no MEAs? Consult the appropriate sectional chart that covers the flight area and pick out the "maximum elevation" figure for the latitude-longitude square you occupy, then add 1,000 feet to comply with FAR minimum altitude regulations (2,000 feet in mountainous terrain—see FAR 91.177). The maximum elevation figures are the big numbers in the center of each latitude-longitude square. You would use the "maximum elevation plus a thousand" figure as a substitute for the MEA.

The likelihood of having to use the two-way radio communications failure procedures is remote, but you have to know them. And "Loss of Communications" is a required task specified in the "Emergency Operations" section of the *Instrument Rating Practical Test Standards*.

In the real world of IFR you can avoid all the complications of lost communications by investing in a hand-held portable transceiver and carrying it on every flight. It will also be invaluable in the next scenario.

COMPLETE ELECTRICAL FAILURE

Single-engine airplanes are vulnerable to complete electrical failure. There is usually only one alternator—driven by one inexpensive V-belt—and only one voltage regulator. So there is no backup if one of these key components fails.

The first step in the event of a complete electrical failure is to turn off all electrical equipment to minimize drain on the battery. Control the airplane by partial panel. At night you will need a flashlight to do this. Careful pilots carry two flashlights for this purpose—one as a backup in case the other starts to fade—plus extra batteries.

If the failure occurs when you are VFR, maintain VFR and land as soon as practicable, just as you would if you had two-way radio communications failure.

If you are in actual IFR conditions when a complete electrical failure occurs, the problem is equally simple, although your work is cut out for you. Find VFR conditions and land as quickly as possible.

When planning for the flight, you filled in the VFR WX AT: block to quickly seek a safe haven in this circumstance (FIG. 17-4). You know where you want to go. If you have been logging the actual time of arrival at each fix along the way, you have an accurate idea of where you were when the failure occurred. Draw a straight line to the new destination from the point you intend to turn. Calculate a new magnetic heading and turn to that heading. Later, with more time, calculate an ETA at the new destination. It's like those "diversions" you worked on so hard during the cross-country phase of private pilot training.

When everything is under control, set the transponder to 7700 and turn it on. Then turn on the transceiver on which you last talked to ATC. You have only a few minutes of battery power left; make sure ATC understands your predicament and intentions, then shut everything off. You can put the remaining battery power to good use at the VFR destination to contact a tower or flash some lights.

Again, a portable transceiver will make life a lot easier. The range of these units is vastly improved if connected to an aircraft antenna; consider installation of an antenna jack.

IFR CROSS-COUNTRY TIPS

Avoid marking up the chart with a highlighter. This can add to confusion on subsequent flights in the same area. I make it a game to see how much of an IFR flight I can complete without consulting the low altitude en route charts except for an amended clearance. A previously marked up chart can be hazardous, especially if you have made a series of flights over the same area since the last chart revision.

After planning the flight, talk yourself through it. Make sure you understand where all those obscure intersections are and note the airports along the way, especially those with instrument approaches.

Carry VFR charts to cover the route. Highlight the IFR route on the VFR charts for quick reference in case VFR landmarks are needed.

Save the flight logs. You might frequently use the same routes, or portions of them, over and over again. ATC often has preferred routings and some of them might be unpublished; old logs will show these preferences and you can plan for them and avoid the frustration of filing for one route and receiving a clearance for another.

Visit the center and the approach control that handles your local area. Call ahead for an appointment; you will find the supervisors and controllers most cooperative in showing you around and answering questions, especially if you can make an appointment for one of their less busy times.

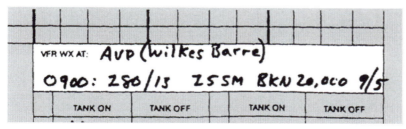

Fig. 17-4. *Flight log showing filled-in block for nearest VFR weather.*

Instructor note. ATC centers and approach control facilities prefer to brief groups of 10–25 people. Visit these facilities in person and find out who can set up a briefing and what are the best dates and times. Post a sign-up sheet in the flight school or FBO and get a group together. Invite everyone. Groups are always more interesting because everyone has different questions and different experiences to bring up. Some of the questions will surprise you!

18
Getting the most out of the instrument written exam

W ITH THE WIDE AVAILABILITY OF EXCELLENT TEST GUIDES THAT
include answers and references, passing the instrument written test is no longer
the nightmare it used to be. But the questions missed most often throw into sharp focus the areas of knowledge causing students the greatest trouble. It is important to master these troublesome subject areas, not only to pass the written test, but also to become
a truly competent instrument pilot.

The most frequently missed questions come from many tricky areas. The questions below get right to the heart of these tricky areas. They are selected from the 930
questions in the FAA's master, computer-based "Knowledge Test Question Bank" for
the instrument rating. (This is the database from which questions are drawn for the
computerized written test.) They serve as a self-diagnostic test to guide you—and your
instructor—to subjects that you might not fully understand. So take this test and see
where you need more work. I don't provide the answers because I want students—and

instructors—to look up the pertinent material and think through the answers for themselves; however, authoritative references are provided to save research time.

Instructor note. I have found that the best time to have a student take the official FAA instrument written test is just after the IFR cross-country. By this time students have had practical experience with most of the subjects covered by the test, such as weather, flight planning, minimums, and ATC procedures. This practical experience will give these specific areas relevance, aiding memory. Use the self-diagnostic test below to help your student polish up any remaining weak areas just before taking the instrument written test.

SELF-DIAGNOSTIC TEST

You should not be satisfied until you score 100 percent on all the questions. (Remember that most of the questions are the result of an accident/incident often resulting in death.) Take the test, review your answers with your instructor, and note the questions you missed. Look up the references for these missed questions and go over this material until you and your instructor are satisfied that you understand all questions and answers. Review the self-diagnostic test just before you take the written exam and study the few remaining subjects you are not sure of. You will be guaranteed to pass the test!

1. Which flight time may be logged as instrument time when on an instrument flight? (FAR 61.51)
   A All of the time the aircraft was not controlled by ground references
   B Only the time you controlled the aircraft solely by reference to flight instruments
   C Only the time you were flying in IFR weather conditions

2. What are the minimum qualifications for a person who occupies the other control seat as a safety pilot during simulated instrument flight? (FAR 91.109)
   A Appropriately rated in the aircraft
   B Private pilot
   C Private pilot with instrument rating

3. A pilot's recent IFR experience expires on July 1 of this year. What is the latest date the pilot can meet the IFR experience requirement without having to take an instrument proficiency check? (FAR 61.57)
   A December 31, this year
   B June 30, next year
   C July 31, this year

4. Which data must be recorded in the aircraft log or other appropriate log by a pilot making a VOR operational check for IFR operations? (FAR 91.171)
   A VOR name or identification
   B Place of operational check, amount of bearing error, date of check, and signature
   C Date of check, VOR name or identification, place of operational check, amount of bearing error

5. If the air temperature is $+8°C$ at an elevation of 1,350 feet and a standard (average) temperature lapse rate exists, what will be the approximate freezing level? (*Aviation Weather,* Chapter 2)
   A 3350 feet MSL
   B 5350 feet MSL
   C 9350 feet MSL

6. If squalls are reported at your destination, what wind conditions should you anticipate? (*Aviation Weather,* Chapter 11)
   A Sudden increase in wind speed of at least 15 knots to a peak of 20 knots or more, lasting for at least 1 minute
   B Peak gusts of at least 35 knots, for a sustained period of 1 minute or longer
   C Rapid variation in wind direction of at least 20° and changes in speed of at least 10 knots between peaks and lulls

7. During the life cycle of a thunderstorm, which stage is characterized predominantly by downdrafts? (*Aviation Weather,* Chapter 11)
   A Cumulus
   B Dissipating
   C Mature

8. Which procedure is recommended if a pilot should unintentionally penetrate embedded thunderstorm activity? (*Aviation Weather,* Chapter 11)
   A The pilot should reverse aircraft heading or proceed toward an area of known VFR conditions
   B Reduce airspeed to maneuvering speed and maintain a constant altitude
   C Set power for recommended turbulence penetration airspeed and attempt to maintain a level-flight attitude

9. In which meteorological environment is aircraft structural icing most likely to have the highest rate of accumulation? (*Aviation Weather,* Chapter 10)
   A Cumulonimbus clouds
   B High humidity and freezing temperature
   C Freezing rain

10. What situation is most conducive to the formation of radiation fog? (*Aviation Weather,* Chapter 12)
    A Warm, moist air over low, flatland areas on clear, calm nights
    B Moist tropical air moving over cold, offshore water
    C The movement of cold air over much warmer water

11. Which primary source should be used to obtain forecast weather information at your destination for the planned ETA? (*Aviation Weather Services,* Section 4)
    A Area forecast
    B Radar summary and weather depiction charts
    C Terminal forecast

12. Which weather conditions should be expected beneath a low-level temperature inversion layer when the relative humidity is high? (*Aviation Weather*, Chapter 2, 12)

    A Smooth air and poor visibility due to fog, haze, or low clouds

    B Light wind shear and poor visibility due to haze and light rain

    C Turbulent air and poor visibility due to fog, low stratus-type clouds, and showery precipitation

13. The reporting station originating this METAR has a field elevation of 620 feet. If the reported sky cover is one continuous layer, what is its thickness? (Tops of OVC are reported at 6,500 feet.) (*Aviation Weather Services*, Section 2)

    METAR KMDW 121856Z AUTO 32005KT 1 1/2SM + RABR OVC007 17/16 AS980

    A 5,180 feet

    B 5,800 feet

    C 5,860 feet

14. What is meant by the entry in the remarks section of the METAR surface report for KBNA? (*Aviation Weather Services*, Section 2)

    METAR KBNA 211250Z 33018KT 290V260 1/2SM R31/2270OFT + SN

    BLSNFG VV008 00/MO3 A2991 RMK RAE42SNB42

    A The wind is variable from 290° to 360°

    B Heavy blowing snow and fog on runway 31

    C Rain ended 42 past the hour, snow began 42 past the hour

15. Hazardous wind shear is commonly encountered near the ground (*Aviation Weather*, Chapter 9)

    A during periods when the wind velocity is stronger than 35 knots

    B during periods when the wind velocity is stronger than 35 knots and near mountain valleys

    C during periods of strong temperature inversion and near thunderstorms

16. While airborne, what is the maximum permissible variation between the two indicated bearings when checking one VOR system against the other? (*Aeronautical Information Manual*, 1-1-4)

    A Plus or minus 4° when set to identical radials of a VOR

    B 4° between the two indicated bearings to a VOR

    C Plus or minus 6° when set to identical radials of a VOR

17. What response is expected when ATC issues an IFR clearance to pilots of airborne aircraft? (*Aeronautical Information Manual*, 4-4-8)

    A Read back the entire clearance as required by the situation

    B Read back those parts containing altitude assignments or vectors and any part requiring verification

    C Read-back should be unsolicited and spontaneous to confirm that the pilot understands all instructions

18. A particular standard instrument departure (SID) requires a minimum climb rate of 210 feet per nautical mile to 8,000 feet. If you climb with a ground speed of 140 knots, what is the rate of climb required in feet/minute? (flight computer)
    A 210
    B 450
    C 490

19. During a flight, the controller advises "traffic 2 o'clock 5 miles southbound." The pilot is holding 20° correction for a crosswind from the right. Where should the pilot look for the traffic? (*Aeronautical Information Manual*, 4-1-14)
    A 40° to the right of the airplane's nose
    B 20° to the right of the airplane's nose
    C Straight ahead

20. When can a VFR-on-top clearance be assigned by ATC? (*Aeronautical Information Manual*, 4-4-7)
    A Only upon request of the pilot when conditions are indicated to be suitable
    B Anytime suitable conditions exist and ATC wishes to expedite traffic flow
    C When VFR conditions exist, but there is a layer of clouds below the MEA

21. What action should you take if the No. 1 VOR receiver malfunctions while operating in controlled airspace under IFR? (Your aircraft is equipped with two VOR receivers: the No. 1 receiver has omni/localizer/glideslope capability, and the No. 2 has only omni. (FAR 91.187)
    A Report the malfunction immediately to ATC
    B Continue the flight as cleared; no report is required
    C Continue the approach and request a VOR or NDB approach

22. You are in IMC and have two-way radio communications failure. If you do not exercise emergency authority, what procedure are you expected to follow? (*Aeronautical Information Manual,* Chapter 6, Section 4)
    A Set the transponder to code 7600, continue flight on assigned route and fly at the last assigned altitude or the MEA, whichever is higher
    B Set transponder to code 7700 for one minute, then to 7600, and fly to an area with VFR weather conditions
    C Set the transponder to 7600 and fly to an area where you can let down in VFR conditions

23. How can an instrument pilot best overcome spatial disorientation? (*Pilot's Handbook of Aeronautical Knowledge,* Chapter 9)
    A Rely on kinesthetic sense
    B Use a very rapid cross-check
    C Read and interpret the flight instruments, and act accordingly

24. (Refer to FIG. 18-1.) What is the magnetic bearing to the station as indicated by the RMI indications in illustration 4? (*Instrument Flying Handbook,* Chapter VIII)
    A 285°
    B 055°
    C 235°

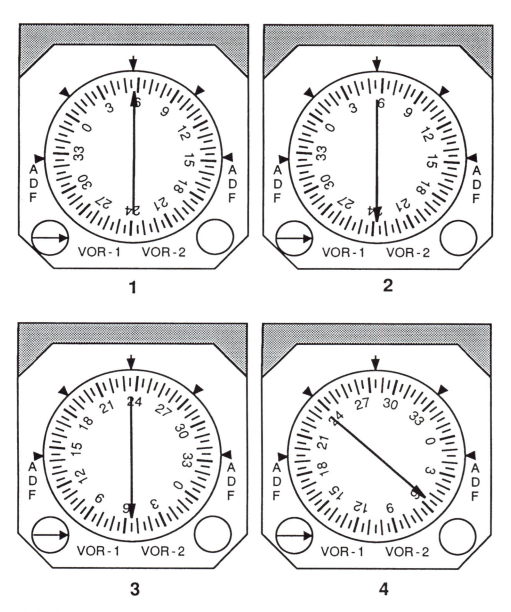

Fig. 18-1.

25. Determine the approximate time and distance to a station if a 5° wingtip bearing change occurs in 1.5 minutes with a true airspeed of 94 knots. (*Instrument Flying Handbook,* Chapter VIII)
    A 16 minutes and 14.3 nm
    B 18 minutes and 28.5 nm
    C 18 minutes and 33.0 nm

26. (Refer to FIG. 18-2.) In which general direction from the VORTAC is the aircraft located? (*Instrument Flying Handbook,* Chapter VIII)
    A Northeast
    B Southwest
    C Southeast

27. A pilot is making an ILS approach and is past the OM to a runway which has a VASI. What action should the pilot take if an electronic glideslope malfunction occurs and the pilot has the VASI in sight? (FAR 91.175)
    A The pilot should inform ATC of the malfunction and then descend immediately to the localizer DH and make a localizer approach
    B The pilot may continue the approach and use the VASI glideslope in place of the electronic glideslope
    C The pilot must request an LOC approach, and may descend below the VASI at the pilot's discretion

28. When landing behind a large jet aircraft, at which point on the runway should you plan to land? (*Aeronautical Information Manual,* Chapter 7, Section 3)
    A If any crosswind, land on the windward side of the runway and prior to the jet's touchdown point
    B At least 1,000 feet beyond the jet's touchdown point
    C Beyond the jet's touchdown point

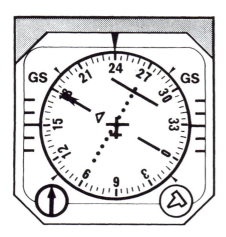

Fig. 18-2.

29. Aircraft approach categories are based on (*Instrument Flying Handbook,* Chapter X)

    A certificated approach speed at maximum gross weight

    B 1.3 times the stall speed in landing configuration at maximum gross landing weight

    C 1.3 times the stall speed at maximum gross weight

30. When passing through an abrupt wind shear which involves a shift from a tailwind to a headwind, what power management would normally be required to maintain a constant indicated airspeed and ILS glideslope? (*Instrument Flying Handbook,* Chapter VIII)

    A Higher than normal power initially, followed by a further increase as the wind shear is encountered, then a decrease

    B Lower than normal power initially, followed by a further decrease as the wind shear is encountered, then an increase

    C Higher than normal power initially, followed by a decrease as the shear is encountered, then an increase

31. What effect will a change in wind direction have upon maintaining a 3° glideslope at a constant true airspeed? (*Instrument Flying Handbook,* Chapter VIII)

    A When ground speed decreases, rate of descent must increase

    B When ground speed increases, rate of descent must increase

    C Rate of descent must be constant to remain on the glideslope

32. You are being vectored to the ILS approach course, but have not been cleared for the approach. It becomes evident that you will pass through the localizer course. What action should be taken? (*Aeronautical Information Manual,* 5-4-3)

    A Turn outbound and make a procedure turn

    B Continue on the assigned heading and query ATC

    C Start a turn to the inbound heading and inquire if you are cleared for the approach

33. Which of these facilities may be substituted for a middle marker during a complete ILS instrument approach? (FAR 91.175)

    A Surveillance and precision radar

    B Compass locator and precision radar

    C A VOR/DME fix

34. The rate of descent on the glideslope is dependent upon (*Instrument Flying Handbook,* Chapter VIII)

    A true airspeed

    B calibrated airspeed

    C groundspeed

35. When is a pilot on an IFR flight plan responsible for avoiding other aircraft? (FAR 91.113)

    A At all times when not in radar contact with ATC

    B When weather conditions permit, regardless of whether operating under IFR or VFR

    C Only when advised by ATC

36. What point at the destination should be used to compute estimated time en route on an IFR flight plan? (FAR 91.169, 91.153)
    A The final approach fix on the expected instrument approach
    B The initial approach fix on the expected instrument approach
    C The point of first intended landing

37. (Refer to FIG. 18-3.) What is the distance (C) from the beginning of the touchdown zone marker to the beginning of the fixed distance marker? (*Aeronautical Information Manual,* 2-3-2)
    A 1,000 feet
    B 500 feet
    C 250 feet

38. What is the first fundamental skill in attitude instrument flying? (*Instrument Flying Handbook,* Chapter V)
    A Aircraft control
    B Instrument cross-check
    C Instrument interpretation

39. As power is reduced to change airspeed from high to low cruise in level flight, which instruments are primary for pitch, bank, and power, respectively? (*Instrument Flying Handbook,* Chapter V)
    A Attitude indicator, heading indicator, and manifold pressure gauge or tachometer
    B Altimeter, attitude indicator, and airspeed indicator
    C Altimeter, heading indicator, and manifold pressure gauge or tachometer

40. While recovering from an unusual flight attitude without the aid of the attitude indicator, approximate level-pitch attitude is reached when the (*Instrument Flying Handbook,* Chapter V)
    A airspeed and altimeter stop their movement and the vertical speed indicator reverses its trend
    B airspeed arrives at cruising speed, the altimeter reverses its trend, and the vertical speed stops its movement
    C altimeter and vertical speed reverse their trend and the airspeed stops its movement

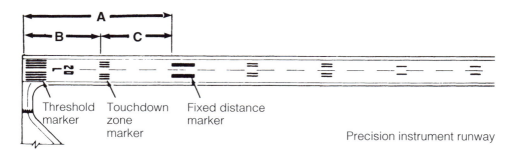

Fig. 18-3.

41. The primary reason why the angle of attack must be increased to maintain a constant altitude during a coordinated turn is because the (*Instrument Flying Handbook,* Chapter III)
   A thrust is acting in a different direction, causing a reduction in airspeed and loss of lift
   B vertical component of lift has decreased as the result of the bank
   C use of ailerons has increased the drag

42. During recoveries from unusual attitudes, level flight is attained the instant (*Instrument Flying Handbook,* Chapter V)
   A the horizon bar on the attitude indicator is exactly overlapped with the miniature airplane
   B a zero rate of climb is indicated on the vertical speed indicator
   C the altimeter and airspeed needles stop prior to reversing their direction of movement

43. (Refer to FIG. 18-4.) What is the flight attitude? One instrument has malfunctioned. (*Instrument Flying Handbook,* Chapter V)
   A Climbing turn to the right
   B Climbing turn to the left
   C Descending turn to the right

Fig. 18-4.

19
Stress can spoil your whole day

FLIGHT INSTRUCTORS MIGHT NOT ALWAYS REALIZE IT, BUT THEY know more about stress than most other pilots. Some have even experienced it! Consider this scenario: ATC says, "Roger your missed approach. You are cleared to the alternate...expect an ILS back course approach to Runway 35 left. Ceiling is now 800 feet overcast, visibility 2 miles in light rain and fog..." The pilot thinks, "A *back course* approach to minimums! I haven't done one of those in years. I thought the weather was supposed to be improving!"

The instrument pilot heading for an alternate to execute an unfamiliar procedure down to minimums is certainly under a lot of stress. So is the white-knuckled student pilot on a second or third flight with thoughts such as *what happens if the engine quits*? Or *please don't let me get sick.* Experienced instrument instructors know also that a student's first time in the clouds for any extended period is very stressful. And we know we might be seeing the effects of stress when a competent instrument student begins getting behind the airplane during routine procedures, or unexpectedly blows an easy approach.

Chapter Nineteen

WHAT IS STRESS?

We are all familiar with stress in daily life, and we often use terms such as "under stress" and "stressed out" to express our feelings when we run into difficulties or have a bad day. Stress is a state of physical, mental, or emotional tension brought on by factors, often beyond our control, that upset our equilibrium.

Stress can come in many forms. Hunger, for example, is a form of stress. When our physical equilibrium is upset by going too long without food, we feel the stress of hunger, which causes us to get something to eat that will relieve the hunger. Anger is another form of stress and is somewhat more complicated than hunger because anger produces the classic "fight or flight" reaction. If your boss unfairly criticizes you during an important meeting it is only natural to feel anger. You will be impelled to "fight or flight" by either responding to the criticism immediately or keeping your thoughts to yourself at the meeting and discussing them later. Some stress can cause mental and emotional tension so severe that a person might never recover from it completely, as in the death of a loved one. On the other hand, stress can be a positive factor, producing physical reactions that can help you cope with the source of the tension. The quick release of adrenaline in a tight situation speeds up the heart and breathing rates so that the body—and especially the brain—receives more oxygen. This gives you an extra boost to help cope with the problem. Navy carrier pilots and other military pilots unconsciously learn to harness stress so that their sensory perceptions become sharper and their thought processes move faster when they are in a tough spot. Or else they don't last long as military pilots.

FLYING STRESS

You hear talk every now and then about someone being a "natural born pilot," but I have never met one. Yes, some people have a greater aptitude for flying than others do. But *nobody* is free of tension while learning to fly. No one is completely comfortable. Instructors and students need to understand this so that the stress of flying can be minimized. Learning will simply not take place if there is too much stress.

Fortunately, causes of flying stress are easy to understand if you think about it a bit. And once you understand the causes you can then take some simple, common-sense steps to minimize their effect. I'm not a psychologist, but over the years I have found the following to be the greatest sources of stress. And I have developed techniques to ease my students' tensions in these situations.

Fear of the unknown

Anytime we face something we have never experienced before, our bodies prepare us for "fight or flight." Flying in actual IFR for the first time sets up an almost classic scenario for fear of the unknown because there is absolutely no way to prepare a person for this experience other than actually doing it.

And doing it is the best way to dispel a student's apprehension about flying in actual IFR. After a few good sessions in actual IFR with an instructor aboard, most students begin to enjoy it and look forward to it. The secret is simple: Expose them to actual IFR on a routine basis whenever the opportunity presents itself. As a designated examiner, I am amazed by the number of students who show up for the checkride who have *never* experienced actual IFR. What do you think will happen when that pilot gets an instrument rating and flies into the clouds for the first time? *World-class stress.*

Other instrument situations evoke fear of the unknown: the first approach to minimums under a hood or in actual IFR, first partial panel work, first experience of vertigo, and any kind of emergency. Again, as in the case of the first actual IFR experience, the way to reduce stress in these situations is for the instructor to expose the student to them routinely during the course. And to keep doing it until the unknown becomes so commonplace that it no longer causes any significant amount of stress.

Instructor note. A good instructor will be alert for other situations that a particular student might be apprehensive about. We're all different and what one person might take in stride might scare the daylights out of someone else the first time it is encountered. The solution, again, is to present these situations routinely and frequently until the student becomes comfortable with them.

Fear of failure

None of us likes to appear foolish or clumsy, and we have a natural tendency to avoid situations that might be embarrassing. And let's face it—some instrument procedures are very intimidating when you first encounter them. This could set up a stressful situation that might make it even more difficult for the student to master the procedure.

The instructor's attitude is very important in overcoming fear of failure. If the instructor makes the student feel like a klutz, the student will perform poorly. In the normal course of instruction, the student will achieve success more often if the instructor is patient and supportive. This attitude will go a long way toward minimizing fear of failure.

Good preflight and postflight briefings are essential for reducing a student's fear of failure. If a student has a clear idea of what's expected in the way of performance on each flight, those intimidating procedures will be a lot easier to handle. But the instructor can't do it all. Students who are conscientious about doing the assigned reading and preparing the background briefings in this book will be much better prepared for new procedures and much less apprehensive about them.

Practice is perhaps the best prescription for fear of failure; once you master something new, you lose your anxiety about it. But it also is important not to get "hung up" on a problem. Sometimes it is better to move on to something else, then return to the problem later when you are feeling more confident.

Instructor note. Sometimes a student will become stressed about one element of a new procedure rather than the whole thing. For example, in an otherwise good NDB approach many students find it difficult to stay on course while tracking outbound. If you spot this type of problem, set up some sessions in an approved instrument simulator. The simulator will allow you to stop and start a procedure, discuss the specific problem, then practice it a few times rather than having to go flying another day to try again.

Fear of catastrophe

Student pilots have no difficulty imagining disaster at every turn. By the time a pilot starts training for an instrument rating many disaster scenarios have been put to rest by practicing or thinking through such emergencies as engine failure, electrical failure, and off-field landings.

Instrument flying gives rise to a whole new set of scenarios: lost communications, partial panel, diversion to an alternate at minimums with low fuel, sudden, unexpected icing in the clouds, and so forth. If these scenarios are not confronted and dealt with, you can end up carrying a pretty heavy load of stress even on routine flights.

Just talking over "worst-case" scenarios with your instructor can help dispel stressful anxiety. For example, it is possible to make a successful landing with 0/0 ceiling and visibility. It's been done in extreme emergencies and there are ways of doing it that work. (Hint: Practice descents at 125 fpm.) But if you never discuss this with your instructor, you'll never be able to dispel the anxiety. Talk about any flying situation that worries you. You will be surprised to find that other pilots, even instructors, have been apprehensive about the same things that bother you and have come up with many imaginative and successful solutions.

Be prepared to respond when your instructor simulates emergencies in flight. On your instrument rating flight test, the examiner might well give you a steady stream of "verbal" emergencies to see how well you respond, such as:

"I see a rapid buildup of ice on the wing."

"You have just lost radio communications."

"The weather is getting worse at the destination."

"You have just lost navigation instruments. Where is the nearest precision approach radar?" Practice and familiarity will ease the anxiety that thinking about these emergencies can cause. This is true, as well, after you obtain the instrument rating. Practice the elements of instrument flying you are *least* familiar with whenever the opportunity arises, with an instructor aboard for retraining if you have lost proficiency or want to try a new or unfamiliar procedure. Then, when you get diverted to that field with an ILS back course approach at minimums, you should have little trouble heading inbound with skill and confidence.

PHYSICAL FACTORS

Altitude, noise, motion, and many other physiological factors produce stress by making it more difficult for your body to function at peak effectiveness. They all grind away as you fly along, and the longer the flight, the greater the stress.

Excellent sections in AIM Chapter 8, "Medical Facts for Pilots," deal with the many physiological factors affecting pilot performance; and I urge you to become thoroughly familiar with this material before the instrument flight test. The discussions of the self-imposed adverse effects of medication and alcohol are particularly important. AIM makes these two points forcefully:

- "The safest rule is not to fly as a crewmember while taking any medication, unless approved to do so by the FAA."
- "An excellent rule is to allow at least 12 to 24 hours between bottle and throttle, depending on the amount of alcoholic beverage consumed."

EFFECTS OF STRESS

Stress can cause numerous problems for a pilot:

- Produce impatience that might deepen almost to the point of paranoia. "They're really out to get me with all these amended clearances."
- Cause enough fatigue over a period of time to produce dangerous lapses of attention. "Um, center, were you talking to me?"
- Slow down thought processes enough to fall "behind the airplane" on a routine approach.
- Interfere with judgment. "No need to file for the alternate. We'll come back around and pick up the lights this time for sure and go on in."
- Ultimately develop into full-blown panic.

Stress is thus a serious problem that affects not only your passengers, but also everyone else sharing your airspace. You should be able to recognize that impatience, lapses of attention, and getting behind the airplane are consequences of stress. Then you can decide on a less stressful course of action, such as returning to your home field if on a local flight, making an intermediate stop on a long cross-country, or voluntarily making a missed approach at the final approach fix or sooner if you are having trouble maintaining courses and altitudes or are unable to keep up with the sequence of events.

The first step is to recognize and accept the fact that you are stressed out, then back off from your current course of action and substitute something less stressful. If you wait too long to do this, panic might take over and rob you of the ability to control the situation. When this happens, the only alternative is to call for help.

Instructor note. Experienced instructors have seen most or all of these problems on training flights, but we have been slow to associate them with stress. Instead we often conclude that the student just hasn't caught on yet, and we repeat the situation in which the problem occurred without considering other alternatives. Either way it's our job to get at the root of the problem and straighten it out regardless of whether the cause is stress, lack of preparation, unfamiliarity, or a combination of factors.

NONFLYING STRESS

Stress has been recognized as a factor in daily life for years and there have been many studies of how stress affects performance, relationships, physical well-being, and mental health. The nonflying stress that is brought to the airport might have more impact on your flying than you think.

A classical tool for measuring stress in daily life is the Holmes/Rahe Life Change Scale (FIG. 19-1) developed several years ago by Dr. Thomas Holmes and Dr. Richard Rahe of the University of Washington.

Start at the top of this list and total the "mean values" of the changes that have occurred in your life over the past year. If a change has occurred more than once, increase the value accordingly. (Add an additional 0 to the dollar amounts in items 20 and 37 to reflect current financial realities.)

Rank	Life Event	Mean Value
1	Death of a spouse	100
2	Divorce	73
3	Marital separation	65
4	Jail term	63
5	Death of close family member	63
6	Personal injury or illness	53
7	Marriage	50
8	Fired at work	47
9	Marital reconciliation	45
10	Retirement	45
11	Changes in family member's health	44
12	Pregnancy	40
13	Sex difficulties	39
14	Gain of new family member	39
15	Business readjustment	39
16	Change in financial state	38
17	Death of close friend	37
18	Change to different line of work	36
19	Change in number of arguments with spouse	35
20	Mortgage over $10,000	31
21	Foreclosure of mortgage or loan	30
22	Change in work responsibilities	29
23	Son or daughter leaving home	29
24	Trouble with in-laws	29
25	Outstanding personal achievement	28
26	Wife begins or stops work	26
27	Begin or end school	26
28	Change in living conditions	25
29	Revision of personal habits	24
30	Trouble with boss	23
31	Change in work hours, conditions	20
32	Change in residence	20
33	Change in schools	20
34	Change in recreation	19
35	Change in church activities	19
36	Change in social activities	18
37	Mortgage or loan under $10,000	17
38	Change in sleeping habits	16
39	Change in number of family get-togethers	15
40	Change in eating habits	15
41	Vacation	13
42	Christmas	12
43	Minor violation of the law	11

TOTAL: _____

NOTE: You can also use the life change scale to project future stress based on expected changes in the upcoming year.

Fig. 19-1. *The Holmes/Rahe Life Change Scale. Reprinted from* Journal of Psychosomatic Research, *Vol. 2, pp. 213–218, T. H. Holmes and R. H. Rahe, "The Social Readjustment Scale," 1967, with permission from Elsevier Science.*

Here is the way the Air Safety Foundation interprets the scores:

- Below 150: Little or no problem—you probably won't have any adverse reactions to the changes in your life.

- 150–199: Mild problem—a 37 percent chance you'll feel the impact of the stress with physical symptoms.

- 200–299: Moderate problem—a 51 percent chance of experiencing a stress related illness or accident.

- 300 and above: Danger! Stress is threatening your well-being. An 80 percent chance of a stress related illness or accident.

The AOPA Air Safety Foundation concludes: "If your score alarms you, *do something about it*. Postpone a move or a job change, or even going on a diet (any change that's under *your* control) even flying, if necessary, until your score settles down. A good person to consult is your aviation medical examiner."

In my opinion it is not necessary to discontinue flying if you recognize that the non-flying stress in your life is on the high side. But be careful about placing yourself in stressful flying situations.

Raise your own personal instrument minimums for a while. One good rule of thumb for reducing the stress of a single-pilot, single-engine IFR flight is do not go if the ceiling is below 1,000 feet anywhere along your route.

Then stick to your new higher minimums even if it means not getting back home on time. "Gethomeitis" is a still a major factor in general-aviation accidents despite years of case histories and warnings about it.

FLIGHT TEST STRESS

I have placed this discussion of stress at this point in the book because you are about to face one of the most stressful events in your pursuit of flying—the instrument rating flight test.

The good news is that if you have absorbed the lessons in this book, you will be so well prepared for the flight test that there will be no surprises. You will have no problem coping with the stress of the checkride. You will be thoroughly familiar with *everything* that will come up, no matter how ornery the examiner might be.

P.S. Get a good night's sleep before the flight test and arrive early! Good luck!

20
How I conduct an instrument flight test

WHEN A CANDIDATE SCHEDULES AN INSTRUMENT FLIGHT TEST, there are a number of things that I, as a designated examiner appointed by the FAA, must check before the test, as all examiners must do.

REQUIRED DOCUMENTS

First of all, is the airplane legal to make this flight? You must bring engine and airframe logs for the airplane to be flown and you must show the entries that indicate the 100-hour (if required) and annual inspections have been done within their deadlines. Also, you should be prepared to show documentation that the pitot static and transponder systems have been tested and if necessary recalibrated within the last 24 months and that the VOR has been checked within the preceding 30 days and properly logged.

You must be able to open the logbooks to the pages with these entries, demonstrate your familiarity with airworthiness directives for the airplane and show that they have all been complied with. Find the entries ahead of time and mark the pages with paper clips to save time on the day of the test—this makes a great first impression on the examiner.

Better yet, make a list of all these airplane requirements along with the hours and/or dates and be able to find the supporting information in the aircraft logs.

Be prepared to show the airplane's current weight and balance data and an operating manual if one is required to be carried on board for that airplane. Prepare a weight and balance calculation for the flight test.

The next question is, are you legally eligible to make this flight?

REQUIREMENTS FOR INSTRUMENT RATING

I will review your application, FAA Form 8710-1, (FIG. 20-1) and pilot logbook to make sure you have satisfied all the requirements specified by the FARs for this flight. To be eligible for the instrument rating, FAR 61.65 requires:

(1) At least 50 hours of cross-country flying as pilot in command, of which at least 10 hours must be in airplanes for an airplane-instrument rating.

(2) A total of 40 hours of actual or simulated instrument time...to include—
   (i) At least 15 hours of instrument flight training from an authorized instructor in the aircraft category for which the instrument rating is being sought;
   (ii) At least 3 hours of instrument training that is appropriate to the instrument rating being sought from an authorized instructor in preparation for the practical test within the 60 days preceding the day of the test;
   (iii) For an instrument-airplane rating, instrument training on cross-country flight procedures specific to airplanes that includes at least one cross-country flight in an airplane that is performed under IFR and consists of—
      (A) A distance of at least 250 nautical miles along airways or ATC-directed routing;
      (B) An instrument approach at each airport;
      (C) Three different kinds of approaches with the use of navigation systems.

Show me your current medical certificate and private or commercial pilot certificate. When you pass the flight test you must surrender the pilot certificate. A temporary certificate will be issued that includes the new instrument rating; you will receive the new permanent certificate by mail within 120 days.

In addition, show me the entry for a current biennial flight review or equivalent. Again, put a paper-clip on the logbook pages that contain the appropriate entries, and you will save time.

I will examine your logbook entries for other items. Sometimes I find people with an entire flight logged as instrument time. The total flight time was 2 hours and they logged 2 hours of simulated instrument time—no, total time is "block-to-block" or Hobbs time. Even if it were low-ceiling instrument weather that day, the entire flight could not possibly be logged as instrument time. Some of this time had to be spent getting clearances, taxiing to the runway, getting off the ground, landing, and taxiing to parking. The maximum instrument time you might reasonably get on a 2-hour instrument flight is about 1.75 hours.

TYPE OR PRINT ALL ENTRIES IN INK

Form Approved OMB No: 2120-0021

U.S. Department of Transportation
Federal Aviation Administration

Airman Certificate and/or Rating Application

I. Application Information □ Student □ Recreational □ Private □ Commercial □ Airline Transport ☒ Instrument
□ Additional Aircraft Rating □ Airplane Single-Engine □ Airplane Multiengine □ Rotorcraft □ Glider □ Lighter-Than-Air
□ Flight Instructor ____ Initial ____ Renewal ____ Reinstatement □ Additional Instructor Rating □ Ground Instructor
□ Medical Flight Test □ Reexamination □ Reissuance ____ Certificate □ Other ____

A. Name (Last, First, Middle)	B. SSN (US Only)	C. Date of Birth Mo. Day Year	D. Place of Birth
DOE, JOHN J.	249-51-4529	02-31-56	NEW YORK, NY

E. Address (Please See Instructions Before Completing)
1010 EAST 86th STREET

F. Nationality (Citizenship) Specify	G. Do you read, speak and understand English?
☒ USA □ Other ____	☒ Yes □ No

City, State, Zip Code
NEW YORK, NY 10021

H. Height 71 In.	I. Weight 175 Lbs	J. Hair BROWN	K. Eyes BROWN	L. Sex ☒ Male □ Female

M. Do you now hold, or have you ever held an FAA Pilot Certificate? ☒ Yes □ No	N. Grade Pilot Certificate PRIVATE	O. Certificate Number 249514529	P. Date Issued 01-08-96

Q. Do you hold a Medical Certificate? ☒ Yes □ No	R. Class of Certificate 3rd CLASS	S. Date Issued 04-31-97	T. Name of Examiner RICHARD R. ROE, MD

U. Have you been convicted for violation of Federal or State statutes relating to narcotic drugs, marijuana, or depressant or stimulant drugs or substances □ Yes ☒ No	V. Date of Final Conviction

W. Glider or Free Balloon Pilots only:	Medical Statement: I have no known physical defect which makes me unable to pilot a glider or free balloon.	Signature	X. Date

II. Certificate or Rating Applied For on Basis of:

☒ A. Completion of Required Test	1. Aircraft to be used (if flight test required) CESSNA 172	2a. Total time in this aircraft 52 hours	2b. Pilot in command 47 hours

□ B. Military Competence Obtained in	1. Service	2. Date Rated	3. Rank or Grade and Service Number
	4. Has flown at least 10 hours as pilot in command during the past 12 months in the following military aircraft.		

□ C. Graduate of Approved Course	1. Name and Location of Training Agency or Training Center		1a Certification Number
	2. Curriculum From Which Graduated		3. Date

□ D. Holder of Foreign License Issued By	1. Country	2. Grade of License	3. Number
	4. Ratings		

□ E. Completion of Air Carrier's Approved Training Program	1. Name of Air Carrier	2. Date	3. Which Curriculum □ Initial □ Upgrade □ Transition

III. Record of Pilot time (Do not write in the shaded areas.)

	Total	Instruction Received	Solo	Pilot in Command	Second in Command	Cross Country Instruction Received	Cross Country Solo	Cross Country Pilot in Command	Instrument	Night Instruction Received	Night Take-off/ Landing	Night Pilot in Command	Night Take-off/ Landing Pilot in Command	Number of Flights	Number of Aero-Tows	Number of Ground Launches	Number of Powered Launches	Number of Free Flights
Airplanes	126	60	20	90		12	10	50	41	7	20	8	4					
Rotorcraft																		
Gliders																		
Lighter than Air																		
Training Device Simulator		8							8									

IV. Have you failed a test for this certificate or rating? □ Yes ☒ No   Within the Past 30 days? □ Yes ☒ No

V. Applicant's Certification — I certify that all statements and answers provided by me on this application form are complete and true to the best of my knowledge, and I agree that they are to be considered as part of the basis for issuance of any FAA certificate to me. I have also read and understand the Privacy Act statement that accompanies this form.

Signature of Applicant *John J. Doe*	Date 02-14-98

FAA Use Only

EMP	REG	D.O.	SEAL	CON	ISS	ACT	LEV	TR	S.H.	SRCH	#RTE		RATING (1)

FAA Form 8710-1 (7-95) Supersedes Previous Edition

NSN: 0052-00-682-5006

Fig. 20-1. *Completed application for instrument flight test, FAA Form 8710-1.*

I also must verify that the long IFR cross-country covered 250 nautical miles and included the required approaches. The entry for that flight should indicate where you went and which instrument approaches you made and at which airports.

ENDORSEMENTS

The next items in the pilot's logbook are the two required endorsements. The first endorsement certifies that your flight instructor has given you the ground instruction required by FAR 61.65 (b)(1) through (10). The second certifies that you have received the flight instruction specified by FAR 61.65 (c)(1) through (5). The recommended form for these endorsements is in Chapter 22.

If this is a retest, the flight instructor must also enter another endorsement to the effect that additional required flight instruction has been given to cover the deficiencies found on the previous flight test. On a retest, you are expected to comply with all other requirements, just as though this was the original test, including a newly completed application (FAA Form 8710-1, FIG. 20-1). You must also present the notice of disapproval from the first flight—the pink slip.

The reverse side of FAA Form 8710-1 must also be signed by your instructor. And the instructor must endorse the written test report in the space provided to attest that this instructor has given instruction on any missed items on the written test. The FAA wants you to be 100 percent in all respects to earn the rating.

I find it helpful to instrument students and instructors alike to have a checklist they can review when the instructor is ready to recommend a student for a flight test (FIG. 20-2). This checklist is based upon the checklist in *Instrument Rating Practical Test Standards*. The checklist in FIG. 20-2 also contains some routine items not covered in this discussion.

Portions of the application (FAA Form 8710-1) should be clarified. The application should be typed using capital letters. Dates should be written with two digits for date, month, and year. For example, February 14, 1998, should be written as 02-14-98. Height should be total inches: 71", not 5'10". Figure 20-1 is a sample of a correctly completed form. Note that the applicant's height is written as 71".

The FAA data entry operators processing applications in Oklahoma City are not pilots and they are not acquainted with our idiosyncrasies. So the information on the application must be entered in the form they are accustomed to and everything must be 100 percent correct or the application will be returned and cause a delay or rejection in issuing the permanent pilot certificate with instrument rating.

Finally, be sure your instructor signs the block at the top of the reverse side of the application. If you have forgotten to sign, it's no problem because you can sign in the examiner's office. But if the instructor has forgotten, you cannot take the flight test.

THE PRACTICAL TEST

The examiner is responsible for evaluating the knowledge and the skill of the applicant in meeting the required standards set forth in the practical test standards. Much of the knowledge can be tested during oral questioning, which might take place prior to or at any time during the flight test.

ACCEPTABLE AIRCRAFT:
- ☐ View-limiting device
- ☐ Aircraft documents:
  - ☐ Airworthiness certificate
  - ☐ Registration certificate
  - ☐ Operating limitations
  - ☐ Weight and balance form
- ☐ Aircraft Maintenance Records: Airworthiness Inspections
- ☐ FCC station license

PERSONAL EQUIPMENT:
- ☐ Identification—Photo/Signature ID
- ☐ Current aeronautical charts
- ☐ Computer and plotter
- ☐ Flight plan form
- ☐ Flight logs
- ☐ Current AIM

PERSONAL RECORDS:
- ☐ Pilot Certificate
- ☐ Medical Certificate
- ☐ Completed FAA Form 8710-1, "Airman Certificate and/or Rating Application"
- ☐ AC Form 8080-2, Airman Written Test Report or Computer Test Report
- ☐ Logbook with Instructor's Endorsements
- ☐ Notice of Disapproval (if applicable)
- ☐ Approved School Graduation Certificate (if applicable)
- ☐ Examiner's Fee

Fig. 20-2. *Checklist for instrument flight test, based upon* Instrument Rating Practical Test Standards.

It is obviously much more practical to conduct some oral questioning on the ground before the flight test rather than in the air. On the other hand, some questioning is actually more pertinent in the air.

It doesn't make much sense to evaluate an applicant's knowledge of weight and balance computations during the actual flight test. On the other hand, I find I can make much better evaluation of an applicant's knowledge of two-way radio communications failure procedures during the flight by asking, "What would you do now if you suddenly lost two-way radio communications?"

After I have reviewed all your required documents and found them to be in order, I will give you several practical problems to solve. One will be a performance problem, probably computing takeoff roll and obstacle clearance distances on a high density altitude day.

When you make your appointment for your flight test, your designated examiner will tell you what destination(s) to plan for, and the examiner will expect you to show up for the test with a complete plan for this flight, including a flight log. You should be ready to get the final weather and file the flight plan. I also ask for a weight and balance calculation. Do yourself a big favor by making blank forms ahead of time for the airplane to be used on the test and photocopying two or three extras. Then all you have to do is fill in the form with the weights and moments of theoretical passengers and baggage the examiner provides and calculate the result. (Be able to do it both ways—the chart method and mathematically.)

You will only have 30 minutes for performing the calculations plus getting a final weather briefing for your preplanned flight, selecting an alternate, and filling out the flight plan form.

Some intelligent advance planning can save you considerable time and anxiety and help you to easily complete all tasks within 30 minutes. Get a full weather briefing for the area within 250 miles before you meet the examiner. Review instrument approaches to airports within that area that might be suitable for use on the flight and study them the night before.

ORAL EXAMINATION

When you present solutions to the performance and weight and balance problems accompanied by the completed flight log and flight plan, I will evaluate your knowledge of these areas.

Other areas of instrument flying in which I will evaluate your knowledge are weather, basic instruments, ATC procedures, FARs, en route and approach charts, and the like. If you have mastered the material covered by Background Briefing 20-20, you should have no trouble with the oral portion of the practical test.

I will also evaluate areas in which you missed written test questions to ensure that you understand them. Even if you scored 100 on the written test, you are going to get some questions on material upon which the test is based.

THE FLIGHT TEST

After all this, I'm sure it will be a positive relief to get into the air! The IFR flight you plan, file, and fly will be to a nearby airport with one of the instrument approaches required for the flight test. During this phase of the flight test I will evaluate your skill in copying clearances and working with ATC during IFR departure, en route, and approach phases. The first instrument approach might be carried through to a landing from a straight-in or a circling approach. Or I might ask you to execute a missed approach, enter the missed approach holding pattern, and then obtain an IFR clearance in the air to a second airport where we will make a second type of instrument approach.

At some point we will cancel IFR (weather permitting) and do some steep turns and unusual attitudes. Count on flying some portion of the flight on partial panel, including an approach.

Expect other simulated emergencies such as electrical failure (simulated by turning off the alternator half of the split power switch) or pitot tube icing (simulated by covering the airspeed indicator).

On occasion I have conducted flight tests in actual IFR flight conditions throughout the entire flight. When the flight plan is filed I will talk to ATC—hopefully with the controller responsible for the sector—and arrange a block of airspace in which to conduct the steep turns and unusual attitudes. I try to get a block of airspace at an assigned altitude ±1,000 feet along a 20-mile length of airway.

It gets to be a little complicated for the examiner, but the applicant should not assume we won't make the flight if the weather is IFR, or that we won't do steep turns, unusual attitudes, minimum controllable airspeed, stalls, and partial panel under actual IFR conditions.

COMMON DEFICIENCIES

As I covered other topics in this book I pointed out several areas in which deficiencies are common on the instrument flight test. I would say that 90 percent of the applicants I see have their paperwork in order when they show up. Few applicants have difficulty completing their flight planning and the other assigned problems within the 30 minutes allowed. I think the good performance reflects the good work of the flight instructors who are apt to be better trained now than in the past.

Holding altitude consistently within the tolerances of the practical standards remains a problem and I see the occasional applicant who goes below the minimums on an instrument approach and does not take prompt corrective action.

That is why I always start my instrument students working toward the goal of 2° of heading, 2 knots of airspeed, and 20 feet in altitude from the beginning. If you fly within these tolerances on an instrument flight test, it will be a big boost to your morale and also demonstrate to the examiner that you are a skillful instrument pilot.

NDB procedures continue to be a problem, and it is not uncommon for an applicant to fly an otherwise acceptable instrument test flight, then fail on an NDB approach. This is definitely an area where flight instructors should devote more time and attention.

Another area that requires more attention is approaches to uncontrolled airports. Sometimes pilots will get so wrapped up in their approaches to an uncontrolled airport that they forget they are about to enter a VFR traffic pattern. They fail to monitor the CTAF and consequently are not sure about which runway is in use, the local altimeter setting if available, and any other traffic. Applicants fail to notify local traffic that they are arriving on the instrument approach.

It's very uncomfortable for an examiner to sit in the right seat while the applicant goes busting into a VFR traffic pattern with a hood on! The examiner can take control when the safety of the flight is at stake and this might well be one of those situations. But most of the time the examiner will just have to sit there and let the applicant make mistakes until it becomes pointless to continue.

It's very important for applicants to realize that they are pilots-in-command on the flight test. The examiner is just an observer. Yes, the examiner is expected to act as a safety pilot while the applicant is under the hood and to take appropriate action if other aircraft, obstructions, or low altitudes are a threat to the safety of the flight, but applicants are expected to demonstrate their ability as pilots-in-command. The applicant must make the decision whether or not to continue an approach to a landing or to execute a missed approach. The examiner can only become involved for safety of the flight and to carry out the requirements of the practical test standards.

YOU PASSED!

Earning an instrument rating is an achievement to be proud of and I'm happy to report that most applicants are successful on their first flight test. Serious pilots recognize the high degree of skill and knowledge required to earn this rating and they respond accordingly. That's certainly one of the reasons I have concentrated on instrument instruction over the years. It's very gratifying to work with highly motivated pilots reaching for another level of excellence.

But the learning process should not end with achievement of the rating. There is always more to learn, and good pilots take advantage of every opportunity to do so. Chapter 21 explores some options and alternatives that you will encounter as you begin flying as a "Proud, Perfect *Instrument* Pilot."

PROFICIENCY

To maintain instrument proficiency you must perform and log:

(1) 6 instrument approaches in the preceding 6 months

(2) Holding procedures

(3) Intercepting and tracking courses through the use of navigation systems. (See FAR 61.57 (c) in Appendix B.)

You cannot legally act as pilot in command under IFR or in weather less than VFR unless you meet these minimums. This could not be more reasonable. Fly 6 approaches in 6 months; while maneuvering for these approaches you also will satisfy the requirement for "intercepting and tracking courses." Throw in one holding pattern—which could, for example, be a course reversal in an NDB approach—and you are legal. These requirements can be logged in actual or simulated conditions or in an approved flight simulator. It should be noted that time in an approved simulator does not count toward a certificate, a rating, or proficiency unless an instructor is present to conduct the simulator training.

You have another 6 months to meet the requirements if someone else flies with you who can act as pilot-in-command while you complete your 6 approaches and other requirements. In actual IFR, this must be someone with an instrument rating who is current. In simulated IFR, this can be anyone who qualifies as an appropriately rated safety pilot according to FAR 91.109(b).

If these minimum requirements lapse, you must pass an *instrument proficiency check* (IPC) that is administered by an instrument instructor (CFI-I), a designated examiner, or someone else designated by the FAA. Look upon the IPC as an opportunity to demonstrate and polish skills in areas such as holding patterns, NDB approaches, or DME arcs that you don't have much opportunity to use in the real world of IFR.

When I give an IPC, I treat it like a miniature instrument flight test. I like to make sure the pilot can plan and file an IFR flight, then go out and fly that flight plan in the airplane that this pilot usually flies IFR, working efficiently with ATC. I always try to have the pilot execute three different approaches: VOR, NDB, and ILS.

Above all, I want to make sure that the pilot is competent and comfortable as pilot-in-command and can confidently handle all the decisions required during an IFR trip. Many sharp pilots take IPCs every six months whether they need them or not in order to maintain a high degree of proficiency.

The biennial flight review (BFR) is another good opportunity to hone your instrument flying skills. You can do an IPC at the same time. This will extend your proficiency requirements for another 6 months as well as provide an opportunity to practice instrument procedures that you haven't used recently. The decision is up to the instructor—are you up to the standard?

21
Moving on–and up

ONE OF THE GREATEST CHALLENGES YOU WILL FACE, ONCE YOU HAVE earned your instrument rating, is how to stay sharp as an instrument pilot. If your career is in flying, you will probably have many regular opportunities to file and fly IFR. But for many pilots, opportunities for solid IFR work do not always come up often enough to stay sharp.

It's not much of a challenge to perform and log the 6 approaches, holding procedures, and course tracking every 6 months to stay legally proficient. This is really a rock-bottom minimum. I highly recommend flying 6 approaches with holding patterns and course tracking routinely every month. If you fly a lot, it's not too hard to give yourself an IFR workout every 4 weeks or so.

But if you don't own your own plane, or if this amount of flight time presents a budget problem, there is now a simple, inexpensive approach to staying sharp. And that is through an IFR proficiency computer program. ASA (Aviation Supplies and Academics), for example, offers its excellent "On Top" IFR proficiency program for less than $400. It is configured for 8 different planes, with the Beech Baron at the top of the line. Its database contains airports and navaids for the United States. On Top has a realistic weather simulation, and you can pause an approach to analyze a problem and check a display to see where you are both horizontally and vertically. There is an excellent On Top demonstration program featuring some interesting approaches to airports in the mountains of Montana. This can be downloaded through the ASA Web site: <www.asa2fly.com/asa> Or call 1-800-ASA-2-FLY for a demonstration disk and further information.

Also offered for less than $400 is the "Pro Trainer" from the Elite line of instrument proficiency trainers and simulators. It comes with a navigational database for the United States and Canada and is configured for the Cessna 172. A yoke is available as an option for the Pro Trainer for a slightly higher price.

Additional information on the Elite line of proficiency trainers and simulators is available on the Elite Web site: <www.flyelite.com>. If you have questions, call 1-800-557-7590. A Pro Trainer demonstration disk can be ordered through the 800 number.

Neither On Top nor Pro Trainer qualifies as PCATDs, thus the time and approaches logged with these two programs cannot be counted toward an instrument rating or toward instrument proficiency. But they are excellent for staying sharp with familiar approaches. You can also use them to rehearse unfamiliar approaches when you plan a trip to someplace you've never been.

TRANSITION TO HIGHER PERFORMANCE

The basic principles of instrument flight don't change when you move up to a high performance airplane or earn a multiengine rating. You still need to scan instruments properly, maintain 2, 2, and 20, and stay ahead of the airplane and the procedures by anticipating the events of the flight while planning and flying.

When you move up to a faster airplane everything will happen much more quickly and you must be prepared for it. The best way to prepare is to get some instrument instruction in that high-performance single or light twin from an instructor who is familiar with the airplane.

One of the most important things you can do when moving up is determine the power settings for the new airplane for the different conditions of flight, then automatically set these power settings whenever a change in flight regime is performed.

Determine the best RPM and manifold pressure settings for fast, normal, and slow cruise; normal climb at 500 fpm; descents at 500 and 1000 fpm; straight-and-level approach cruise with one increment of flaps down; and approach descent, with one increment of flaps and landing gear down. An actual IFR flight is not the time to experiment or discover new knowledge.

Always refer to these standard settings when changing from one flight condition to another. You can set up the new flight regime very quickly with standard power settings every time. This will avoid throttle jockeying and other forms of overcontrolling. Go right to these standard power settings whenever you make a transition; you can always fine-tune them or make minor adjustments if you don't get the exact desired airspeed or rate of climb or descent.

Write down these settings on a card and carry the card on a clipboard or kneepad where you can refer to it quickly. Better still make up a placard with the power settings and place it on the instrument panel near the tachometer and manifold pressure gauge. A taped-on paper placard will do nicely and you can replace it easily as you fine-tune the power settings with experience. Then consider having a permanent placard made up for you by a local office supply store.

With a multiengine airplane, take advantage of every opportunity to practice single-engine instrument approaches. Check the operating manual for a zero-thrust power setting to simulate the drag of a feathered engine. If you can't find the zero-thrust setting in the operating manual, a multiengine instructor can help determine one with a little experimentation. With a zero-thrust setting, you can always have the power of the wind-milling engine available during an approach in case of a go-around. If the propeller is feathered, a go-around might be impossible.

Consider the engine out scenario at altitude on an IFR flight plan. If you lose an engine in actual IFR, inform ATC immediately and plan to divert to the closest available airport where the weather is VFR or comfortably above IFR minimums.

You might have to descend to a lower altitude immediately. You might have no choice because the airplane will drift down to its single-engine service ceiling no matter what you do. Check the operating manual; the single-engine ceiling might be lower than you think it is.

Light twins frequently come with features you might not be familiar with such as three-axis autopilots, flight directors, and horizontal situation indicators (HSIs). Get checked out by an instructor who is proficient with equipment that is new to you, then become proficient with it yourself by using it whenever possible.

MOVING UP

After earning the instrument rating, the next move depends on your goals in aviation. Are you going to pursue a career in aviation working toward an ATP and a job with an airline, or will you—like most of us—remain in general aviation, honing your skills and advancing your knowledge at every opportunity so that you can do whatever kind of flying you wish with confidence and pleasure?

When you earned your instrument rating you just took out some life insurance in the form of greatly increased competency. You are now a safer pilot. Much the same can be said of the commercial pilot certificate, the next step up the ladder, whether or not you intend to pursue aviation as a future career. I highly recommend obtaining a commercial certificate.

To earn a commercial you have to demonstrate instrument competency up to commercial standards, along with improved performance in other areas: precision landings, more demanding maneuvers, emergencies, etc. For the normally competent pilot, the commercial is a breeze compared to the instrument rating. It is well worth the extra effort and is certainly looked upon kindly by insurance companies. (In the long run, if you are an aircraft owner, the reduced premiums might go a long way toward paying for the cost of the training.)

Why not make your next step after the commercial the instrument flight instructor rating? A surprising number of pilots are certificated as flight instructor-instrument without the airplane endorsement. Presumably they intend only to teach instruments, which is not such a bad idea either.

On the other hand, a better idea might be to do the extra work to become certificated as a flight instructor—airplane first, then add the "double I" CFII—after that. It is all up to your hopes and desires. It is all up to you.

A further step up could be a multiengine rating; however, this rating is not very useful unless you expect to own a multiengine airplane or work in the industry where you can do some multiengine flying. Just remember that if you want to pilot a multiengine airplane on IFR, you will have to demonstrate multiengine instrument competency on your flight test. You will then have to log all the required instrument approaches in a multiengine airplane (or an approved simulator) in order to maintain your instrument proficiency, or pass an instrument proficiency check in a multiengine aircraft. Otherwise, you will have to be satisfied to have a multiengine rating with a VFR restriction.

The ultimate pilot's license is, of course, the airline transport pilot certificate with a type rating as a finishing touch. The ATP is the advanced degree of a pilot's education, and, if you have the time and the money, the high professional standards it requires will make you a much better pilot, regardless of whether or not you intend to pursue a career in aviation.

The designated examiner on an ATP flight test will expect airline standards or better in basic instrument work, ATC procedures, and emergencies. Some ATP checkrides are virtually one emergency after another, starting with an engine failure on takeoff just as you are about to enter the clouds! If you have been trained from the beginning of your instrument flying career in the "2, 2, and 20 club" and the other fundamentals in this book, you should be able to take the ATP in stride—plus a lot of hard work. The ATP is the equivalent "Ph.D." of aviation and it is a substantial achievement in which you can take great pride.

A type rating is restricted to one type of highly complex airplane, such as the B-727. Type ratings are usually obtained after an ATP-rated pilot hires on as a corporate, commuter, or long-haul pilot, and the type ratings are for the airplanes operated by that company. Type ratings are limited to only those airplanes in which you have been qualified. (I have seen ATP certificates with more than 10 different types of aircraft listed.)

As you think ahead to a high-performance single or a light twin, you should investigate some optional equipment that will make IFR flying much easier. This equipment is available for most single-engine airplanes. If you are buying or renting, fly an airplane that has the optional features you want and check them out.

But first, you should have some personal items regardless of your aircraft.

PERSONAL EQUIPMENT

For all IFR flights, night or day, you should carry two flashlights. At night without a flashlight, you cannot conduct preflight inspections and read checklists before starting the engine. In the daytime while flying in the clouds, it can get awfully dark in the cockpit when the power fails. A backup flashlight is essential for obvious reasons. When are the batteries of the first flashlight most likely to turn up dead? When you want to use the flashlight, of course! These are obvious points, yet I am constantly amazed at how many students show up ill-prepared in this regard.

I highly recommend a stop watch or digital timer for timing approaches and holding patterns. The clock on the instrument panel is not adequate for this purpose because it is easy to get confused about how many minutes have passed on a long final approach segment. I see more and more timers included as features in battery powered flight calculators. The electronic calculator is a good option itself because it is much easier to use in a crowded cockpit than the manual E6B. Prices for these dual-purpose calculators are reasonable. Why not get both in one compact unit? It is your choice.

I have also mentioned the desirability of getting a battery powered, hand-held communications transceiver. The portable radio is good insurance if you lose power on an IFR flight and you need to stay in touch with ATC, as well as a very useful aid for copying ATIS and clearances before starting the engine.

The more advanced models feature VOR radials and CDIs and have cables and connectors for hooking up to the aircraft's outside antennas. Usually an avionics shop is required to install a proper jack for connecting the transceiver. There are many models and a wide range of prices to choose from. If you do get a transceiver with VOR capability, practice some VOR using only the hand-held transceiver with other VOR receivers turned off—with a safety pilot on board, of course. And don't forget to carry a fresh set of batteries in the flight case for the transceiver. The last thing you want to cope with is a power failure and a dead or dying hand-held transceiver.

I urge you to get a good comfortable headset for IFR operations. This is a great aid in reducing cockpit confusion because it eliminates groping for a microphone. Why is it that ATC *always* calls when you need both hands to fly—and sometimes your teeth to hold a pencil or a flashlight?

Another important feature of headsets is that they reduce the noise that bombards the ears in most lightplane cockpits and can cause hearing loss. This is not just a "nice-to-have" feature. Cockpit noise is a major cause of fatigue, especially on long flights. Anything you can do to reduce fatigue will help you make a better, less stressful approach at the end of that long flight.

Dozens of makes and models of headsets are available. I suggest that you find an FBO or an aviation supply store that has a wide selection that you can try on. Select a headset that feels light and comfortable. Tight, heavy sets might eventually feel like a vise!

GOOD OPTIONS FOR IFR

And now we are going to talk about some more "goodies." Price is often the determining factor when selecting new equipment. But before committing to new equipment you should also ask yourself:

- How much will this reduce my workload?
- Is it user-friendly, given my level of competence?
- Will this improve my safety?
- How much training will I need to fully utilize this equipment?

Engine monitors

Limited advances have been made in engine design and performance, but great strides have been made in performance monitoring equipment. When pilots identified a need, someone came along with a solution. These solutions include such advances as carburetor air temperature gauges, carburetor ice detectors, EGT/CHT monitors for either one cylinder or all four or six cylinders, battery voltage indicators, cabin pressurization gauges, temperature monitors for turbocharged engines, and more.

Any or all of these deserve consideration by the IFR pilot because they can give information on the reliability of the powerplant during the course of an IFR flight. Although these additional gauges add to the pilot's workload and scan requirements, they can prove very comforting during any flight, whether VFR or IFR, because they can warn of potential trouble before it becomes a problem. Do you ever have that nagging feeling, "How goes it?"

Avionics

"Whatever man can dream, man can achieve." Nowhere is this more true than in the field of avionics and related equipment.

Distance measuring equipment (DME) is no longer really an option. While DME is not required by law below 24,000 feet, it is almost impossible to navigate safely and legally without it in Class B airspace or within the "mode C veil" that extends out to 30 miles from a primary Class B airport. The same is true for Class C airspace. In fact, in every situation that requires a transponder, DME will help you navigate with the precision that ATC expects.

DME is a great aid in visualizing position at all times. Close to a VOR station DME helps you hold a steady reference course until station passage instead of chasing the needle, as you might be tempted to do if you didn't know how close you were. DME holding patterns are easier because you don't have to add or subtract time on the outbound leg to account for the wind. And DME provides access to many approaches that would otherwise be unavailable. One of the great benefits of DME is that you can learn to use it with no special training. Simply follow the directions in the owner's manual and read the distances directly off the dial. One point to remember is that DME reports *slant range* to a station. Directly over the DME facility, for example, the nautical mile indication on the DME display will equal the altitude in nautical miles.

Radio magnetic indicators (RMI) have been around for many years, but very few newly-rated instrument pilots know about them or understand their many uses. The RMI instrument (FIG. 21-1) consists of a gyro-stabilized heading indicator slaved to the aircraft's magnetic compass system. This combination provides the readouts of a magnetic compass stabilized by a gyroscope so that there are no lag-lead or acceleration-deceleration errors to account for—a big improvement over the basic magnetic compass with its wild gyrations and the conventional heading indicator that must be reset frequently.

Most RMIs have a double-barred bearing indicator needle that shows the magnetic bearing to a selected VOR station. The single-barred needle displays magnetic bearings

Fig. 21-1. *Typical radio magnetic indicator (RMI).*

to a selected NDB station. The beauty of this system is that you can read the magnetic bearings to and from a station directly from the pointers without having to add relative bearing to heading.

Some knowledge of the RMI is necessary to answer a few questions on current FAA written exams. Good pilot training is called for in solving the mysteries of the RMI. Make it a point to fly with an instructor in a plane equipped with an RMI and practice some NDB and VOR approaches with this instrument, cross-checking with other radio navigation instruments. You will quickly see how much easier the RMI is to work with than the conventional ADF needle/gyro heading indicator combination.

Instructor note. Interpret the RMI much the same as you would an old-fashioned ADF with "a rotatable" azimuth card. The slaved azimuth of the RMI always gives the magnetic bearing to or from a station. What could be simpler? The VOR needle on an RMI is interpreted in the same manner as an ADF. With a little practice and ingenuity on your part you will find this to be a truly amazing navigation tool, very accurate, and a great aid developing situational awareness and helping answer the question "Where am I?"

Horizontal situation indicators (HSIs) give another dimension to answering the question "Where am I?" The HSI is a combination of two instruments: the azimuth card of the RMI and the needles of the VOR/ILS indicator (FIG. 21-2). On the HSI, your airplane appears in miniature in the center of the instrument. The VOR/ILS localizer is shown in relation to the miniature airplane. This is all on the face of a heading indicator slaved to a remote compass—a marriage of the heading indicator and VOR/ILS (with glideslope), which is frequently interconnected with an autopilot with altitude hold and the capability of following the glideslope all the way to decision height and beyond.

The next step in sophistication comes when command bars are added to the attitude indicator giving climb, descent, and turn commands telling you what to do to make VOR

Fig. 21-2. *Typical horizontal situation in-dicator (HSI).*

and ILS altitude and heading corrections. When an attitude indicator has command bars it is called a **flight director**.

Automatic pilot with an approach coupler is invaluable on long IFR flights or when fatigue is a factor. With an autopilot you can give yourself a break. The autopilot can do a major portion of the work if you are rerouted and have to prepare new flight log entries or if you simply want to eat a sandwich. Because fatigue is at its greatest at the end of an instrument flight, the approach coupler will help reduce the anxiety that often accompanies fatigue and ease the workload.

With a fully coupled autopilot engaged, you just sit back and monitor the progress of the flight and the proficiency of the autopilot, stay ever alert for a malfunction that can turn a routine flight into near panic unless you have been thoroughly trained in autopilot emergency procedures.

Instructor note. Be sure *you* are trained by a competent instructor before you get in over your head! This can lead to disaster with you on board (suddenly as pilot-in-command). Or worse yet, you could unwittingly send an unsafe pilot out into the "wild blue" all alone or with passengers without proper preparation in all facets of an airplane's equipment. Who would really be at fault here?

Instill in the student's mind the importance of hand-flying as much of every flight as humanly possible! Autopilots make human pilots lazy. (Otherwise, kiddingly suggest that the student purchase an extra logbook to log the autopilot's flying time.)

While autopilots can be a great aid on a long flight when fatigue begins to set in and the going gets rough, they can, if depended upon too much, become the greatest de-stroyer of pilot skills known to aviation. I have known pilots who engaged the autopilot as soon as airborne, even before retracting the landing gear, letting "Old George" (the autopilot) do all the work right on down to a fully coupled, hands-off approach to deci-sion height at the end of the flight. A pilot who does this becomes nothing more than a sophisticated passenger.

Recall how busy you were on a recent IFR flight. It is easy to become overloaded by too many complications and distractions too fast. It is amazing how much work you can handle with proper training and thorough planning until a minor emergency surfaces and you begin piloting on "overload." That is the time you will wish that you had had much more rigorous training, practice, and drill in the autopilot and any other special equipment that might be on board.

Where is the good and bad weather? Sometimes it is easy enough to look out the window and analyze the hue of the clouds then check the seat of the pants to determine the level of turbulence. Our ability to analyze weather in flight has been considerably upgraded and supplemented by Stormscopes and airborne color radar. Both systems benefit from good pilot training regarding the use of the equipment and proper interpretation of the available information. For example, radar can be tilted down 15° to give outlines of shorelines, rivers, lakes, and even bridges.

Stormscope can help you work with ATC to deviate around thunderstorms because ATC radar is not designed to detect thunderstorms, only precipitation. A Stormscope detects and shows electrical activity. This is especially important when coping with embedded thunderstorms that hide along weather fronts. You can't see them, ATC can't see them, but a Stormscope can.

Airborne weather radar is also effective in spotting thunderstorms. But it is more costly than Stormscope and is impractical for many general aviation airplanes.

GPS (Global Positioning System) has been discussed already in connection with en route navigation and approaches. While GPS is clearly the system of the future, the panel-mounted units required for IFR use are still costly and time-consuming to learn.

Not quite ready to step up to a panel-mounted system? Why not consider buying an inexpensive hand-held unit to become acquainted with GPS? There are many excellent models on the market these days. With them you can master features such as satellite acquisition, flight planning and waypoint designation, moving map displays, and many other features. Then, when you are ready to move up to a panel-mounted system, your learning time will be reduced substantially.

GPS hand-held units are acceptable for VFR navigation, and they provide superb situational awareness with their moving maps and the wealth of information these maps make available. They tend to run down batteries quickly, however; so a cigarette lighter adapter is a good idea for longer flights. A suction cup mounted interior antenna with an extension cord will allow you to position the antenna for maximum satellite reception.

All these options are wonderful but you must always remember to aviate, navigate, and communicate without autopilot, Stormscope, GPS, moving map displays, or other advanced equipment.

You must be able to fly hands on, even at the end of a long, fatiguing flight with a lot of weather and turbulence all the way. Many aids can make the flight simpler, easier, more precise, and with less effort. But they are not substitutes for mastering instrument flying.

Only you can do that.

22
Instrument rating syllabus

Flight lesson 1: Introduction to IFR

WITH INSTRUCTOR ASSISTANCE THE STUDENT WILL PLAN, FILE, AND fly a short instrument cross-country flight to a destination 51 to 75 nautical miles distant and return. The student will learn how to plan an IFR flight, prepare a flight log, obtain a thorough weather briefing, file the IFR flight plans, obtain clearances, and conduct the flight.

The instructor will demonstrate an unhooded instrument approach to a full stop landing at the destination. A new clearance will be obtained for the return flight. Ideally, the student will be able to fly 60 percent to 90 percent of the flight with some coaching from the instructor.

Assigned reading:

Chapter 3—Preparing for an Instrument Flight

Chapter 4—Weather/Whether to Fly?

Chapter 5—How to Get a Good Weather Briefing

Chapter Twenty-Two

Chapter 6—Airplane, Equipment, and Instrument Checks

Chapter 7—Clearances and Communications

Preflight briefing
Review:

1. Communications procedures and frequencies

Introduce:

1. IFR flight planning
2. IFR weather briefings
3. Filing IFR flight plans
4. Copying clearances
5. Instrument preflight checks
6. Departure, en route, and approach procedures

Completion standards:
The lesson is complete when the student has a sufficient overview of the planning, filing, and conduct of an IFR flight to begin planning and filing IFR flight plans with a minimum of assistance from the instructor. The student will meet private pilot standards for holding headings within ±10°, airspeeds within ±10 knots, and altitudes within ±100 feet.
Postflight critique

Background briefing 1-2: Introduction to basic instruments

After Flight Lesson 1, the instructor should plan to give several hours of ground instruction covering basic instruments, their purpose, interpretation, and appropriate pilot actions. The emphasis in this briefing is on the relationships and interactions of the control, primary, and support instruments and how to scan them properly using the heading, altitude, and attitude scanning pattern. Use FIG. 8-5, Basic Instruments, and FIG. 12-3, Partial Panel Instruments, as guides to control, primary, and support instruments.

The briefing is complete when the student knows what the instruments show during straight and level flight, turns, climbs, and descents; how control, primary, and support instruments relate to each other during these fundamental maneuvers; and how to scan the instruments properly.
References:

Chapter 6—Airplane, Instrument, and Equipment Checks

Chapter 8—Basic Instruments

Chapter 9—Turns, Climbs, and Descents

Chapter 12—Stalls, Unusual Attitudes, and Partial Panel

FAA Instrument Flying Handbook. Chapter II, Instrument Flying: Coping with Illusions in Flight and Chapter IV, Basic Flight Instruments.

Questions:

1. What is the difference between a control instrument and a primary instrument?

2. Give your definition of a primary instrument.

3. Under what conditions would the needle of a primary instrument be moving?

4. What is the primary instrument for:
   Bank during straight and level flight?
   Bank during a standard rate turn?
   Bank during a climbing turn?
   Bank during a straight ahead descent?
   Pitch during level flight?
   Pitch during a constant speed climb?
   Pitch during a constant rate descent?
   Pitch during an ILS approach?
   Power during the transition from a climb to straight and level?
   Power during a standard rate turn?

5. Describe at least six errors in the use of the heading indicator.

6. Describe at least six common errors in the use of pitch instruments.

7. Describe four common errors in power management.

8. Describe four common errors in the use of trim.

9. The vacuum pump has failed. Describe what happens and what instruments are affected.

10. What causes vertigo? How can you remedy this problem?

11. What causes incorrect airspeed indications? What are preventative measures?

12. Describe the recovery from nose low and nose high unusual attitudes.

13. What indications confirm that you have indeed recovered from an unusual attitude?

14. Describe VOR accuracy requirements prescribed by the FARs and how the regulations may be satisfied.

15. Make a list of the items added to a VFR checklist for the instrument pilot.

16. What are the fuel requirements for an instrument flight?

17. List the equipment checks required by the FARs and when they must be done.

18. What erroneous indications can the attitude indicator show and why?

19. When do you reset the attitude indicator?

20. Describe the errors of the magnetic compass and how to compensate for them.

21. Describe the errors of the VSI and how to compensate.

22. Describe the errors of the fuel gauges and how to compensate.

Flight lesson 2:
Maneuvering Solely by Reference to Instruments—Part I

The student will plan, file, obtain the clearance, and depart IFR for the same destination as in Flight Lesson 1. Approximately 20 nautical miles into the flight, the instructor will direct the student to cancel IFR and continue VFR for training in maintaining heading and altitude solely by reference to instruments during straight and level flight, climbs, turns, speed transitions, and descents. The student will monitor a practice IFR approach conducted by the instructor and try to fly as much of it as possible on return to the airport.

Assigned reading:

Chapter 8—Basic Instruments

Chapter 9—Turns, Climbs, Descents

Review Chapters 3 through 7

Preflight Briefing
Review:

1. Flight planning, weather briefing, filing flight plans, copying clearances

2. Preflight instrument checks

3. IFR communications

4. Departure and en route procedures

Introduce:

1. Precision straight and level flight

2. Speed transitions

3. Standard rate turns

4. Minimum controllable airspeed

Supplementary Exercises:

1. Pattern A

2. Oboe Pattern

Completion standards:
The lesson is complete when the student can maintain headings within $\pm 2°$, airspeed within ± 2 knots, altitudes within ± 20 feet for periods of 30 seconds to one minute throughout all maneuvers introduced in this lesson. The student should also begin to develop a general understanding of the instrument approach procedure.

Postflight critique

Flight lesson 3:
Maneuvering Solely by Reference to Instruments—Part II

The student will plan, file, obtain a clearance, and depart on a short IFR cross-country flight to a destination other than that selected for the first two flight lessons. When well-established on the en route portion of the IFR cross-country, the instructor will direct the student to cancel IFR. The student will practice precise heading, altitude, and airspeed control in straight and level flight, climbs, descents, turns, and speed transitions. The student will learn precise heading, altitude, and airspeed control in constant airspeed and constant rate climbs and descents. The student will monitor a practice IFR approach conducted by the instructor on return to the airport.

Assigned reading:

Review Chapters 3 through 9

Preflight Briefing
Review:

1. Flight planning, weather briefing, filing flight plans, and copying clearances
2. Preflight instrument checks
3. IFR communications
4. Departure and en route procedures
5. Straight and level, standard rate turns, and speed transitions
6. Minimum controllable airspeed

Introduce:

1. Constant airspeed climbs
2. Constant rate climbs
3. Constant airspeed descents
4. Constant rate descents

Supplementary exercises:

1. Step climbs and descents
2. Vertical S
3. Pattern B
4. Pattern C

Completion standards:
The lesson is complete when the student can maintain headings within ±2°, airspeed within ±2 knots, and altitudes within ±20 feet throughout all maneuvers in this lesson for periods of one or two minutes.

Postflight critique

Flight lesson 4: VOR tracking and bracketing

The student will plan, file, obtain a clearance, and depart on an IFR cross-country flight to a destination suggested by the instructor. The student will practice precise control in straight and level flight, climbs, descents, turns and speed transitions as they occur during the IFR cross-country phase of the flight. When directed by the instructor, the student will cancel IFR.

The student will practice flight at minimum controllable airspeed and practice establishing position in relation to a VOR station. The student will learn to intercept, bracket, and track VOR courses and radials with the needle held within the center circle at all times. The student will monitor an unhooded approach conducted by the instructor on return to the home airport.

Assigned reading:

Chapter 10—VOR Procedures

Preflight Briefing
Review:

1. Constant airspeed climbs and descents

2. Minimum controllable airspeed

Introduce:

1. VOR orientation

2. VOR tracking and bracketing

Supplementary exercises:

1. 16-point orientation exercise

2. VOR time-distance checks

3. Pattern A around a VOR

4. Pattern B around a VOR

Completion standards

The lesson is complete when the student can (1) maintain the course deviation indicator needle within one-half of its full deflection throughout all the maneuvers in this lesson, except for station passage, and (2) maintain desired headings within ±2°, airspeed within ±2 knots, and altitudes within ±100 feet during the VOR work and then ±20 feet for extended periods throughout the rest of the flight.

Postflight critique

Flight lesson 5: VOR holding patterns

The student will file and depart on an IFR cross-country flight to a destination suggested by the instructor using VOR stations as outbound fixes so the student may practice VOR tracking and bracketing; when directed by the instructor, the student will cancel IFR. The student will learn to enter a holding pattern at a VOR station or VOR fix using direct, tear-drop, and parallel methods. The student will learn how to correct for the wind to maintain standard and nonstandard holding patterns. On return to the home field, the student will monitor a practice instrument approach conducted by the instructor.

Assigned reading:

Chapter 11—Holding Patterns
Review Chapter 10

Preflight Briefing
Review:

1. VOR tracking and bracketing, inbound and outbound

Introduce:

1. Holding pattern entry
2. Holding patterns

Completion standards

The lesson is complete when the student can (1) maintain the course deviation indicator needle within the center circle bull's-eye on all inbound courses, (2) maintain inbound legs of holding patterns within ±15 seconds of the desired one-minute length, and (3) maintain headings within ±2°, airspeed within ±2 knots, and altitudes within ±100 feet during the VOR work and ±20 feet for extended periods during the rest of the flight.

Postflight critique

Flight lesson 6: Unusual attitudes, partial panel

The student will file and depart on an IFR flight to a destination suggested by the instructor. When directed by the instructor, the student will cancel IFR. The student will practice turns to headings with the magnetic compass. The student will learn power-off stalls and steep turns and will learn to recover from unusual attitudes. The student will learn precise control of the aircraft under partial panel conditions with the attitude indicator and directional indicator covered to simulate a vacuum system failure. The student will fly a no-gyro instrument approach on return to the airport.

Assigned reading:

Chapter 12—Stalls, Unusual Attitudes, and Partial Panel

Preflight Briefing
Review:

1. Turning to a heading with magnetic compass

Introduce:

1. Power-off stalls
2. Steep turns
3. Critical attitude recovery
4. Partial panel

Completion standards:

The lesson is complete when the student can (1) recover from power-off stalls with a loss of altitude of 50 feet or less, (2) maintain altitude within ±100 feet, airspeed within ±10 knots, desired angle of bank within ±5°, and roll out within ±10° of the specified heading during 45° banked turns in either direction, (3) recover to straight and level flight without the use of the attitude indicator after an unusual attitude, (4) consistently maintain headings within ±2°, airspeed within ±2 knots, and altitudes within ±20 feet while maneuvering on partial panel.

The student will demonstrate magnetic compass errors, recognize imminent stalls, and use correct control pressures and movements in proper sequence in unusual attitude recovery in order to complete the lesson.

Postflight critique

Flight lesson 7: ADF orientation, tracking, and bracketing

The student will file and depart on an IFR cross-country flight to a destination selected by the instructor along a route in the vicinity of a strong NDB or standard AM broadcast station. The student will practice control by partial panel in straight and level flight, climbs, turns, descents, and speed transitions as they occur during the IFR cross-country phase of the flight.

The instructor will direct the student to cancel IFR in the vicinity of the NDB. The student will practice steep (45° bank) turns under the hood. The student will learn to orient the airplane around an NDB and to intercept, bracket, and track inbound and outbound bearings. Pattern A may be used to practice interception, tracking, and bracketing.

Assigned reading:

Chapter 13—The NDB Unmasked

Review:

1. Partial panel
2. Steep turns

Introduce:

1. ADF orientation
2. ADF tracking and bracketing
3. Pattern A around an NDB
4. Pattern B around an NDB

Supplementary exercise:

1. ADF time-distance checks

Completion standards:

The lesson is complete when the student maintains ±100 feet while performing steep turns and is able to predetermine magnetic bearing to an NDB within 10 seconds, then turn to the station and intercept a bearing and track within ±5° of the course.

Postflight critique

Flight lesson 8: ADF holding

The student will file and depart on an IFR cross-country flight to a destination selected by the instructor along a route in the vicinity of a strong NDB. When directed by the instructor, the student will cancel IFR and practice ADF orientation, tracking, and bracketing. The student will learn to enter a holding pattern at an NDB by direct, teardrop, and parallel methods. The student will learn how to correct for the wind to maintain holding patterns.

Assigned reading:

Review Chapter 13

Preflight Briefing
Review:

1. Partial panel
2. ADF orientation
3. ADF bracketing and tracking

Introduce:

1. ADF holding patterns

Completion standards
The lesson is complete when the student demonstrates proficiency in predetermining bearing to an NDB, entering holding patterns by the correct method, and intercepting, bracketing, and tracking a magnetic bearing to and from the NDB within ±5°.

Postflight critique

Background briefing 8-9: Instrument approach procedures

Immediately after completing Background Briefing 1-2, the student should commence work on Background Briefing 8-9, writing as many answers as possible. The student should allow adequate time to prepare for this briefing because it covers a large body of information of great importance, especially on the instrument flight test.

The briefing covers weather minimums, approach charts, alternate airports, communications and clearances, ADF, VOR, and ILS procedures, holding patterns, straight-in and circling approaches, at uncontrolled airports, and missed approaches.

The briefing is complete when the student can talk through ADF, VOR, and ILS approaches—and the appropriate missed approach procedures—at nearby airports.

References:

Chapter 14—Approaches I: Approaches Basics and NDB Approaches

Chapter Twenty-Two

Chapter 15—Approaches II: VOR, DME, and GPS

Chapter 16—Approaches III: ILS, Localizer, and Radar

Questions:

1. Instrument approach procedures are based on criteria established in what publication of the U.S. government?

2. What publication serves as the instrument pilot's reference for transitioning from the en route phase to the landing phase in instrument conditions?

3. Basic weather minimums are prescribed for what two broad categories of approaches?

4. Name six factors that change the published minimums.

5. What does the phrase "vectors to the final approach course" mean?

6. What is the lowest forecast ceiling permitted for an instrument approach at your home airport?

7. How is the highest obstruction at an airport depicted on an approach chart?

8. What is the determining factor in whether or not a legal approach may be attempted?

9. Takeoff weather minimums are found in what publications?

10. Alternate airport weather minimums are found in what publications?

11. When is an alternate airport required? When is an airport authorized as an alternate?

12. In case of communications failure, when and where can an instrument approach be commenced?

13. What are the obstacle clearance altitudes on:
    VOR approaches?
    Localizer approaches?
    ASR approaches?
    NDB approaches?
    DF approaches?
    VOR approaches with FAF?

14. What is the maximum permissible distance from the airport during a circling approach in a Cessna 172?

15. Describe the procedure for executing an early missed approach.

16. Describe holding pattern protection and variations in holding patterns that a pilot might encounter; describe the reasons for these variations.

17. Describe five different acceptable procedures when executing a circling approach.

18. Position reports are not required when in radar contact, except in five specific instances. What are they?

19. When can you descend below the glideslope on an ILS approach?

20. When can you descend below the MDA on a nonprecision approach?

21. Can a takeoff be legally and safely executed when the current METAR weather is 1/8SM FG VV006?

22. What are you giving up during a takeoff at 1/8SM FG VV006?

23. What are your personal weather minimums for takeoff and for landing? Why?

24. Describe the clearance delivery procedures at your home airport.

25. Describe at least six variations in procedures in receiving IFR clearances.

26. When issued a cruise clearance, how do you get clearance to commence the approach?

27. When in VFR conditions executing an approach, why would you want to cancel IFR?

28. Describe the method of activating pilot-controlled lighting.

29. Describe the purposes of holding patterns depicted by a heavy solid line, a light solid line, and a light dotted line.

30. When stabilized on an ILS approach, you find you are flying 15 knots too fast. What is one popular method of correcting this excess speed?

31. What is the recommended procedure when intercepting the ILS glideslope in a fixed gear, single-engine airplane? In a high performance retractable?

32. When landing at an airport without a control tower, when should you attempt contact on the CTAF? Where do you get the local altimeter setting, and what does it buy you?

33. Where and when should you use that old memory jogger checklist, time, turn, twist, throttle, talk? Explain its use and significance.

34. Must you time an ILS approach? Why? Why not?

35. What special actions must you take when flying into a Class B or Class C airspace? Into a restricted area or an MOA?

36. Where and how can you tell quickly if your destination airport has an approved instrument approach procedure?

37. Talk your way through several NDB, VOR, DME, ILS, and LOC/BC approaches, from the feeder fix through the missed approach.

Flight lesson 9: NDB approaches—I

The student will file and fly a short IFR cross-country flight to a nearby destination with a published NDB approach. The student will fly the NDB approach, conduct a missed approach, cancel IFR, then make additional approaches as directed by the instructor. ADF holding patterns in both the approach and the missed approach procedure will be included, if possible.

Assigned reading:

Chapter 14—Approaches I: Approach Basics and NDB Approaches

Review Chapter 13
Preflight Briefing
Review:

1. Partial panel

Introduce:

1. NDB approaches
2. Missed approaches

Completion standards:
The lesson is complete when the student demonstrates an understanding of the NDB approach, tracks the inbound and outbound bearings within ±5°, and maintains altitudes within ±100 feet to MDA. The MDA must be maintained to +100/−0 feet.
Postflight critique

Flight lesson 10: NDB approaches—II

The student will file and fly a short IFR cross-country to a different destination with a published NDB approach. The student will fly the NDB approach, conduct a missed approach, cancel IFR, then make additional NDB approaches, with ADF holding patterns, as directed by the instructor. At least one of the additional NDB approaches will be made with partial panel.
Assigned reading:

Review Chapters 13 and 14

Preflight Briefing
Introduce:

1. No-gyro approaches

Review:

1. Partial panel
2. NDB approaches
3. Missed approaches
4. ADF holding patterns

Completion standards:
The lesson is complete when the student demonstrates an understanding of the NDB approach, tracks the inbound and outbound bearings within ±5°, and maintains altitudes within ±100 feet to MDA. The MDA must be maintained to +100/−0 feet.
Postflight critique

Flight lesson 11: VOR approaches—I

The student will file and fly a short IFR cross-country flight to the nearest destination with a published VOR approach. The student will fly the VOR approach, conduct a missed approach, cancel IFR, then make additional VOR approaches as directed by the instructor. VOR holding patterns in either the approach or the missed approach procedure will be included.

Assigned reading:

Chapter 15—Aproaches II: VOR, DME, and GPS

Review Chapters 10 and 14

Preflight Briefing
Review:

1. Partial panel
2. Missed approaches
3. VOR holding patterns

Introduce:

1. VOR approaches

Completion standards:
The lesson is complete when the student demonstrates an understanding of the VOR approach, tracks VOR radials within $\pm 5°$, maintains altitudes within ± 100 feet to MDA, then $+100/-0$ feet, and maintains the desired airspeed within ± 10 knots.
Postflight critique

Flight lesson 12: VOR approaches—II

The student will file and fly a short IFR cross-country to a different destination with a published VOR approach. The student will fly the VOR approach, conduct a missed approach, cancel IFR, then make additional VOR approaches, with VOR holding patterns, as directed by the instructor. At least one of the additional VOR approaches will be made on partial panel. The student will practice recovery from unusual attitudes.

Assigned reading:

Review Background Briefing 8-9

Preflight Briefing
Review:

1. Partial panel
2. VOR approaches
3. Missed approaches
4. VOR holding patterns
5. Critical attitude recovery

Completion standards:

The lesson is complete when the student demonstrates competence in performing VOR approaches and generally tracks VOR radials within $\pm 2°$, maintains altitudes within ± 50 feet to MDA, then $+50/-0$ feet, promptly executes the missed approach, and properly enters a missed approach holding pattern.

Postflight critique

Flight lesson 13: VOR approaches—III

This repeats Flight Lesson 12. The student will file and fly a short IFR cross-country to a different destination with a published VOR approach. The student will fly the VOR approach, conduct a missed approach, cancel IFR, then make additional VOR approaches, with VOR holding patterns, as directed by the instructor. At least one of the additional VOR approaches will be made on partial panel. The student will practice recovery from unusual attitudes.

Assigned reading:

Review chapters and briefings as directed by instructor.

Preflight Briefing
Review:

1. Partial panel

2. VOR approaches

3. Missed approaches

4. VOR holding patterns

5. Critical attitude recovery

Supplementary exercise:

1. DME arc approaches

Completion standards:

The lesson is complete when the student demonstrates competence under full and partial panel conditions and generally tracks VOR radials within $\pm 2°$, maintains altitudes within ± 50 feet to MDA, then $+50/-0$ feet, maintains airspeed within ± 5 knots, promptly executes the missed approach, and properly enters a missed approach holding pattern.

Postflight critique

Flight lesson 14: ILS approaches—I

The student will file and depart on an IFR cross-country flight to a nearby destination with an ILS approach. The student will fly the ILS approach, conduct a missed approach, cancel IFR, then make additional ILS approaches as directed by the instructor. Holding patterns in either the approach or the missed approach procedure will be included.

Assigned reading:

Chapter 16—Approaches III: ILS, Localizer, and Radar

Preflight Briefing
Review:

1. Missed approaches

Introduce:

1. ILS approaches

Completion standards:
The lesson is complete when the student demonstrates an understanding of the ILS approach and maintains altitudes within ±100 feet, tracks the localizer and glideslope without exceeding full scale deflections, maintains the desired airspeed within ±10 knots, and takes prompt action at DH.
Postflight critique

Flight lesson 15: ILS approaches—II

The student will file and fly a short IFR cross-country to a different destination with published ILS and ADF approaches. The student will fly the ILS approach, conduct a missed approach, cancel IFR, then make additional ILS approaches, with holding patterns, as directed by the instructor. At least one of the additional approaches will be made on partial panel.

One of the ILS approaches will be conducted if possible, without the glideslope or using only localizer minimums to simulate the loss of the glideslope receiver. The student will also make one ADF approach to simulate total failure of the aircraft's ILS receiver.

Assigned reading:

Review Chapter 16

Preflight Briefing
Review:

1. Partial panel
2. ILS approaches
3. ADF approaches
4. Missed approaches

Introduce:

1. Loss of radio navigation equipment
2. Localizer approaches

Supplementary exercise:

1. Instrument takeoff

Completion standards:

The lesson is complete when the student maintains altitudes within ±50 feet, tracks localizer and glideslope within $\frac{1}{2}$ scale deflections, maintains desired airspeed within ±5 knots, recognizes and copes with instrument failures such as loss of glideslope, ILS/localizer, and attitude indicator.

Postflight critique

Flight lesson 16:
ILS back course, localizer, LDF, SDF, and radar approaches

The student will file and depart on an IFR cross-country to a destination with a published ILS back course approach. The student will fly the ILS back course approach, conduct a missed approach, cancel IFR, then make additional back course approaches, with holding patterns, as directed by the instructor. At least one approach will be made on partial panel.

If an airport with a radar approach is available within the local flying area, a second IFR cross-country should be filed and flown, and several practice radar approaches should be made. If radar approaches are not available, the instructor should simulate the radar approaches by providing vectors and other standard radar approach instructions to a nearby airport.

If airports with SDF and LDA approaches are available within the local flying area, additional IFR cross-country flights should be filed and flown to these airports, time permitting, for practice with these distinctive approaches. Otherwise, they must be covered in ground instruction.

Assigned reading:

Review chapters and briefings as directed by instructor.

Preflight Briefing
Review:

1. Partial panel
2. Loss of radio navigation equipment
3. ILS approaches
4. Missed approaches

Introduce:

1. ILS back course approaches
2. Localizer, LDA, and SDF approaches
3. Radar approaches

Completion standards:

The lesson is complete when the student demonstrates competence in ILS back course approaches and tracks the localizer within $\frac{1}{2}$ scale deflections, and competence in following directions on radar and no-gyro approaches, maintaining headings within $\pm2°$ and altitudes within ±50 feet.

Postflight critique

Background briefing 16-17: IFR cross-country procedures

Upon completion of Background Briefing 8-9 the student should commence working on Background Briefing 16-17, again writing as many of the answers as possible. Considerable emphasis is placed on planning an IFR cross-country flight and its ramifications.

As a designated FAA flight test examiner, I see cross-country planning surfacing again and again as one of the weak areas. Some candidates even expect to conduct their instrument flight test with only a low altitude en route chart, an approach chart or two, and little or no orderliness. A well-organized flight log, thoroughly worked out, is the key to mastering instrument flying.

This briefing is based on the planning for a 200 nautical mile (nm) IFR cross-country. To obtain maximum benefit from the training, it should be a different flight from the 250 nm IFR cross-country flight required by FAR 61.65 (d)(iii), which is conducted in Flight Lesson 17.

The briefing is complete when the student is ready in all respects to plan and conduct the 250 nm IFR cross-country required by FAR 61.65 (d)(iii).

References:

Chapters 3—Preparing for an Instrument Flight

Chapter 17—Putting It All Together: The Long IFR Cross-Country

AIM and A/FD—relevant sections

En route low altitude chart legend

Questions:

1. What are four sources of information for determining the IFR route for your flight plan?

2. Plan a 200 nm IFR cross-country flight and work out all the details on your flight log. (Each section of this flight log should be thoroughly explored with your instructor to ensure that you have a complete understanding of the use and value of the flight log.)

3. You have just picked up your clearance and you find that your routing for a segment of the flight plan has been substantially changed. You have at least five courses of action. What are they? When and why would each option be appropriate?

4. Explain the purpose of the TEC and when you would use it.

5. Explain the "preferred route" system and how it operates.

6. Give the two main reasons to note the time when you start a takeoff roll on an IFR flight.

7. Give at least two methods for maintaining a record of changes in assigned altitude. Why is this necessary?

8. How are you guaranteed terrain clearance when flying direct (VOR to VOR or off airways)? Who is ultimately responsible?

9. What do the following give you? Explain. MEA, MOCA, MCA, MRA, MAA, MSA.

10. Review all the symbols on the en route low altitude chart legend. Ask your instructor to explain any that you don't understand.

11. What is the advantage in recording the time when you reach each fix during an IFR cross-country?

12. What advantage is there in "visualizing" where you are at all times?

13. What is the purpose in writing your clearance limit?

14. Explain when, how, and why you might use the frequency 122.0 MHz.

15. How might you use the frequency 121.5 MHz? Why? When?

Flight lesson 17: Long IFR cross-country flight

This flight will satisfy the requirements of FAR 61.65 (d)(iii) for one 250-mile cross-country flight in simulated or actual IFR conditions, on federal airways, or as routed by ATC, including three different kinds of approaches.

The student will file and fly IFR flight plans for each leg of the 250-mile cross-country. Each approach will be made to a full-stop landing, and the student will refile an IFR flight plan after each full-stop landing en route. One approach and landing will be made at an uncontrolled airport.

Assigned reading:

Chapter 17—Putting It All Together: The Long IFR Cross-Country

Preflight Briefing
Review:

1. IFR departure and en route procedures

2. ADF approaches

3. VOR approaches

4. ILS approaches

Introduce:

1. Lost communications procedures

2. IFR departures from uncontrolled airports

Completion standards:

The lesson is complete when the student demonstrates competence in "putting it all together" and generally maintains airspeed within ±2 knots, headings and VOR radials within $\pm 2°$, and altitudes within ± 20 feet. The approaches shall meet the FAA practical test standards as stated in the completion standards for Flight Lessons 10, 13, and 15.

Deficiencies on this flight will not normally require repetition of this lesson, but will be corrected with extra work in Flight Lessons 18 and 19.

Postflight critique

Flight lesson 18: Progress check

This flight lesson will be conducted by another instrument flight instructor. The student will file and depart on an IFR cross-country flight to a destination selected by the instructor. When directed by the instructor, the student will cancel IFR and practice maneuvers chosen by the instructor to determine the student's proficiency in carrying out the tasks required by *Instrument Rating Practical Test Standards*. At least one of the instrument approaches will be on partial panel.

Assigned reading:

As directed by the instructor.

Preflight Briefing
Review:

1. Flight planning
2. Obtaining and analyzing weather information
3. Filing an IFR flight plan
4. IFR departure and en route procedures
5. Partial panel
6. Lost communications procedures
7. Loss of radio navigation equipment
8. ADF, VOR, and ILS approaches
9. Holding patterns
10. Missed approaches

Completion standards:

The lesson is complete when the instructor determines what deficiencies, if any, require additional practice.

Postflight critique

Flight lesson 19: Flight test preparation

The student will file and depart on an IFR cross-country flight to a destination selected by the instructor. When directed by the instructor, the student will cancel IFR and practice any maneuvers that may require further attention to attain the proficiency required by

Instrument Pilot Practical Test Standards. At least one of the approaches will be on partial panel.

Assigned reading:

As directed by the instructor.

Preflight Briefing
Review:

1. Flight planning
2. Obtaining and analyzing weather information
3. Filing IFR flight plans
4. IFR departure and en route procedures
5. Partial panel
6. Lost communications procedures
7. Loss of radio navigation equipment
8. ADF, VOR, and ILS approaches
9. Holding patterns
10. Missed approaches

Completion standards:
The lesson is complete when the student has corrected any deficiencies noted in Flight Lessons 17 and 18.

Postflight critique

Background briefing 19-20:
Preparation for the Instrument Flight Test Oral Exam

This very important final background briefing covers material the student can expect on the oral examination that the designated examiner will give prior to the flight test. Consult Appendix B for FARs that deal specifically with IFR training and IFR flight. A selection of AIM "Pilot/Controller Glossary" items relating to IFR flight may be found in Appendix C.

The briefing is complete when the student can promptly and accurately answer the questions and work out the problems that can be expected on the oral examination prior to the flight test.

References:

Chapter 19—How I Conduct an Instrument Flight Test

Questions:
(Appropriate reference at end of each question.)

1. What are the IFR currency requirements? (FAR 61.57)

2. If instrument currency expires, how do you become current? (FAR 61.57)

3. What is an "appropriate" safety pilot? (FAR 91.109)

4. What is legally considered "instrument flight time?" (FAR 61.51)

5. Who has direct responsibility and final authority for the operation of an aircraft? (FAR 91.3)

6. What certificates must pilots carry in their personal possession for flight? (FAR 61.3)

7. What documents must be on board the aircraft for an IFR flight? (FAR 91.9, FAR 91.203).

8. What is the fuel requirement for flight in IFR conditions? For VFR flight? (FAR 91.167, FAR 91.151)

9. Under what conditions must you list an alternate when filing IFR? (FAR 91.169)

10. What are alternate airport weather minimums? (FAR 91.169)

11. What restrictions apply regarding the operation of portable electronic devices on board an aircraft on an IFR flight? (FAR 91.21)

12. What are the four methods of checking VOR accuracy and the required records? (FAR 91.171)

13. How often must VOR accuracy be checked for IFR operations? (FAR 91.171)

14. At what point can you cancel an IFR flight plan? (AIM 5-1-13)

15. How frequently should you check your altimeter setting? (91.121)

16. What are the minimum weather conditions for IFR takeoff? (FAR 91.175)

17. What additional instruments and equipment are required for IFR over VFR? (FAR 91.205)

18. Explain DH versus MDA. (FAA *Instrument Flying Handbook*, FAR 1)

19. Explain MEA, MOCA, MRA, MAA. (FAA *Instrument Flying Handbook*, FAR 1)

20. How should you navigate your course on an IFR flight? (FAR 91.181)

21. Name the components of the ILS system. (AIM 1-1-10)

22. Can anything be substituted for an outer marker on an ILS approach? (FAR 91.175)

23. When is a procedure turn prohibited on an instrument approach? (FAR 91.175, Instrument Approach Procedures legend)

24. When may you descend below DH or MDA? (FAR 91.175)

25. Explain the terms "straight in" versus "circling" minimums. (AIM "Pilot/Controller Glossary")

26. How do you determine the minimum safe altitude on a "direct" off-airway flight? (FAR 91.177)

27. When must the pitot-static system and altimeter be inspected for IFR operations? (FAR 91.411)

28. How often must the transponder be inspected? (FAR 91.413)

29. Give the appropriate cruising altitudes when operating IFR below 18,000 feet. (FAR 91.179)

30. Describe clearance to "VFR on top." (AIM 4-4-7)

31. Do the FARs require an alternate static source? (FAR 91.205)

32. How will the alternate static source affect the instruments? (FAA *Instrument Flying Handbook*)

33. When should pitot heat be used? When is it recommended? (Chapter 6 of this book: "Airplane, Instrument, and Equipment Checks")

34. Explain HAA and HAT. (AIM "Pilot/Controller Glossary")

35. Explain "maintain" versus "cruise" in an IFR assignment. (AIM "Pilot/Controller Glossary")

36. List the inspections required on an aircraft to be operated IFR. (FAR 91.409, FAR 91.411, FAR 91.413)

37. Describe the operations and limitations of the gyroscopic instruments. (FAA *Instrument Flying Handbook*)

38. Discuss the purpose and use of SIDs and STARs. (AIM 5-2-6, AIM 5-4-1, "Pilot/Controller Glossary")

39. Describe a contact approach. (AIM 5-4-22, AIM 5-5-3, "Pilot/Controller Glossary")

40. Describe a visual approach. (AIM 5-5-11, "Pilot/Controller Glossary")

41. Describe minimums as determined by aircraft approach category. How do you know in which category your aircraft is classified? (Instrument approach procedure chart)

42. In a radar environment, what radio reports are expected from you without being requested by ATC? (AIM 5-3-2)

43. Outline your actions if you lose radio communications with ATC. (FAR 91.185, AIM 6-4-1)

44. What would you do if *all* radio equipment failed?

45. How can you determine where restricted areas are located along your route and what are your actions? (En route low altitude chart legend)

46. How can you identify the boundaries between ATC centers? (En route low altitude chart legend)

47. What does the symbol "x" on a flag on an en route chart indicate to the IFR pilot? (En route low altitude chart legend)

48. Describe mandatory changeover points on an airway. What is their purpose? (AIM 5-3-6, En route low altitude chart legend.)

49. Describe the different altitudes shown on an airway. (En route low altitude chart legend)

50. What does the symbol "r" on a flag on an en route chart indicate to the IFR pilot? (En route low altitude chart legend)

51. Describe the mileage markings on an airway. (En route low altitude chart legend)

52. Describe the correct ways to identify intersections. (En route low altitude chart legend)

53. Explain the purpose of aural signals carried by VORs and NDBs and when they are used. (FAA *Instrument Flying Handbook*)

54. What is the significance of a "T" bar on an airway at an intersection versus the absence of such a "T" bar on an airway at an intersection. (En route low altitude chart legend)

55. When can you use DME to identify an intersection? (En route low altitude chart legend)

56. Is there any significance to a solid triangle in the center of a VORTAC? (En route low altitude chart legend)

57. What is the purpose of "Flight Watch?" What is the frequency? (AIM 7-1-4)

58. How can you contact the nearest flight watch as indicated on the chart? (En route low altitude chart legend)

59. What is the difference between holding patterns depicted in fine lines versus those depicted in dark lines on approach charts? (Instrument approach procedure chart)

60. Describe the intent of the circle on approach charts. (Instrument approach procedure chart)

61. What are the different ways you might identify an outer marker. (Instrument approach procedure chart)

62. How do you identify the highest obstruction on an approach chart? (Instrument approach procedure chart)

63. Describe the holding pattern entry on the missed approach at several nearby NDBs. Show examples of where direct, parallel, and tear drop entries must be used. (Instrument approach procedure charts, AIM 5-3-7)

64. Describe the same entries prescribed in No. 63 on several nearby VOR approaches. (Instrument approach procedure charts, AIM 5-3-7)

65. Explain feeder routes and show how they are used on several nearby approaches. (AIM "Pilot/Controller Glossary," Instrument approach procedure charts)

66. Describe how an NDB is depicted on an approach chart and on a low altitude en route chart. (Instrument approach procedure chart, en route low altitude chart)

67. What is the significance of altitudes associated with feeder routes? Show one on a nearby approach. (Instrument approach procedure charts)

68. What is the significance of altitudes marked on the profiles of nearby VOR, VOR/DME, NDB, and ILS approaches. (Instrument approach procedure charts)

69. In case of communications failure, when and where can you begin an approach? (FAR 91.185, AIM 6-3-1)

70. Describe the use of the magnetic compass and common errors associated with this instrument. (FAA *Instrument Flying Handbook*)

71. On an instrument approach, describe the criteria for making a missed approach. (FAR 91.175)

Flight lesson 20: Flight test recommendation flight

This flight is a dress rehearsal for the FAA flight test; your instructor will play the role of the FAA examiner. The instructor will direct the student to plan and fly a short IFR cross-country flight. The student will carry out all the tasks specified in the Instrument Pilot Practical Flight Test Standards.

Assigned reading:

Chapter 19—Stress Can Spoil Your Whole Day

Review Chapter 20

Preflight Briefing
Review:

1. Flight planning
2. Obtaining and analyzing weather information
3. Filing IFR flight plans
4. IFR departure and en route procedures
5. Partial panel
6. Lost communications procedures
7. Loss of radio navigation equipment
8. ADF, VOR, and ILS approaches
9. Holding patterns
10. Missed approaches

Completion standards

The lesson is complete when the instructor is confident the student will pass the FAA instrument flight test and so indicates by endorsing the student's logbook and signing the "Airman Certificate and/or Rating Application," FAA Form 8710-1 (7-95).

Postflight critique

Flight instructor endorsements

Here is wording approved by the FAA for the various endorsements flight instructors must make in student logbooks to sign a student off for an instrument flight test:

Aeronautical Knowledge:
I have given Mr./Ms. _____ the required ground training on the aeronautical areas required by FAR 61.65 (b) and certify that he/she is prepared for the instrument rating knowledge test.
(signed by instructor)
(flight instructor certificate no. and exp. date)
(date of endorsement)

Flight Proficiency:
I have given Mr./Ms. _____ the training in an (airplane/training device) required by FAR 61.65 (c) and certify that (he/she) is prepared for the instrument rating practical test.
(signed by instructor)
(flight instructor certificate no. and exp. date)
(date of endorsement)

Instructors will also review written test questions that were missed, as listed on the written test report, and sign and date the report in the space provided to indicate that this review has been completed.

Finally, the instructor must sign and date the "Instructor's Recommendation" portion of the "Airman Certificate and/or Rating Application" (FAA Form 8710-1) (7-95) (*see* FIG. 20-1 after checking to make sure that the applicant has supplied all the required information and filled out the applicant portion of the form correctly.

Appendix A
The instrument pilot's professional library

IN AVIATION'S EARLIER DAYS IT WAS EASY TO REMEMBER JUST ABOUT everything you needed to know to become a safe, competent pilot. Now FARs run to thousands of pages, the AIM contains almost 400 pages, and aviation weather requires not one but two books to explain it all. The list of detailed material you need to master goes on and on.

It is all but impossible to absorb and remember all this detail. But it is possible—and easy—to look up hard-to-remember details if you have assembled a professional pilot's library for your own personal use. This is particularly important for the instrument pilot—and the instrument instructor.

Recently the electronic media have added new and exciting dimensions to the resources available to pilots. In the following bibliography you will find both electronic and print sources, with electronic sources listed first. The sources range from the essential to the nonessential but great to have available. All have been carefully checked by the authors of *Mastering Instrument Flying*.

With these sources as the foundation, your professional library will quickly become something you will turn to again and again to refresh your memory about something you haven't encountered recently, to broaden your knowledge of subjects such as weather, or to assist in flight planning.

INTERNET RESOURCES

To make the most of the hundreds of excellent aviation-oriented programs and Web sites now available, you must have a modem-equipped desktop computer with at least Windows 95. (See page 42 for additional requirements.) With this basic equipment a fascinating new world will open up for you that was barely conceivable less than ten years ago.

The quickest way to become familiar with what's out there in aviation is to get a copy of *200 Best Aviation Web Sites...and 100 More that Are Worth Bookmarking* by John A. Merry, published by McGraw-Hill. This excellent guide is widely available through bookstores. It can also be ordered directly through the publisher's Web site:

www.books.mcgraw-hill.com

Following are some electronic sources we recommend that were especially helpful in preparing the 3rd edition of *Mastering Instrument Flying*. Web sites are subject to constant change, and some even shut down completely just as you become accustomed to using them! If you are inexperienced at such things as installing or downloading programs, have a knowledgeable friend help you. This can save you a lot of time and frustration.

- *AccuWeather*—<www.accuweather.com> AccuWeather, the largest Internet weather resource, will customize a free weather page for you. Fee-paying subscribers can create customized pages from more than 35,000 types of weather and climate data with graphics of all kinds. A 30-day free trial of the main, "premier" service is available.

- *AOPA*—<www.aopa.org> For members, the Aircraft Owners and Pilot's Association offers a full menu of aviation weather, with graphics, tailored especially for general aviation pilots. Membership in AOPA offers a wide range of other benefits, including a subscription to *AOPA Pilot* magazine with news and articles of direct interest to general aviation pilots.

- *DUATS*—Two important things to know about the weather and flight planning services offered by the two DUATS providers are (1) they are free of charge to pilots with current medicals (the FAA contracts for this service), and (2) they constitute "official" weather briefings since you are logged on as a certificated pilot with your aircraft N number whenever you get a DUATS weather briefing.

DTC DUATS:

Information and customer service: 1-800-243-3828

Data line: 1-800-245-3828

Internet address: <www.duat.com>

Graphics: Free graphics. Obtain disk through customer service number.

GTE DUATS:

Information and customer service: 1-800-345-3828

Data line: 1-800-767-9989

Internet address: <www.skycentral.gte.com>

Graphics: Free Cirrus graphics. Download through www.skycentral.gte.com or obtain disk through customer service number.

- *The Weather Channel*—<www.weather.com> Easily browsable free weather reports and graphics similar to what you see on TV, but without all the commercial and promotional interruptions. Nor do you have to wait for "weather on the eights," etc., to get what you want. Free, customized information is available for your location.

PUBLICATIONS RESOURCES

An instrument pilot's professional publications library will contain all the information on which written, oral, and flight tests are based. You must read these publications and understand what they cover, especially on such subjects as air traffic control procedures, instrument construction and behavior, theory of flight, weather, and weather reports.

The following 6 basic books and 2 subscriptions are available through the Government Printing Office (GPO). There are also many excellent reprints of most of these available at FBOs or offered through catalogs. The prices are subject to change:

- *Aeronautical Information Manual: Official Guide to Basic Flight Information and ATC Procedures.* Contains fundamentals required to fly in the U.S. National Airspace System. It also contains items of interest to pilots concerning health and medical facts, factors affecting flight safety, a pilot/controller glossary of terms used in the air traffic control system, and information on safety, accident, and hazard reporting. Subscription consists of basic manual and changes issued every 112 days. Annual subscription $72, stock number SN 950-074-00000-1.

- *Aviation Weather* AC 00-6. Provides an up-to-date and expanded text for pilots and other flight operations personnel whose interest in meteorology applies to flying. $8.50, stock number SN 050-007-00283-1.

- *Aviation Weather Services* AC 00-45. Supplements *Aviation Weather* AC 00-6 in that it explains the weather service in general and the use and interpretation of reports, forecasts, weather maps, and prognostic charts in detail. $12, stock number SN 050-007-01082-6.

- *Computerized Testing Supplement for Instrument Rating* FAA-CT-8O8O-3B. Computerized testing designees will be required to use this supplement to administer those computer-assisted airman knowledge test questions that reference figures. $16, stock number SN 050-007-01088-5.

- *FAA Aviation News.* A bimonthly report issued by the FAA giving air traffic control and safety news, current interpretations of FARs, and general information. Annual subscription $15, stock number SN 750-002-00000-5.

- *Instrument Flying Handbook* AC 61-27. Provides the pilot with basic information needed to acquire an FAA instrument rating. It is designed for the reader who holds at least a private pilot certificate and is knowledgeable in all areas covered in the *Pilot's Handbook of Aeronautical Knowledge*. $8.50, stock number SN 050-007-00585-7.

- *Instrument Rating Practical Test Standards* FAA-S 8081-4C. The standards contained in this practical test book are to be used by FAA inspectors and designated pilot examiners when conducting airman practical tests. Instructors are expected to use this book when preparing applicants for practical tests. $2.50, stock number SN 050-007-01062-1.

- *Pilot's Handbook of Aeronautical Knowledge* AC 61-23. Provides basic knowledge that is essential for pilots. Introduces pilots to the broad spectrum of knowledge that will be needed as they progress in their pilot training. Except for the Code of Federal Regulations pertinent to civil aviation, most of the knowledge areas applicable to pilot certification are presented. This handbook is useful to beginning pilots as well as those pursuing more advanced pilot certificates. $13, stock number SN 050-011-00078-0.

Order the publications above from:
U.S. Government Printing Office (GPO)
Washington, D.C. 20402
202-512-1800

GPO bookstores are also located in several large metropolitan areas; if you call the GPO in Washington, ask about a bookstore in your region.

You may copy the form in FIG. A-1 and use it for purchasing single publications. Copy and use the form in FIG. A-2 for subscriptions. You can also purchase reprints of many of these publications at FBOs and through many aviation supply outlets.

The FAA's master guide to its publications is *Advisory Circular Checklist* AC 00-2, which includes the status of all FAA publications, is updated annually in August. The current update of AC 00-2 will be sent to you free of charge on request. Free advisory circulars can be obtained by requesting them through the form on FIG. A-3:

While you're at it, order some other free ACs. Be persistent; if something you want is reported as out of stock or being revised, try again at a later date. Copy the order form on FIG. A-3 and use it for free additional publications. The form will speed up the process. When the ACs come in, put them in a ring binder.

Here are the free ACs that every instrument pilot should be familiar with:

- *Thunderstorms* AC 00-24. Describes the aviation hazards of thunderstorms and offers guidance to help prevent accidents caused by thunderstorms.

Superintendent of Documents **Publications** Order Form

Order Processing Code:
*

☐ **YES,** please send me the following publications:

Qty.	Title / Stock Number	Price Each	Total Price
1	Catalog—Bestselling Government Books		
	S/N 021-602-00001-9	**FREE**	**FREE**
	S/N	$	
	S/N	$	
	S/N	$	

The total cost of my order is $_____. International customers please add 25%. Prices include regular domestic postage and handling and are subject to change.

_____
(Company or Personal Name)   (Please type or print)

_____
(Additional address/attention line)

_____
(Street address)

_____
(City, State, ZIP Code)

_____
(Daytime phone including area code)

_____
(Purchase Order No.)

May we make your name/address available to other mailers? YES ☐ NO ☐

Charge your order. It's Easy!

MasterCard

VISA

P3S

Please Choose Method of Payment:

☐ Check Payable to the Superintendent of Documents

☐ GPO Deposit Account ☐☐☐☐☐☐—☐

☐ VISA or MasterCard Account

☐☐☐☐☐☐☐☐☐☐☐☐☐☐☐☐☐☐☐☐

☐☐☐☐ (Credit card expiration date)

To fax your orders
(202) 512–2250

_____
(Authorizing Signature)

Thank you for your order!

Mail To: New Orders, Superintendent of Documents
P.O. Box 371954, Pittsburgh, PA 15250–7954

Fig. A-1. *Form for ordering nonsubscription publications from the Government Printing Office.*

Appendix A

Superintendent of Documents **Subscriptions** Order Form

Order Processing Code:

*

☐ **YES,** please send me the following publications:

Qty.	List ID	Title	Price Each	Total Price

The total cost of my order is $_____. International customers please add 25%. Prices include regular domestic postage and handling and are subject to change.

_____
(Company or Personal Name)   (Please type or print)

_____
(Additional address/attention line)

_____
(Street address)

_____
(City, State, ZIP Code)

_____
(Daytime phone including area code)

_____
(Purchase Order No.)   YES   NO
May we make your name/address available to other mailers? ☐   ☐

Charge your order. It's Easy!

MasterCard

VISA

P3S

Please Choose Method of Payment:

☐ Check Payable to the Superintendent of Documents

☐ GPO Deposit Account ☐☐☐☐☐☐☐ – ☐

☐ VISA or MasterCard Account

☐☐☐☐☐☐☐☐☐☐☐☐☐☐☐☐☐☐☐☐

☐☐☐☐ (Credit card expiration date)

To fax your orders (202) 512–2250

_____
(Authorizing Signature)

Mail To: New Orders, Superintendent of Documents
          P.O. Box 371954, Pittsburgh, PA 15250-7954

Thank you for your order!

Fig. A-2. *Form for ordering subscriptions from the Government Printing Office.*

314

ORDER BLANK [Free Publications]   DATE____ /____ /____

For Faster Service Use A Self-Addressed Mailing Label When
Not Using This Blank. Please Print Or Type All Information.

Mail To: U.S. Department of Transportation
             Subsequent Distribution Office, SVC-121.23
             Ardmore East Business Center
             3341 Q 75th Ave
             Landover, MD 20785

Help Line: 301-322-4961 _____ FAX REQUEST TO 301-386-5394 DOT Warehouse

NUMBER	TITLE	QUANTITY

SVC-121.23
Request Filled By:_____   Date ____ /____ /____

1. Out of Stock [reorder in   days]* *   3. Cancelled, no replacement

2. Being revised   4. Cancelled by _____[enclosed]

     5. Other:_____

* * IF YOU DO NOT RECEIVE DESIRED PUBLICATION(S) AFTER YOUR <u>SECOND</u>
REQUEST PLEASE CALL FAA'S TOLL-FREE CONSUMER HOTLINE: 1-800 FAA-SURE.

    TO COMPLETE ORDER: Enter Name and Address.   <u>DO NOT DETACH.</u>

NAME
_____

STREET ADDRESS
_____

CITY   **STATE**   **ZIP CODE**
_____

Fig. A-3. *Form for ordering free FAA advisory circulars.*

- *Status and Availability of Military Handbooks and ANC Bulletins for Aircraft* AC 20-3. Announces the status and availability of military handbooks and ANC bulletins prepared jointly by the FAA, U.S. Navy, and U.S. Air Force.

- *Aids Authorized for Use by Airmen Written Test Applicants* AC 60-11. Clarifies FAA policy concerning aids that applicants may use when taking airman written tests.

- *Aeronautical Decision Making* AC 60-22. Provides introductory material, background information, and reference material on aeronautical decision making. Provides a systematic approach to risk assessment and stress management in aviation, illustrates how personal attitudes can influence decision making and how these attitudes can be modified to enhance safety in the cockpit.

- *Qualification and Approval of Personal Computer-Based Aviation Training Devices* AC 61-126. Provides information and guidance to potential training device manufacturers and aviation training consumers concerning a means, acceptable to the Administrator, by which personal computer-based aviation training devices (PCATD) may be qualified and approved for flight training toward satisfying the instrument rating training under the provisions of FAR parts 61 and 141.

- *Traffic Advisory Practices at Nontower Airports* AC 90-42. Contains good operating practices and procedures for use when approaching and departing airports without an operating control tower and airports that have control towers operating part time. Includes changes in radio frequencies and phraseology.

- *Gyroscopic Instruments—Good Operating Practices* AC 91-46. Issued to reemphasize to general aviation instrument-rated pilots the need to determine the proper operation of gyroscopic instruments, the importance of instrument cross-checks, and proficiency in partial panel operations.

- *Runway Visual Range* AC 97-1. Describes RVR measuring equipment and its operating use.

The FAA maintains a mailing list of people to whom they routinely send updated or reissued free ACs. Use the form shown in FIG. A-4 to have your name added to this list. Note that the form also has a checklist for other services, such as change of address. Again, using this form for the listed services will speed up the process.

For obvious reasons, the use of obsolete charts or publications for navigation is dangerous. Aeronautical information changes rapidly, and it is vitally important that pilots check the effective dates on each aeronautical chart and publication to be used. Obsolete charts should always be discarded and replaced by current editions.

Because you can't always rely upon an FBO or flight school to have exactly what you need, you should subscribe to the *Airport/Facility Directory* (the green book), a set of en route charts that covers the region in which you fly, and current booklets of approach, SID, and STAR charts. For prices, subscription rates, and ordering instructions, get a copy of the latest free catalog, *Aeronautical Charts and Related Products*, published by the National Oceanic and Atmospheric Administration (NOAA) by writing to:

REQUEST FOR MAILING LIST ACTION (Advisory Circulars)

Number	Subject		Number	Subject
				Air Carriers and Commercial Operators of Large Aircraft _____
00	GENERAL _____		125	Certification and Operations: Airplanes Having a Seating Capacity of 20 or More Passengers or a Maximum Payload Capacity of 6,000 Pounds or More _____
1	Definitions and Abbreviations _____		127	Certification and Operations of Scheduled Air Carriers with Helicopters _____
10	PROCEDURAL RULES _____		129	Operations of Foreign Air Carriers _____
11	General Rule-Making Procedures _____		133	Rotorcraft External-Load Operations _____
13	Investigation and Enforcement Procedures _____		135	Air Taxi Operators and Commercial Operators _____
			137	Agricultural Aircraft Operations _____
20	AIRCRAFT _____		139	Certification and Operations: Land Airports Serving CAB-Certificated Air Carriers _____
21	Certification Procedures for Products and Parts _____			
13	Investigation and Enforcement Procedures _____		140	SCHOOLS AND OTHER CERTIFICATED AGENCIES _____
23	Airworthiness Standards: Normal, Utility,and Acrobatic Category Airplanes _____			
25	Airworthiness Standards: Transport Category Airplanes _____		141	Pilot Schools _____
27	Airworthiness Standards: Normal Category Rotorcraft _____		143	Ground Instructors _____
29	Airworthiness Standards: Transport Category Rotorcraft _____		145	Repair Stations _____
31	Airworthiness Standards: Manned Free Balloons _____		147	Aviation Maintenance Technician Schools _____
33	Airworthiness Standards: Aircraft Engines _____			
34	Fuel Venting and Exhaust Emission Requirements for Turbine Engine Powered Airplaness _____		150	AIRPORT NOISE COMPATIBILITY PLANNING _____
35	Airworthiness Standards: Propellers _____		151	Federal Aid to Airports _____
36	Noise Standards: Aircraft Type and Airworthiness Certification _____		152	Airport Aid Program _____
			155	Release of Airport Property from Surplus Property Disposal Restrictions _____
39	Airworthiness Directives _____		156	State Block Grant Pilot Program _____
43	Maintenance, Preventive Maintenance, Rebuilding and Alteration _____		157	Notice of Construction, Alteration, Activation, and Deactivation of Airports _____
45	Identification and Registration Marking _____		158	Passenger Facilities Charges _____
47	Aircraft Registration _____		159	National Capital Airports _____
49	Recording of Aircraft Titles and Security Documents _____		159/10	Washington National Airport _____
			159/20	Dulles International Airport _____
60	AIRMEN _____		161	Notice and Approval of Airport Noise and Access Restrictions _____
61	Certification: Pilots and Flight Instructors _____		169	Expenditures of Federal Funds for Nonmilitary Airports or Air Navigational Facilities Thereon _____
63	Certification: Flight Crewmembers Other Than Pilots _____			
65	Certification: Airmen Other Than Flight Crewmembers _____		170	NAVIGATIONAL FACILITIES _____
67	Medical Standards and Certification _____			
			170	Establishment and Discontinuance Criteria for Airport Traffic Control Tower Facilities _____
70	AIRSPACE _____		171	Non-Federal Navigation Facilities _____
71	Designation of Federal Airways, Area Low Routes, Controlled Airspace,and Reporting Points _____		180	ADMINISTRATIVE REGULATIONS _____
73	Special Use Airspace _____		183	Representatives of the Administrator _____
75	Establishment of Jet Routes and Area High Routes _____		185	Testimony by Employees and Production of Records in Legal Proceedings _____
77	Objects Affecting Navigable Airspace _____		187	Fees _____
			189	Use of Federal Aviation Administration Communication System _____
90	AIR TRAFFIC AND GENERAL OPERATING RULES _____			
91	General Operating and Flight Rules _____		190	WITHHOLDING SECURITY INFORMATION; WAR RISK INSURANCE; AIRCRAFT LOAN GUARANTEE PROGRAM _____
93	Special Air Traffic Rules and Airport Traffic Patterns _____			
95	IFR Altitudes _____		191	Withholding Security Information from Disclosure Under the Air Transportation Security Act of 1974 _____
97	Standard Instrument Approach Procedures _____			
99	Security Control of Air Traffic _____		210	FLIGHT INFORMATION _____
101	Moored Balloons, Kites, Unmanned Rockets and Unmanned Free Balloons _____		211	Aeronautical Charts and Flight Information Publications _____
103	Ultralight Vehicles _____		212	Publication Specification: Charts and Publications _____
105	Parachute Jumping _____			
107	Airport Security _____			
108	Airplane Operators Security _____			
109	Indirect Air Carrier Security _____			
120	AIR CARRIERS, AIR TRAVEL CLUBS,AND OPERATORS FOR COMPENSATION OR HIRE CERTIFICATION AND OPERATIONS _____			
121	Certification and Operations: Domestic Flag, and Supplemental			

Note: Advisory Circulars numbered 150/5000—150/5049 will be considered in Airports/General category and will be distributed to subscribers of each of the five listings below.

150/5000/150-5099	Airport Planning./Environment _____
150/5100/150/5199	Airport Grants/Assistance _____
150/5200/150/5299	Airport Safety/Certification/Operations _____
150/5300/150/5399	Airport Design/Construction/ Maintenance _____
150/5340/150/5354	Airport Visual/Navigation Equipment _____

DATE: _____/_____/_____

[ ] ADD MY NAME TO THE LIST.
[ ] ADD MY NAME FOR AC CHECKLIST.
[ ] REMOVE MY NAME FROM LISTS CHECKED ABOVE.
[ ] CHANGE OF ADDRESS AS INDICATED BELOW:

CUSTOMER'S NAME AND ADDRESS (attach label from back cover, if available)

_____

_____

_____

_____ Zip _____

MAIL TO:
DOT, SVC-121-.21
DISTRIBUTION REQUIREMENTS SECTION
WASHINGTON, DC 20590

Fig. A-4. *Form for adding name to FAA mailing list and for making mailing list changes.*

Appendix A

NOAA Distribution Division, N/A CC3
National Ocean Service
Riverdale, Maryland 20737-1199

The NOAA distribution branch can also be reached at the following telephone numbers:

1-800-638-8972 or 301-436-6993

Information on Canadian charts and publications can be obtained from:

Canada Map Office
Department of Energy, Mines, and Resources
130 Bentley Ave.
Nepean (Ottawa) Ontario
K1A 0E9 Canada
613-952-7000

Information on Jeppesen charts can be obtained by contacting:

Jeppesen Sanderson
55 Inverness Drive East
Englewood, Colorado 80112-5498
1-800-621-5377

Instrument flight instructors and instrument pilots who wish to delve deeper into the real world of instrument flying should also obtain these publications:

- *NOAA Aeronautical Chart User's Guide.* This is a guide to the wealth of information provided on NOAA's aeronautical charts, both VFR and IFR. It includes a discussion of IFR chart terms and symbols and complete, illustrated lists of all symbols used on all NOAA charts. $8, see listing in the booklet *Aeronautical Charts and Related Products* for ordering instructions.

- *Air Traffic Control Handbook* 7110.65. Prescribes air traffic control procedures and phraseology for use by air traffic control personnel. Controllers are required to be familiar with all provisions of this handbook. Annual subscription $64. GPO stock number SN 950-002-00000-0.

- *Contractions* 7340.1. Gives approved word and phrase contractions used by personnel connected with air traffic control, communications, weather, charting, and associated services. Annual subscription $40. GPO stock number SN 950-003-00000-6.

- *Flight Services* 7110.10. Prescribes procedures and phraseology for use by personnel providing flight assistance services. Annual subscription $40, GPO stock number SN 950-032-00000-6.

- *United States Standard for Terminal Instrument Procedures (TERPS)* 8260.3B. Contains criteria that shall be used to formulate, review, approve, and publish procedures for instrument approach and departure of aircraft to and from civil and military airports. Be sure to ask for changes when ordering. $21. GPO stock number SN 050-007-01006-1.

Appendix B
FAR excerpts

FEDERAL AVIATION REGULATION PARTS 1, 61, AND 91 are fundamental regulations for general aviation, and every pilot should be familiar with all sections and subsections. The excerpts reproduced here include major changes made to FAR Part 61 in 1997. Excerpted here are only those FARs pertinent to instrument flight.

PART 1—DEFINITIONS AND ABBREVIATIONS

1.1 General Definitions

Air traffic means aircraft operating in the air or on an airport surface, exclusive of loading ramps and parking areas.

Air traffic clearance means an authorization by air traffic control, for the purpose of preventing collision between known aircraft, for an aircraft to proceed under specified traffic conditions within controlled airspace.

Air traffic control means a service operated by appropriate authority to promote the safe, orderly, and expeditious flow of air traffic.

Alternate airport means an airport at which an aircraft may land if a landing at the intended airport becomes inadvisable.

Area navigation (RNAV) means a method of navigation that permits aircraft operations on any desired course within the coverage of station-referenced navigation signals or within the limits of self-contained system capability.

Appendix B

Area navigation high route means an area navigation route within the airspace extending upward from, and including, 18,000 feet MSL to Flight Level 450.

Area navigation low route means an area navigation route within the airspace extending upward from 1,200 feet above the surface of the earth to, but not including 18,000 feet MSL.

Category II operation, with respect to the operation of aircraft, means a straight-in ILS approach to the runway of an airport under a Category II ILS instrument approach procedure issued by the Administrator or other appropriate authority.

Category III operations, with respect to the operation of aircraft, means an ILS approach to, and landing on, the runway of an airport using a Category III ILS instrument approach procedure issued by the Administrator or other appropriate authority.

Ceiling means the height above the earth's surface of the lowest layer of clouds or obscuring phenomena that is reported as "broken," "overcast," or "obscuration," and not classified as "thin" or "partial."

Decision height, with respect to the operation of aircraft, means the height at which a decision must be made, during an ILS or PAR instrument approach, to either continue the approach or to execute a missed approach.

Flight level means a level of constant atmospheric pressure related to a reference datum of 29.92 inches of mercury. Each is stated in three digits that represent hundreds of feet. For example, flight level 250 represents a barometric altimeter indication of 25,000 feet: flight level 255, an indication of 25,500 feet.

Flight plan means specified information, relating to the intended flight of an aircraft, that is filed orally or in writing with air traffic control.

Flight time means pilot time that commences when an aircraft first moves under its own power for the purpose of flight and ends when the aircraft comes to rest after landing.

Flight visibility means the average forward horizontal distance, from the cockpit of an aircraft in flight, at which prominent unlighted objects may be seen and identified by day and prominent lighted objects may be seen and identified by night.

IFR conditions means weather conditions below the minimum for flight under visual flight rules.

IFR over-the-top, with respect to the operation of aircraft, means the operation of an aircraft over-the-top on an IFR flight plan when cleared by air traffic control to maintain "VFR conditions" or "VFR conditions on top."

Minimum descent altitude means the lowest altitude, expressed in feet above mean sea level, to which descent is authorized on final approach or during circle-to-land maneuvering in execution of a standard instrument approach procedure, where no electronic glideslope is provided.

Night means the time between the end of evening civil twilight and the beginning of morning civil twilight, as published in the American Air Almanac, converted to local time.

Nonprecision approach procedure means a standard instrument approach procedure in which no electronic glideslope is provided.

Over-the-top means above the layer of clouds or other obscuring phenomena forming the ceiling.

Pilot in command means the person who:

(1) Has final authority and responsibility for the operation and safety of the flight.

(2) Has been designated as pilot in command before or during the flight.

Positive control means control of all air traffic, within designated airspace, by air traffic control.

Powered-lift means a heavier-than-air aircraft capable of vertical takeoff, vertical landing, and low speed flight that depends principally on engine-driven lift devices or engine thrust for lift during these flight regimes and on nonrotating airfoil(s) for lift during horizontal flight.

Precision approach procedure means a standard instrument approach procedure in which an electronic glideslope is provided.

RNAV way point (W/P) means a predetermined geographical position used for route or instrument approach definition or progress reporting purposes that is defined relative to a VORTAC station position.

Route segment means a part of a route. Each end of that part is identified by: (A) A continental or insular geographical location; or (B) A point at which a definite radio fix can be established.

True airspeed means the airspeed of an aircraft relative to undisturbed air.

VFR over-the-top, with respect to the operation of aircraft, means the operation of an aircraft over-the-top under VFR when it is not being operated on an IFR flight plan.

1.2 Abbreviations and Symbols

AGL	above ground level.
ALS	approach light system.
ASR	airport surveillance radar.
ATC	air traffic control.
DH	decision height.
DME	distance measuring equipment compatible with TACAN.
FM	fan marker.
GS	glideslope.
HIRL	high-intensity runway light system.
IAS	indicated airspeed.
IFR	instrument flight rules.
ILS	instrument landing system.
IM	ILS inner marker.
INT	intersection.
LDA	localizer-type directional aid.
LMM	compass locator at middle marker.
LOC	ILS localizer.
LOM	compass locator at outer marker.
MAA	maximum authorized IFR altitude.

Appendix B

MALS	medium intensity approach light system.
MALSR	medium intensity approach light system with runway alignment indicator lights.
MCA	minimum crossing altitude.
MDA	minimum descent altitude.
MEA	minimum en route IFR altitude.
MM	ILS middle marker.
MOCA	minimum obstruction clearance altitude.
MRA	minimum reception altitude.
MSL	mean sea level.
NDB(ADF)	nondirectional beacon (automatic direction finder).
NOPT	no procedure turn required.
OM	ILS outer marker.
PAR	precision approach radar.
RAIL	runway alignment indicator light system.
RBN	radio beacon.
RCLM	runway centerline marking.
RCLS	runway centerline light system.
REIL	runway end identification lights.
RVR	runway visual range as measured in the touchdown zone area.
SALS	short approach light system.
SSALS	simplified short approach light system.
SSALSR	simplified short approach light system with runway alignment indicator lights.
TACAN	ultra-high frequency tactical air navigational aid.
TAS	true airspeed.
TDZL	touchdown zone lights.
TVOR	very high frequency terminal omnirange station.
VFR	visual flight rules.
VHF	very high frequency.
VOR	very high frequency omnirange station.
VORTAC	collocated VOR and TACAN.

PART 61—CERTIFICATION: PILOTS AND FLIGHT INSTRUCTORS

61.1 Applicability

(a) This part prescribes:

(1) The requirements of issuing pilot, flight instructor, and ground instructor certificates and ratings, the conditions under which those certificates and ratings are necessary, and the privileges and limitations of those certificates and ratings.

61.3 Requirements for Certificates, Rating, and Authorizations

(c) *Medical certificate.*

(1) A person may not act as pilot in command or in any other capacity as a required pilot flight crewmember, under a certificate issued to that person under this part, unless that person has a current and appropriate medical certificate that has been issued under part 67 of this chapter.

(e) *Instrument rating.* No person may act as pilot in command of a civil aircraft under IFR or in weather conditions less than the minimums prescribed for VFR unless that person holds:

(1) The appropriate aircraft category, class, and type (if necessary), and instrument rating on that person's pilot certificate for any airplane, helicopter, or powered-lift being flown.

61.4 Qualification and approval of flight simulators and flight training devices.

(a) Except as provided in paragraph (b) or (c) of this section, each flight simulator and flight training device used for training, and for which an airman is to receive credit to satisfy any training, testing, or checking requirement under this chapter, must be qualified and approved by the Administrator for—

(1) The training, testing, and checking for which it is used;

(2) Each particular maneuver, procedure, or crewmember function performed; and

(3) The representation of the specific category and class of aircraft, particular variation within the type of aircraft, or set of aircraft for certain flight training devices.

(b) Any device used for flight training, testing, or checking that has been determined to be acceptable to or approved by the Administrator prior to August 1, 1996, which can be shown to function as originally designed, is considered to be a flight training device, provided it is used for the same purposes for which it was originally accepted or approved and only to the extent of such acceptance or approval.

(c) The Administrator may approve a device other than a flight simulator or flight training device for specific purposes.

61.43 Practical Tests: General Procedures

(a) Except as noted in paragraph (b) of this section, the ability of an applicant for a certificate or rating issued under this part to perform the required tasks on the practical test is based on that applicant's ability to safely:

(1) Perform the tasks specified in the areas of operation for the certificate or rating sought within the approved standards;

(2) Demonstrate mastery of the aircraft with the successful outcome of each task performed never seriously in doubt;

(3) Demonstrate satisfactory proficiency and competency within the approved standards;

(4) Demonstrate sound judgment; and

(5) Demonstrate single-pilot competence if the aircraft is type certificated for single-pilot operations.

(b) If an applicant does not demonstrate single pilot proficiency, as required in paragraph (a) (5) of this section, a limitation of "Second in Command Required" will be placed on the applicant's airman certificate. The limitation may be removed if the applicant passes the appropriate practical test by demonstrating single-pilot competency in the aircraft in which single-pilot privileges are sought.

(c) If an applicant fails any area of operation, that applicant fails the practical test.

(d) An applicant is not eligible for a certificate or rating sought until all the areas of operation are passed.

(e) The examiner or applicant may discontinue a practical test at any time:

(1) When the applicant fails one or more of the areas of operation; or

(2) Due to inclement weather conditions, aircraft airworthiness, or any safety-of-flight concern.

(f) If a practical test is discontinued, the applicant is entitled credit for those areas of operation that were passed by the applicant, but only if the applicant:

(1) Passes the remainder of the practical test within the 60-day period after the date the practical test was discontinued;

(2) Presents to the examiner for the retest the original notice of disapproval form or the letter of discontinuance form, as appropriate;

(3) Satisfactorily accomplishes any additional training needed and obtains the appropriate instructor endorsements, if additional training is required; and

(4) Presents to the examiner for the retest a properly completed and signed application.

61.51 Pilot Logbooks

(a) *Training time and aeronautical experience.* Each person must document and record the following time in a manner acceptable to the Administrator:

(1) Training and aeronautical experience used to meet the requirements for a certificate, rating, or flight review of this part.

(2) The aeronautical experience required for meeting the recent flight experience requirements of this part.

(b) *Logbook entries.* For the purposes of meeting the requirements of paragraph (a) of this section, each person must enter the following information for each flight or lesson logged:

(1) General—

(i) Date.

(ii) Total flight time or lesson time.

(iii) Location where the aircraft departed and arrived, or for lessons in a flight simulator or flight training device, the location where the lesson occurred.

(iv) Type and identification of aircraft, flight simulator, or flight training device, as appropriate.

(v) The name of the safety pilot, if required by 91.109 (b) of this chapter.

(2) Type of pilot experience or training—

(i) Solo.

(ii) Pilot in Command.

(iii) Second in command,

(iv) Flight and ground training received by an authorized instructor.

(v) Training received in a flight simulator or flight training device from an authorized instructor.

(3) Conditions of flight—

(i) Day or night.

(ii) Actual instrument.

(iii) Simulated instrument conditions in flight, a flight simulator, or a flight raining device.

(c) *Logging of pilot time*. The pilot time described in this section may be used to:

(1) Apply for a certificate or rating issued under this part; or

(2) Satisfy the recent flight experience requirements of this part.

(e) *Logging pilot in command flight time*.

(1) A recreational, private, or commercial pilot may log pilot-in-command time only for that flight time during which that person—

(i) Is the sole manipulator of the controls of an aircraft for which the pilot is rated.

(3) An authorized instructor may log as pilot-in-command time all flight time while acting as an authorized instructor.

(g) *Logging instrument flight time*.

(1) A person may log instrument time only for that flight time when the person operates the aircraft solely by reference to instruments under actual or simulated instrument flight conditions.

(2) An authorized instructor may log instrument time when conducting instrument flight instruction in actual instrument flight conditions.

(3) For the purposes of logging instrument time to meet the requirements of 61.57 (c) of this part, the following information must be recorded in the person's logbook—

(i) The location and type of each instrument approach accomplished; and

(ii) The name of the safety pilot, if required.

61.57 Recent Flight Experience: Pilot in Command

(c) *Instrument experience*. No person may act as pilot in command under IFR or in weather conditions less than the minimums prescribed for VFR, unless within the preceding 6 calendar months, that person has:

(1) For the purpose of obtaining instrument experience in an aircraft (other than a glider) performed and logged under actual or simulated instrument conditions, either in flight in the appropriate category of aircraft for the instrument privileges sought or in a flight simulator or flight training device that is representative of the aircraft category for the instrument privileges sought—

(i) At least six instrument approaches;

(ii) Holding procedures; and

(iii) Intercepting and tracking courses through the use of navigation systems.

(d) *Instrument proficiency check.* A person who does not meet the instrument experience requirements of paragraph (c) of this section within the prescribed time, or within 6 calendar months after the prescribed time, may not serve as pilot in command under IFR or in weather conditions less than the minimums prescribed for VFR until that person passes an instrument proficiency check consisting of a representative number of tasks required by the instrument rating practical test.

(1) The instrument proficiency check must be—

(i) In an aircraft that is appropriate to the aircraft category;

(ii) For other than a glider, in a flight simulator or flight training device that is representative of the aircraft category; or

(iii) For a glider, in a single-engine aircraft or a glider.

(2) The instrument proficiency check must be given by—

(i) An examiner;

(ii) A person authorized by the U.S. Armed Forces to conduct instrument flight tests, provided the person being tested is a member of the U.S. Armed Forces;

(iii) A company check pilot who is authorized to conduct instrument flight tests under part 121, 125, or 135 of this chapter, and provided that both the check pilot and the pilot being tested are employees of that operator;

(iv) An authorized instructor; or

(v) A person authorized by the Administrator to conduct instrument practical tests.

61.65 Instrument Rating Requirements

(a) *General.* A person who applies for an instrument rating must:

(1) Hold at least a current private pilot certificate with an appropriate airplane, helicopter, or powered-lift rating appropriate to the instrument rating being sought;

(2) Be able to read, speak, write, and understand the English language;

(3) Receive and log ground training from an authorized or instructor or accomplish a home-study course of training on the aeronautical knowledge areas of paragraph (b) of this section that apply to the instrument rating being sought;

(4) Receive a logbook or training record endorsement from an authorized instructor certifying that the person is prepared to take the required knowledge test;

(5) Receive and log training on the areas of operation of paragraph (c) of this section from an authorized in instructor in an aircraft, flight simulator, or flight raining device that represents an airplane, helicopter, or powered-lift appropriate to the instrument rating being sought;

(6) Receive a logbook or training record endorsement from an authorized instructor certifying that the person is prepared to take the required practical test;

(7) Pass the required knowledge test on the aeronautical areas of paragraph (b) of this section; however, an applicant is not required to take another knowledge test when that person already holds an instrument rating;

(8) Pass the required practical test on the areas of operation of paragraph (c) of this section in—

(i) An airplane, helicopter, or powered-lift appropriate to the rating being sought; or

(ii) A flight simulator or a flight training device appropriate to the rating sought and approved for the specific maneuver or procedure performed. If a flight training device is used for the practical test, the instrument approach procedures conducted in that flight training device are limited to one precision and one nonprecision approach, provided the flight training device is approved for the procedure performed.

(b) *Aeronautical knowledge.* A person who applies for an instrument rating must have received and logged ground training from an authorized instructor or accomplished a home-study course on the following areas that apply to the instrument rating being sought:

(1) Federal Aviation Regulations of this chapter that apply to flight operations under IFR;

(2) Appropriate information that applies to flight operations under IFR in the "Aeronautical Information Manual;"

(3) Air traffic control system and procedures for instrument flight operations;

(4) IFR navigation and approaches by use of navigation systems;

(5) Use of IFR en route and instrument approach procedure charts;

(6) Procurement and use of aviation weather reports and forecasts and the elements of forecasting weather trends based on that information and personal observation of weather conditions;

(7) Safe and efficient operation of aircraft under instrument flight rules and conditions;

(8) Recognition of critical weather situations and windshear avoidance;

(9) Aeronautical decision making and judgment; and

(1) Crew resource management, including crew communications and coordination.

(c) *Flight proficiency.* A person who applies for an instrument rating must receive and log training from an authorized instructor in an aircraft, or in a flight simulator or flight training device, in accordance with paragraph (e) of this section, that includes the following areas of operation:

(1) Preflight preparation;

(2) Preflight procedures;

(3) Air traffic control clearances and procedures;

(4) Flight by reference to instruments;

(5) Navigation systems;

(6) Instrument approach procedures;

(7) Emergency operations; and

(8) Postflight procedures.

(d) *Aeronautical experience.* A person who applies for an instrument rating must have logged the following:

(1) At least 50 hours of cross-country flight time as pilot in command, of which at least 10 hours must be in airplanes for an instrument-airplane rating; and

(2) A total of 40 hours of actual or simulated instrument time on the areas of operation of this section, to include—

(i) At least 15 hours of instrument flight training from an authorized flight instructor in the aircraft category for which the instrument rating is being sought;

(ii) At least 3 hours of instrument training that is appropriate to the instrument rating being sought from an authorized flight instructor in preparation for the practical test within 60 preceding the date of the test;

(iii) For an instrument-airplane rating, instrument training on cross-country flight procedures specific to airplanes that includes at least one cross-country flight that is performed under IFR, and consists of—

(A) A distance of at least 250 nautical miles along airways or ATC-directed routing;

(B) An instrument approach at each airport; and

(C) Three different kinds of approaches with the use of navigation systems;

(iv) For an instrument helicopter rating, instrument training specific to helicopters on cross-country procedures that includes at least one cross-country flight in a helicopter that is performed under IFR, and consists of—

(A) A distance of at least 100 nautical miles along airways or ATC-directed routing;

(B) An instrument approach at each airport; and

(C) Three different kinds of approaches with the use of navigation systems; and

(v) For an instrument-powered-lift rating, instrument training specific to a powered-lift on cross-country flight procedures that includes at least one cross-country flight in a powered-lift that is performed under LFR and consists of—

(A) A distance of at least 250 nautical miles along airways or ATC-directed routing;

(B) An instrument approach at each airport; and

(C) Three different kinds of approaches with the use of navigation systems.

(e) *Use of flight simulators or flight training devices.* If the instrument training was provided by an authorized instructor in a flight simulator or flight training device—

(1) A maximum of 30 hours may be performed in that flight simulator or flight training device if the training was accomplished in accordance with Part 142 of this chapter; or

(2) A maximum of 20 hours may be performed in that flight simulator or flight training device if the training was not accomplished in accordance with Part 142 of this chapter.

PART 91—GENERAL OPERATING AND FLIGHT RULES

91.109 Flight Instruction; Simulated Instrument Flight and Certain Flight Tests

(b) No person may operate a civil aircraft in simulated instrument flight unless—

(1) The other control seat is occupied by a safety pilot who possesses at least a private pilot license with category and class ratings appropriate to the aircraft being flown.

91.123 Compliance with ATC Clearances and Instructions

(a) When an ATC clearance has been obtained, no pilot in command may deviate from that clearance unless an amended clearance is obtained, an emergency exists, or the deviation is in response to a traffic alert and collision avoidance system resolution advisory. However, except in Class A airspace, a pilot may cancel an IFR flight plan if the operation is being conducted in VFR weather conditions. When a pilot is uncertain of an ATC clearance, that pilot shall immediately request clarification from ATC.

(b) Except in an emergency, no person may operate an aircraft contrary to an ATC instruction in an area in which air traffic control is exercised.

(c) Each pilot in command who, in an emergency, or in response to a traffic alert and collision avoidance system resolution advisory, deviates from an ATC clearance or instruction shall notify ATC of that deviation as soon as possible.

(d) Each pilot in command who (though not deviating from a rule of this subpart) is given priority by ATC in an emergency, shall submit a detailed report of that emergency within 48 hours to the manager of that facility, if requested by ATC.

(e) Unless otherwise authorized by ATC, no person operating an aircraft may operate that aircraft according to any clearance or instruction that has been issued to the pilot of another aircraft for radar air traffic control purposes.

91.167 Fuel Requirements for Flight in IFR Conditions

(a) Except as provided in paragraph (b) of this section, no person may operate a civil aircraft in IFR conditions unless it carries enough fuel (considering weather reports and forecasts, and weather conditions) to—

(1) Complete the flight to the first airport of intended landing;

(2) Fly from that airport to the alternate; and

(3) Fly after that for 45 minutes at normal cruising speed.

(b) Paragraph (a) (2) of this section does not apply if:

(1) Part 97 of this subchapter prescribes a standard instrument approach procedure for the first airport of intended landing; and

(2) For at least 1 hour before and 1 hour after the estimated time of arrival at the airport, the weather reports or forecasts or any combination of them, indicate—

(i) The ceiling will be at least 2,000 feet above the airport elevation; and

(ii)   Visibility will be at least 3 miles.

91.169 IFR Flight Plan: Information Required

(a) *Information required.* Unless otherwise authorized by ATC, each person filing an IFR flight shall include in it the following information:

(1) Information required under (1.153 (a).

(2) An alternate airport, except as provided in paragraph (b) of this section.

(b) *Exceptions to applicability of paragraph (a) (2) of this section.* Paragraph (a) (2) of this section does not apply if part 97 of this chapter prescribes a standard instrument approach procedure for the first airport of intended landing and, for at least 1 hour before and 1 hour after the estimated time of arrival, the weather reports or forecasts, or any combination of them, indicate—

(1) The ceiling will be at least 2,000 feet above the airport elevation; and

(2) The visibility will be at least 3 statute miles.

(c) *IFR alternate weather minimums.* Unless otherwise authorized by Administrator, no person may include an alternate airport in an IFR flight plan unless current weather forecasts indicate that, at the estimated time of arrival at the alternate airport, the ceiling and visibility at that airport will be at or above the following alternate airport weather minimums:

(1) If an instrument approach procedure has been published in part 97 of this chapter for that airport, the alternate minimums specified in that procedure or, if none are specified, the following minimums:

(i) Precision approach procedure: Ceiling 600 feet, visibility 2 statute miles.

(ii) Nonprecision approach procedure: Ceiling 800 feet and visibility 2 statute miles.

(2) If no instrument approach procedure has been published in part 97 of this chapter for that airport, ceiling and visibility minimums are those allowing descent from the MEA, approach, and landing under basic VFR.

(d) *Cancellation.* When a flight plan has been activated, the pilot in command, upon canceling or completing the flight under the flight plan, shall notify an FAA Flight Service Station or ATC facility.

91.171 VOR Equipment Check for IFR Operations

(a) No person may operate a civil aircraft under IFR using the VOR system of radio navigation unless the VOR equipment of that aircraft—

(1) Is maintained, checked, and inspected under an approved procedure; or

(2) Has been operationally checked within the preceding 30 days, and was found to be within the limits of the permissible indicated bearing error set forth in paragraph (b) or (c) of this section.

(b) Except as provided in paragraph (c) of this section, each person conducting a VOR check under subparagraph (a) (2) of this section, shall—

(1) Use, at the airport of intended departure, an FAA operated or approved test signal or a test signal radiated by a certificated and appropriately rated radio repair station, or outside the United States, a test signal operated or approved by appropriate authority, to check the VOR equipment (the maximum permissible indicated bearing error is plus or minus 4 degrees); or

(2) Use at the airport of intended departure, a point on the airport surface designated as a VOR system checkpoint by the Administrator or, outside the United States, by appropriate authority (the maximum permissible bearing error is plus or minus 4 degrees); or

(3) If neither a test signal nor a designated checkpoint on the surface is available, use an airborne checkpoint designated by the Administrator, or outside the United States, by appropriate authority (the maximum permissible bearing error is plus or minus 6 degrees); or

(4) If no check signal or point is available, while in flight—

(i) Select a VOR radial that lies along the centerline of an established VOR airway;

(ii) Select a prominent ground point along the selected radial preferably more than 20 nautical miles from the VOR ground facility and maneuver the aircraft directly over the point at a reasonably low altitude; and

(iii) Note the VOR bearing indicated by the receiver when over the ground point (the maximum permissible variation between the published radial and the indicated bearing is 6 degrees).

(c) If dual system VOR (units independent of each other except for the antenna) is installed in the aircraft, the person checking the equipment may check one system against the other in place of the check procedures specified in paragraph (b) of this section. Both systems shall be tuned to the same VOR ground facility and note the indicated bearings to that station. The maximum permissible variation between the two indicated bearings is 4 degrees.

(d) Each person making the VOR operational check as specified in paragraph (b) or (c) of this section shall enter the date, place, bearing error, and sign the aircraft log or other record. In addition, if a test signal radiated by a repair station, as specified in paragraph (b) (1) of this section, is used, an entry must be made in the aircraft log or other record by the repair station certificate holder or the certificate holder's representative certifying to the bearing transmitted by the repair station for the check and the date of transmission.

91.173 ATC Clearance and Flight Plan Required

No person may operate an aircraft in controlled airspace under IFR unless that person—
(a) Has filed an IFR flight plan; and
(b) Has received an appropriate ATC clearance.

91.175 Takeoff and Landing under IFR

(a) *Instrument approaches to civil airports.*
Unless otherwise authorized by the Administrator, when an instrument letdown to a civil airport is necessary, each person operating an aircraft, except a military aircraft of the United States, shall use a standard instrument approach procedure prescribed for the airport in Part 97 of this chapter.

(b) *Authorized DH or MDA.* For the purpose of this section, when the approach procedure being used provides for and requires use of a DH or MDA, the authorized DH or MDA is the highest of the following:
(1) The DH or MDA prescribed by the approach procedure;
(2) The DH or MDA prescribed for the pilot in command;
(3) The DH or MDA for which the aircraft is equipped.

(c) *Operation below DH or MDA.* Where a DH or MDA is applicable, no pilot may operate an aircraft, except a military aircraft of the United States, at any airport below the authorized MDA or continue an approach below the authorized DH unless—
(1) The aircraft is continuously in a position from which a descent to a landing on the intended runway can be made at a normal rate of descent using normal maneuvers, and for operations conducted under Part 121 or Part 135 unless that descent rate will allow touchdown to occur within the touchdown zone of the runway of intended landing;
(2) The flight visibility is not less than the visibility prescribed in the standard instrument approach being used; and
(3) Except for a Category II or Category III approach where any necessary visual reference requirements are specified by the Administrator, at least one of the following visual references for the intended runway is distinctly visible and identifiable to the pilot:

Appendix B

(i) The approach light system, except that the pilot may not descend below 100 feet above the touchdown zone elevation using the approach lights as a reference unless the red terminating bars or the red side row bars are also distinctly visible and identifiable.

(ii) The threshold.

(iii) The threshold markings.

(iv) The threshold lights.

(v) The runway end identifier lights.

(vi) The visual approach slope indicator.

(vii) The touchdown zone or touchdown zone markings.

(viii) The touchdown zone lights.

(ix) The runway or runway markings.

(d) *Landing.* No pilot operating an aircraft, except a military aircraft of the United States, may land that aircraft when the flight visibility is less than the visibility prescribed in the standard instrument approach procedure being used.

(e) *Missed approach procedures.* Each pilot operating an aircraft, except a military aircraft of the United States, shall immediately execute an appropriate missed approach procedure when either of the following conditions exist:

(1) Whenever the requirements of paragraph (c) of this section are not met at either of the following times:

(i) When the aircraft is being operated below MDA; or

(ii) Upon arrival at the missed approach point, including a DH where a DH is specified and its use is required, and at any time after that until touchdown.

(2) Whenever an identifiable part of the airport is not distinctly visible to the pilot during a circling maneuver at or above MDA, unless the inability to see an identifiable part of the airport results only from a normal bank of the aircraft during the circling approach.

(f) *Civil airport takeoff minimums.* Unless otherwise authorized by the Administrator, no pilot operating an aircraft under Part 121, 125, 127, 129, or 135 of this chapter may take off from a civil airport under IFR unless weather conditions are at or above the weather minimums for IFR takeoff prescribed for that airport under Part 97 of this chapter. If takeoff minimums are not prescribed under Part 97 of this chapter for a particular airport, the following minimums apply to takeoffs under IFR for aircraft operating under those Parts:

(1) For aircraft, other than helicopters, having two engines or less—1 statute mile visibility.

(2) For aircraft having more than two engines—$\frac{1}{2}$ statute mile visibility.

(3) For helicopters—$\frac{1}{2}$ statute mile visibility.

(g) *Military airports.* Unless otherwise prescribed by the Administrator, each person operating a civil aircraft under IFR into or out of a military airport shall comply with the instrument approach procedures and the takeoff and landing minimum prescribed by the military authority having jurisdiction of that airport.

(h) *Comparable values of RVR and ground visibility.*

(1) Except for Category II or Category III minimums, if RVR minimums for takeoff or landing are prescribed in an instrument approach procedure, but RVR is not

reported for the runway of intended operation, the RVR minimum shall be converted to ground visibility in accordance with the table in paragraph (h) (2) of this section and shall be the visibility minimum for takeoff or landing on that runway.

(2)

RVR (feet)	Visibility (statute miles)
1,600	¼
2,400	½
3,200	⅝
4,000	¾
4,500	⅞
5,000	1
6,000	1-¼

(i) *Operations on unpublished routes and use of radar in instrument approach procedures.* When radar is approved at certain locations for ATC purposes, it may be used not only for surveillance and precision radar approaches, as applicable, but also may be used in conjunction with instrument approach procedures predicted on other types of radio navigational aids. Radar vectors may be authorized to provide course guidance through the segments of an approach to the final course or fix. When operating on an unpublished route or while being radar vectored, the pilot, when an approach clearance is received, shall, in addition to complying with 91.177, maintain the last altitude assigned to that pilot until the aircraft is established on a segment of a published route or instrument approach procedure unless a different altitude is assigned by ATC. After the aircraft is so established, published altitudes apply to descent within each succeeding route or approach segment unless a different altitude is assigned by ATC. Upon reaching the final approach course or fix, the pilot may either complete the instrument approach in accordance with a procedure approved for the facility or continue a surveillance or precision radar approach to a landing.

(j) *Limitation on procedure turns.* In the case of a radar vector to a final approach course or fix, a timed approach from a holding fix, or an approach for which the procedure specifies "No PT," no pilot may make a procedure turn unless cleared to do so by ATC.

(k) *ILS components.* The basic ground components of an ILS are the localizer, glideslope, outer marker, middle marker, and, when installed for use with Category II or Category III instrument approach procedures, an inner marker. A compass locator or precision radar may be substituted for the outer or middle marker. DME, VOR, or nondirectional beacon fixes authorized in the standard instrument procedure or surveillance radar may be substituted for the outer marker. Applicability of, and substitution for, the inner marker for Category II or III approaches is determined by the appropriate Part 97 approach procedure, letter of authorization, or operations specification pertinent to the operations.

91.177 Minimum Altitudes for IFR Operations

(a) *Operation of aircraft at minimum altitudes.*

Except when necessary for takeoff or landing, or unless otherwise authorized by the Administrator, no person may operate an aircraft under IFR below—

(1) The applicable minimum altitudes prescribed in Parts 95 and 97 of this chapter; or

(2) If no applicable minimum altitude is prescribed in those Parts—

(i) In the case of operations over an area designated as a mountainous area in Part 95, an altitude of 2,000 feet above the highest obstacle within a horizontal distance of 4 nautical miles from the course to be flown; or

(ii) In any other case, an altitude of 1,000 feet above the highest obstacle within a horizontal distance of 4 nautical miles from the course to be flown. However, if both a MEA and a MOCA are prescribed for a particular route or route segment, a person may operate an aircraft below the MEA down to, but not below, the MOCA, when within 22 nautical miles of the VOR concerned (based on the pilot's reasonable estimate of that distance).

(b) *Climb.* Climb to a higher minimum IFR altitude shall begin immediately after passing the point beyond which that minimum altitude applies, except that, when ground obstructions intervene, the point beyond which the higher minimum altitude applies shall be crossed at or above the applicable MCA.

91.179 IFR Cruising Altitude or Flight Level

(a) *In controlled airspace.* Each person operating an aircraft under IFR in level cruising flight in controlled airspace shall maintain the altitude or flight level assigned that aircraft by ATC. However, if the ATC clearance assigns "VFR conditions on top," that person shall maintain altitude or flight level as prescribed by 91.159.

(b) *In uncontrolled airspace.* Except while holding in a holding pattern of 2 minutes or less, or while turning, each person operating an aircraft under IFR in level cruising flight in uncontrolled airspace, shall maintain an appropriate altitude as follows:

(1) When operating below 18,000 feet MSL and—

(i) On a magnetic course of zero degrees through 179 degrees, any odd thousand foot MSL altitude (such as 3,000, 5,000, or 7,000; or

(ii) On a magnetic course of 180 degrees through 359 degrees, any even thousand foot MSL altitude (such as 2,000, 4,000, or 6,000).

91.181 Course to Be Flown

Unless otherwise authorized by ATC, no person may operate an aircraft within controlled airspace, under IFR, except as follows:

(a) On a Federal airway, along the centerline of that airway.

(b) On any other route, along the direct course between the navigational aids or fixes defining that route. However, this section does not prohibit maneuvering the aircraft to pass well clear of other air traffic or the maneuvering of the aircraft in VFR conditions to clear the intended flight path both before and during climb or descent.

91.183 IFR Radio Communications

The pilot in command of each aircraft operated under IFR in controlled airspace shall have a continuous watch maintained on the appropriate frequency and shall report by radio as soon as possible—

(a) The time and altitude of passing each designated reporting point, or the reporting points specified by ATC, except that while the aircraft is under radar control, only the passing of those reporting points specifically requested by ATC need be reported;

(b) Any unforecast weather conditions encountered; and

(c) Any other information relating to the safety of flight.

91.185 IFR Operations: Two-Way Radio Communications Failure

(a) *General.* Unless otherwise authorized by ATC, each pilot who has two-way radio communications failure when operating under IFR shall comply with the rules of this section.

(b) *VFR conditions.* If the failure occurs in VFR conditions, or if VFR conditions are encountered after the failure, each pilot shall continue the flight under VFR and land as soon as practicable.

(c) *IFR conditions.* If the failure occurs in IFR conditions, or if paragraph (b) of this section cannot be complied with, each pilot shall continue the flight according to the following:

(1) *Route.*

(i) By the route assigned in the last ATC clearance received;

(ii) If being radar vectored, by the direct route from the point of radio failure to the fix, route, or airway specified in the vector clearance;

(iii) In the absence of an assigned route, by the route that ATC has advised may be expected in a further clearance; or

(iv) In the absence of an assigned route or a route that ATC has advised may be expected in a further clearance, by the route filed in the flight plan.

(2) *Altitude.* At the highest of the following altitudes or flight levels for the route segment being flown:

(i) The altitude or flight level assigned in the last ATC clearance received;

(ii) The minimum altitude (converted, if appropriate, to minimum flight level as prescribed in 91.121 (c) for IFR operations; or

(iii) The altitude or flight level ATC has advised may be expected in a further clearance.

(3) *Leave clearance limit.*

(i) When the clearance limit is a fix from which an approach begins, commence descent or descent and approach as close as possible to the expect further clearance time if one has been received, or if one has not been received, as close as possible to the estimated time of arrival as calculated from the filed or amended (with ATC) estimated time en route.

(ii) If the clearance limit is not a fix from which an approach begins, leave the clearance limit at the expect further clearance time if one has been received, or if none has been received, upon arrival over the clearance limit, and proceed to a fix from which an approach begins and commence descent or descent and approach as close as possible to the estimated time of arrival as calculated from the filed or amended (with ATC) estimated time en route.

91.187 Operation under IFR in Controlled Airspace; Malfunction Reports

(a) The pilot in command of each aircraft operated in controlled airspace under IFR, shall report as soon as practical to ATC any malfunctions of navigational, approach, or communications equipment occurring in flight.

(b) In each report required by paragraph (a) of this section, the pilot in command shall include the—

(1) Aircraft identification;

(2) Equipment affected;

(3) Degree to which the capability of the pilot to operate under IFR in the ATC system is impaired; and

(4) Nature and extent of assistance he desires from ATC.

91.205 Powered Civil Aircraft with Standard Category U.S. Airworthiness Certificates; Instruments and Equipment

(b) *Visual flight rules (day).* For VFR flight during the day the following instruments and equipment are required:

(1) Airspeed indicator.

(2) Altimeter.

(3) Magnetic direction indicator.

(4) Tachometer for each engine.

(5) Oil pressure gauge for each engine using pressure system.

(6) Temperature gauge for each liquid-cooled engine.

(7) Oil temperature gauge for each air-cooled engine.

(8) Manifold pressure gauge for each altitude engine.

(9) Fuel gauge indicating the quantity of fuel in each tank.

(10) Landing gear position indicator, if the aircraft has a retractable landing gear.

(11) For small civil aircraft certificated after March 11, 1996, in accordance with Part 23 of this chapter, an approved aviation red or aviation white anticollision light system. In the event of the failure of any light of the anticollision light system, operation of the aircraft may continue to a location where repairs or replacement can be made.

(12) If the aircraft is operated for hire over water and beyond the power-off gliding distance from shore, approved flotation gear readily available to each occupant and at least one pyrotechnic signaling device.

(13) An approved safety belt with an approved metal-to-metal latching device for each occupant 2 years of age or older.

(14) For small civil airplanes manufactured after July 18, 1978, an approved shoulder harness for each front seat.

(15) An emergency locator transmitter, if required by FAR 91.207.

(c) *Visual flight rules (night).* For VFR flight at night the following instruments and equipment are required:

(1) Instruments and equipment specified in paragraph (b) of this section.

(2) Approved position lights.

(3) An approved aviation red or aviation white anticollision light system on all U.S. registered civil aircraft.

(4) If the aircraft is operated for hire, one electric landing light.

(5) An adequate source of electrical energy for all installed electrical and radio equipment.

(6) One spare set of fuses, or three spare fuses of each kind required that are accessible to the pilot in flight.

(d) *Instrument flight rules.* For IFR flight the following instruments and equipment are required:

(1) Instruments and equipment specified in paragraph (b) of this section and for night flight, instruments and equipment specified in paragraph (c) of this section.

(2) Two-way radio communications system and navigational equipment appropriate to the ground facilities to be used.

(3) Gyroscopic rate-of-turn indicator.

(4) Slip-skid indicator.

(5) Sensitive altimeter adjustable for barometric pressure.

(6) A clock displaying hours, minutes, and seconds with a sweep-second pointer or digital presentation.

(7) Generator or alternator of adequate capacity.

(8) Gyroscopic bank and pitch indicator (artificial horizon).

(9) Gyroscopic direction indicator (directional gyro or equivalent).

(e) *Flight at and above 24,000 feet MSL.* If VOR navigational equipment is required under paragraph (d) (2) of this section, no person may operate a U.S. registered civil aircraft within the 50 states, and the District of Columbia, at or above 24,000 feet MSL unless that aircraft is equipped with approved distance measuring equipment (DME).

Appendix C
Glossary

The terms in this glossary are excerpted from the "Pilot/Controller Glossary" that appears at the end of each edition of the *Aeronautical Information Manual* (AIM). The glossary was compiled by the FAA to promote a common understanding of the terms used in the air traffic control system. It includes those terms that are intended for pilot and controller communications. Be familiar with *all* the terms and definitions in the complete glossary. A competent instrument pilot knows the excerpted definitions and checks new editions of the glossary to review any redefinitions and learn new definitions. (Selected definitions in this book's glossary are edited for style and presentation, but the meaning remains unchanged.)

abbreviated IFR flight plans An authorization by ATC requiring pilots to submit only that information needed for the purpose of ATC. It includes only a small portion of the usual IFR flight plan information. In certain instances, this may be only aircraft identification, location, and pilot request. Other information may be requested if needed by ATC for separation/control purposes. It is frequently used by aircraft that are airborne and desire an instrument approach or by aircraft that are on the ground and desire a climb to VFR-on-top. (Refer to AIM).

advisory frequency The appropriate frequency to be used for airport advisory service. (Refer to Advisory Circular No. 90-42 and AIM).

affirmative Yes.

airport advisory area The area within 10 miles of an airport without a control tower or where the tower is not in operation, and on which a Flight Service Station is located. (Refer to AIM.)

airport lighting Various lighting aids that may be installed on an airport. Types of lighting include:

a. Approach Light System (ALS). An airport lighting facility which provides visual guidance to landing aircraft by radiating light beams in a directional pattern by which the pilot aligns the aircraft with the extended centerline of the runway on his final approach for landing. Condenser-Discharge Sequential Flashing Lights/Sequenced Flashing Lights might be installed in conjunction with the ALS at some airports. Types of Approach Light Systems are:

1. ALSF-1—Approach Light System with Sequenced Flashing Lights in ILS Cat-III configuration.

2. ALSF-2—Approach Light System with Sequenced Flashing Lights in ILS Cat-II configuration., The ALSF-2 may operate as an SSALR when weather conditions permit.

3. SSALF—Simplified Short Approach Light System with Sequenced Flashing Lights.

4. SSALR—Simplified Short Approach Light System with Runway Alignment Indicator Lights.

5. MALSF—Medium Intensity Approach Light System with Sequenced Flashing Lights.

6. MALSR—Medium Intensity Approach Light System with Runway Alignment Indicator Lights . . . .

7. LDIN—Lead-in-light system: Consists of one or more series of flashing lights installed in or near ground level that provides positive visual guidance along an approach path, either curving or straight, where special problems exist with hazardous terrain, obstructions, or noise abatement procedures.

8. RAIL—Runway Alignment Indicator Lights: Sequenced flashing lights that are installed only in combination with other light systems.

9. ODALS—Omidirectional Approach Lighting System consists of seven omnidirectional flashing lights located in the approach area of a nonprecision runway. Five lights are located on the runway centerline extended with the first light located 300 feet from the threshold and extending at equal intervals up to 1,500 feet from the threshold. The other two lights are located, one on each side of the runway threshold, at a lateral distance of 40 feet from the runway edge, or 75 feet from the runway edge when installed on a runway equipped with VASI...

b. Runway Lights/Runway Edge Lights. Lights having a prescribed angle of emission used to define the lateral limits of a runway. Runway lights are uniformly spaced at intervals of approximately 200 feet, and the intensity may be controlled or preset.

c. Touchdown Zone Lighting. Two rows of transverse light bars located symmetrically about the runway centerline normally at 100 foot intervals. The basic system extends 3,000 along the runway.

d. Runway Centerline Lighting. Flush centerline lights spaced at 50-foot intervals beginning 75 feet from the landing threshold and extending to within 75 feet of the opposite end of the runway.

e. Threshold Lights. Fixed green lights arranged symmetrically left and right of the runway centerline, identifying the runway threshold.

f. Runway End Identifier Lights (REIL). Two synchronized flashing lights, one on each side of the runway threshold, which provide rapid and positive identification of the approach end of a particular runway.

g. Visual Approach Slope Indicator (VASI)—An airport lighting facility providing vertical approach slope guidance to aircraft during approach to landing by radiating a directional pattern of high intensity red and white focused light beams which indicate to the pilot that he is "on path" if he sees red/white, "above path" if white/white, and "below path" if red/red. Some airports serving large aircraft have three-bar VASIs that provide two visual glide paths to the same runway.

Airport Surveillance Radar/ASR Approach control radar used to detect and display an aircraft's position in the terminal area. ASR provides range and azimuth information but does not provide elevation data. Coverage of the ASR can extend up to 60 miles.

Air Route Surveillance Radar/ARSR Air Route Traffic Control Center (ARTCC) radar used primarily to detect and display an aircraft's position while en route between terminal areas. The ARSR enables controllers to provide radar air traffic control service when aircraft are within the ARSR coverage. In some instances, ARSR may enable an ARTCC to provide terminal radar services similar to but usually more limited than those provided by a radar approach control.

Air Route Traffic Control Center/ARTCC A facility established to provide air traffic control service to aircraft operating on IFR flight plans within controlled airspace and principally during the en route phase of flight. When equipment capabilities and controller workload permit, certain advisory/assistance services may be provided to VFR aircraft. (Refer to AIM.)

Air Traffic Control/ATC A service operated by appropriate authority to promote the safe, orderly and expeditious flow of air traffic.

alternate airport An airport at which an aircraft may land if a landing at the intended airport becomes inadvisable.

approach clearance Authorization by ATC for a pilot to conduct an instrument approach. The type of instrument approach for which a clearance and other pertinent information is provided in the approach clearance when required. (Refer to AIM and FAR Part 91.)

approach control facility A terminal ATC facility that provides approach control service in a terminal area.

ATC clears Used to prefix an ATC clearance when it is relayed to an aircraft by other than an air traffic controller.

ATC instructions Directives issued by air traffic control for the purpose of requiring a pilot to take specific actions: e.g., "Turn left heading two five zero," "Go around," "Clear the runway." (Refer to FAR Part 91.)

automatic altitude reporting That function of a transponder which responds to Mode C interrogations by transmitting the aircraft's altitude in 100-foot increments.

automatic direction finder/ADF An aircraft radio navigation system that senses and indicates the direction to a L/MF nondirectional radio beacon (NDB) ground transmitter. Direction is indicated to the pilot as a magnetic bearing or as a relative bearing to the longitudinal axis of the aircraft depending on the type of indicator installed in the aircraft. In certain applications, such as military, ADF operations may be based on airborne and ground transmitters in the VHF/UHF frequency spectrum.

automatic terminal information service/ATIS The continuous broadcast of recorded noncontrol information in selected terminal areas. Its purpose is to improve controller effectiveness and to relieve frequency congestion by automating the repetitive transmission of essential but routine information: e.g., "Los Angeles information Alfa. One three zero zero Coordinated Universal Time. Weather, measured ceiling two thousand overcast, visibility three, haze, smoke, temperature seven one, dew point five seven, wind two five zero at five, altimeter two niner niner six. I-L-S Runway Two Five Left approach in use, Runway Two Five Right closed, advise you have Alfa." (Refer to AIM.)

Aviation Weather Service A service provided by the National Weather Service (NWS) and FAA that collects and disseminates pertinent weather information for pilots, aircraft operators, and ATC. Available aviation weather reports and forecasts are displayed at each NWS office and FAA FSS. (Refer to AIM.)

center weather advisory/CWA An unscheduled weather advisory issued by Center Weather Service Unit meteorologists for ATC use to alert pilots of existing or anticipated adverse weather conditions within the next 2 hours. A CWA may modify or redefine a SIGMET. (Refer to AIM.)

circle-to-land maneuver/circling maneuver a maneuver initiated by the pilot to align the aircraft with a runway for landing when a straight-in landing from an instrument approach is not possible or is not desirable. This maneuver is made only after ATC authorization has been obtained and the pilot has established required visual reference to the airport. (Refer to AIM.)

clearance limit The fix, point, or location to which an aircraft is cleared when issued an air traffic clearance.

clearance void if not off by (time) Used by ATC to advise an aircraft that the departure clearance is automatically canceled if takeoff is not made prior to a specified time. The pilot must obtain a new clearance or cancel the IFR flight plan if not off by the specified time.

cleared as filed Means the aircraft is cleared to proceed in accordance with the route of flight filed in the flight plan. This clearance does not include the altitude, SID, or SID transition. (Refer to AIM.)

cleared for (type of) approach ATC authorization for an aircraft to execute a specific instrument approach procedure to an airport: e.g., "Cleared for ILS Runway Three Six Approach." (Refer to AIM, FAR Part 91.)

cleared approach ATC authorization for an aircraft to execute any standard or special instrument approach procedure for that airport. Normally, an aircraft will be cleared for a specific instrument approach procedure. (Refer to AIM, FAR Part 91.)

codes The number assigned to a particular multiple pulse reply signal transmitted by a transponder.

common traffic advisory frequency/CTAF A frequency designed for the purpose of carrying out airport advisory practices while operating to or from an uncontrolled airport. The CTAF may be a UNICOM, multicom, FSS, or tower frequency and is identified in appropriate aeronautical publications. (Refer to AC 90-42C.)

compass locator A low power, low or medium frequency (L/MF) radio beacon installed at the site of the outer or middle marker of an instrument landing system (ILS). It can be used for navigation at distances of approximately 15 miles or as authorized in the approach procedure.

    a. Outer compass locator/LOM—A compass locator installed at the site of the outer marker of an instrument landing system.

    b. Middle compass locator/LMM—A compass locator installed at the site of the middle marker of an instrument landing system.

contact approach An approach wherein an aircraft on an IFR flight plan, having an air traffic control authorization, operating clear of clouds with at least 1 mile flight visibility and a reasonable expectation of continuing to the destination airport in those conditions, may deviate from the instrument approach procedure and proceed to the destination airport by visual reference to the surface. This approach will only be authorized when requested by the pilot and the reported ground visibility at the destination airport is at least 1 statute mile. (Refer to AIM.)

convective SIGMET/WST/convective significant meteorological information A weather advisory concerning convective weather significant to the safety of all aircraft. Convective SIGMETs are issued for tornadoes, lines of thunderstorms, embedded thunderstorms of any intensity level, areas of thunderstorms greater than or equal to VIP level 4 with an aerial coverage of $^4/_{10}$ (40 percent) or more, and hail $^3/_4$ inch or greater. (Refer to AIM.)

cruise Used in an ATC clearance to authorize a pilot to conduct flight at any altitude from the minimum IFR altitude up to and including the altitude specified in the clearance. The pilot may level off at any intermediate altitude within this block of airspace. Climb/descent within the block is to be made at the discretion of the pilot; however, once the pilot starts descent and verbally reports leaving an altitude in the block, the pilot may not return to that altitude without additional ATC clearance. Further, it is approval for the pilot to proceed to and make an approach at destination airport.

decision height/DH With respect to the operation of aircraft, means the height at which a decision must be made during an ILS or PAR instrument approach to either continue the approach or to execute a missed approach.

delay indefinite (reason if known) expect further clearance (time) Used by ATC to inform a pilot when an accurate estimate of the delay time and the reason for the delay cannot immediately be determined: e.g., a disabled aircraft on the runway, terminal or center area saturation, weather below landing minimums, etc.

departure control A function of an approach control facility providing air traffic control service for departing IFR and, under certain conditions, VFR aircraft. (Refer to AIM.)

deviations

    a. A departure from a current clearance, such as an off-course maneuver to avoid weather or turbulence.

    b. Where specifically authorized in the FARs and requested by the pilot, ATC may permit pilots to so deviate from certain regulations. (Refer to AIM.)

Direct Straight line flight between two navigational aids, fixes, points, or any combination thereof. When used by pilots in describing off-airway routes, points defining direct route segments become compulsory reporting points unless the aircraft is under radar contact.

displaced threshold A threshold that is located at a point on the runway other than the designated beginning of the runway.

distance measuring equipment/DME Equipment (airborne and ground) used to measure, in nautical miles, the slant range distance of an aircraft from the DME navigational aid.

emergency locator transmitter/ELT A radio transmitter attached to the aircraft structure that operates from its own power source on 121.5 MHz and 243.0 MHz. It aids in locating downed aircraft by radiating a downward sweeping audio tone, 2–4 times per second. It is designed to function without human action after an accident. (Refer to FAR Part 91, AIM.)

en route air traffic control services Air traffic control service provided aircraft on IFR flight plans, generally by centers, when these aircraft are operating between departure and destination terminal areas. When equipment, capabilities, and controller workload permit, certain advisory/assistance services may be provided to VFR aircraft. (Refer to AIM.)

en route flight advisory service/EFAS A weather service specifically designed to provide, upon pilot request, timely weather information pertinent to the type of flight, intended route of flight, and altitude. The FSSs providing this service are listed in the Airport/Facility Directory. (Refer to AIM.)

expect (altitude) at (time) or (fix) Used under certain conditions to provide a pilot with an altitude to be used in the event of two-way communications failure. It also provides altitude information to assist the pilot in planning. (Refer to AIM.)

expected departure clearance time/EDCT The runway release time assigned to an aircraft in a controlled departure time program and shown on the flight progress strip as an EDCT.

expect further clearance (time)/EFC The time a pilot can expect to receive clearance beyond a clearance limit.

feeder route A route depicted on instrument approach procedure charts to designate routes for aircraft to proceed from the en route structure to the initial approach fix (IAF).

final approach fix/FAF The fix from which the final approach (IFR) to an airport is executed and which identifies the beginning of the final approach segment. It is designated on government charts by the Maltese cross symbol for nonprecision approaches and the lightning bolt symbol for precision approaches; or when ATC directs a lower-than-published glideslope/path intercept altitude, it is the resultant actual point of the glideslope/path intercept.

final approach-IFR The flight path of an aircraft which is inbound to an airport on a final instrument approach course, beginning at the final approach fix or point and extending to the airport or the point where a circle-to-land maneuver or a missed approach is executed.

Flight Service Station/FSS Air traffic facilities which provide pilot briefing, en route communications and VFR search and rescue services, assist lost aircraft and aircraft in emergency situations, relay ATC clearances, originate Notices to Airmen, broadcast aviation weather and NAS information, receive and process IFR flight plans, and monitor navaids. In addition, at selected locations, FSSs provide en route flight advisory service (Flight Watch), take weather observations, issue airport advisories, and advise customs and immigration of transborder flights. (Refer to AIM.)

Flight Standards District Office/FSDO An FAA field office serving an assigned geographical area and staffed with flight standards personnel who serve the aviation industry and the general public on matters relating to the certification and operation of air carrier and general aviation aircraft. Activities include general surveillance of operational safety, certification of airmen and aircraft, accident prevention, investigation, enforcement, etc.

flight test A flight for the purpose of:
  a. Investigating the operation/flight characteristics of an aircraft or aircraft component.
  b. Evaluating an applicant for a pilot certificate or rating.

flight watch A shortened term for use in air-ground contacts to identify the FSS providing en route flight advisory service; e.g., "Oakland Flight Watch."

gate hold procedures Procedures at selected airports to hold aircraft at the gate or other ground location whenever departure delays exceed or are anticipated to exceed 15 minutes. The sequence for departure will be maintained in accordance with initial call-up unless modified by flow control restrictions. Pilots should monitor the ground control/clearance delivery frequency for engine startup advisories or new proposed start time if the delay changes.

glideslope/glidepath Provides vertical guidance for aircraft during approach and landing. The glideslope/glidepath is based on the following:
  a. Electronic components emitting signals which provide vertical guidance by reference to airborne instruments during instrument approaches such as ILS/MLS.
  b. Visual ground aids, such as VASI, which provide vertical guidance for a VFR approach or for the visual portion of an instrument approach and landing.

c. **PAR.** Used by ATC to inform an aircraft making a PAR approach of its vertical position (elevation) relative to the descent profile.

glideslope/glidepath intercept altitude The minimum altitude to intercept the glideslope/path on a precision approach. The intersection of the published intercept altitude with the glideslope/path, designated on government charts by the lightning bolt symbol, is the precision FAF; however, when ATC directs a lower altitude, the resultant lower intercept position is then the FAF.

global positioning system (GPS) A space-base radio positioning, navigation, and time-transfer system. The system provides highly accurate position and velocity information, and precise time, on a continuous global basis, to an unlimited number of properly equipped users. The system is unaffected by weather, and provides a worldwide common grid reference system. The GPS concept is predicated upon accurate and continuous knowledge of the spatial position of each satellite in the system with respect to time and distance from a transmitting satellite to the user. The GPS receiver automatically selects appropriate signals from the satellites in view and translates these into a three-dimensional position, velocity, and time. System accuracy for civil users is normally 100 meters horizontally.

ground delay The amount of delay attributed to ATC, encountered prior to departure, usually associated with a CDT program.

ground speed The speed of an aircraft relative to the surface of the earth.

handoff An action taken to transfer the radar identification of an aircraft from one controller to another if the aircraft will enter the receiving controller's airspace and radio communications with the aircraft will be transferred.

hazardous inflight weather advisory service (HIWAS) Continuous recorded hazardous inflight weather forecasts broadcast to airborne pilots over selected VOR outlets defined as a HIWAS broadcast area.

height above airport/HAA The height of the minimum descent altitude above the published airport elevation. This is published in conjunction with circling minimums.

height above landing/HAL The height above a designated helicopter landing area used for helicopter instrument approach procedures. (Refer to FAR Part 97.)

height above touchdown/HAT The height of the decision height or minimum descent altitude above the highest runway elevation in the touchdown zone (first 3,000 feet of the runway). HAT is published on instrument approach charts in conjunction with all straight-in minimums. (*See* decision height, minimum descent altitude.)

hold/holding procedure A predetermined maneuver which keeps aircraft within a specified airspace while awaiting further clearance from air traffic control. Also used during ground operations to keep aircraft within a specified area or at a specified point while awaiting further clearance from air traffic control. (Refer to AIM.)

holding fix A specified fix identifiable to a pilot by navaids or visual reference to the ground used as a reference point in establishing and maintaining the position of an aircraft while holding. (Refer to AIM.)

hold for release Used by ATC to delay an aircraft for traffic management reasons; i.e., weather, traffic volume. etc. Hold for release instructions (including departure delay information) are used to inform a pilot or a controller (either directly or through an authorized relay) that a departure clearance is not valid until a release time or additional instructions have been received.

homing Flight toward a NAVAID, without correcting for wind, by adjusting the aircraft heading to maintain a relative bearing of zero degrees.

IFR aircraft/IFR flight An aircraft conducting flight in accordance with instrument flight rules.

IFR conditions Weather conditions below the minimum for flight under visual flight rules.

IFR takeoff minimums and departure procedures FAR Part 91 prescribes standard takeoff rules for certain civil users. At some airports, obstructions or other factors require the establishment of nonstandard takeoff minimums, departure procedures, or both to assist pilots in avoiding obstacles during climb to the minimum en route altitude. Those airports are listed in NOS/DOD instrument approach charts (IAPs) under a section entitled "IFR Takeoff Minimums and Departure Procedures." The NOS/DOD IAP chart legend illustrates the symbol used to alert the pilot to nonstandard takeoff minimums and departure procedures. When departing IFR from such airports or from any airports where there are no departure procedures, SIDs, or ATC facilities available, pilots should advise ATC of any departure limitations. Controllers may query a pilot to determine acceptable departure directions, turns, or headings after takeoff. Pilots should be familiar with the departure procedures and must assure that their aircraft can meet or exceed any specified climb gradients.

ILS Categories

1. ILS Category I. An ILS approach procedure which provides for approach to a height above touchdown of not less than 200 feet and with runway visual range of not less than 1800 feet.

2. ILS Category II. An ILS approach procedure which provides for approach to a height above touchdown of not less than 100 feet and with runway visual range of not less than 1200 feet.

3. ILS Category III:
    a. IIIA—An ILS approach procedure that provides for approach without a decision height minimum and with runway visual range of not less than 700 feet.
    b. IIIB—An ILS approach procedure that provides for approach without a decision height minimum and with runway visual range of not less than 150 feet.
    c. IIIC—An ILS approach procedure that provides for approach without a decision height minimum and without runway visual range minimum.

initial approach fix/IAF The fixes depicted on instrument approach procedure charts that identify the beginning of the initial approach segment(s).

instrument approach procedure/IAP/instrument approach A series of predetermined maneuvers for the orderly transfer of an aircraft under instrument flight conditions from the beginning of the initial approach to a landing or to a point from which a landing may be made visually. It is prescribed and approved for a specific airport by competent authority. (Refer to FAR Part 91, AIM.)

    a. U.S. civil standard instrument approach procedures are approved by the FAA as prescribed under FAR Part 97 and are available for public use.

    b. U.S. military standard instrument approach procedures are approved and published by the Department of Defense.

    c. Special instrument approach procedures are approved by the FAA for individual operators but are not published in FAR Part 97 for public use.

instrument flight rules/IFR Rules governing the procedures for conducting instrument flight. Also a term used by pilots and controllers to indicate type of flight plan. (Refer to AIM.)

Instrument Landing System/ILS A precision instrument approach system which normally consists of the following electronic components and visual aids:

    a. localizer (*See* localizer.)

    b. glideslope (*See* glideslope.)

    c. outer marker (*See* outer marker.)

    d. middle marker (*See* middle marker.)

    e. approach lights (*See* airport lighting.)

(Refer to FAR Part 91, AIM.)

instrument meteorological conditions/IMC Meteorological conditions expressed in terms of visibility, distance from cloud, and ceiling less than the minimum specified for visual meteorological conditions.

instrument runway A runway equipped with electronic and visual navigation aids for which a precision or nonprecision approach procedure having straight-in landing minimums has been approved.

intersection

    a. A point defined by any combination of courses, radials, or bearings of two or more navigational aids.

    b. Used to describe the point where two runways, a runway and a taxiway, or two taxiways cross or meet.

jet route A route designed to serve aircraft operations from 18,000 feet MSL up to and including flight level 450. The routes are referred to as "J" routes with numbering to identify the designated route: e.g., J105. (Refer to FAR Part 71.)

jet stream A migrating stream of high-speed winds present at high altitudes.

known traffic With respect to ATC clearances, means aircraft whose altitude, position, and intentions are known to ATC.

landing minimums The minimum visibility prescribed for landing a civil aircraft while using an instrument approach procedure. The minimum applies with other limitations set forth in FAR Part 91 with respect to the minimum descent altitude (MDA) or decision height (DH) prescribed in the instrument approach procedures as follows:

    a. Straight-in landing minimums. A statement of MDS and visibility, or DH and visibility, required for a straight-in landing on a specified runway, or

    b. Circling minimums. A statement of MDA and visibility required for the circle-to-land maneuver.

Note: Descent below the established MDA or DH is not authorized during an approach unless the aircraft is in a position from which a normal approach to the runway of intended landing can be made and adequate visual reference to required visual cues is maintained. (Refer to FAR Part 91.)

localizer The component of an ILS which provides course guidance to the runway.

localizer type directional aid/LDA A NAVAID used for nonprecision instrument approaches with utility and accuracy comparable to a localizer but which is not part of a complete ILS and is not aligned with the runway. (Refer to AIM.)

localizer usable distance The maximum distance from the localizer transmitter at a specified altitude, as verified by flight inspection, at which reliable course information is continuously received. (Refer to AIM.)

loran An electronic navigational system by which hyperbolic lines of position are determined by measuring the difference in the time of reception of synchronized pulse signals from two fixed transmitters. Loran A operates in the 1750-1950 kHz frequency band. Loran C and D operate in the 100-110 kHz frequency band. (Refer to AIM.)

low-altitude airway structure/federal airways The network of airways serving aircraft operations up to but not including 18,000 feet MSL. (Refer to AIM.)

low-altitude alert system/LAAS An automated function of the TPX-42 that alerts the controller when a Mode C transponder-equipped aircraft on an IFR flight plan is below a predetermined minimum safe altitude. If requested by the pilot, LAAS monitoring is also available to VFR Mode C transponder-equipped aircraft.

low approach An approach over an airport or runway following an instrument approach or a VFR approach including the go-around maneuver where the pilot intentionally does not make contact with the runway. (Refer to AIM.)

marker beacon An electronic navigation facility transmitting a 75 MHz vertical fan or bone shaped radiation pattern. Marker beacons are identified by their modulation frequency and keying code, and when received by compatible airborne equipment, indicate to the pilot, both aurally and visually, that the aircraft is passing over the facility. (Refer to AIM.)

maximum authorized altitude/MAA A published altitude representing the maximum usable altitude or flight level for an airspace structure or route segment. It is the highest altitude on a Federal airway, jet route, area navigation low or high route, or other direct route for which an MEA is designated in FAR Part 95 at which adequate reception of navigation aid signals is assured.

microburst A small downburst with outbursts of damaging winds extending 2.5 miles or less. In spite of its small horizontal scale, an intense microburst could induce winds as high as 150 knots. (Refer to AIM)

middle marker/MM A marker beacon that defines a point along the glideslope of an ILS normally located at or near the point of decision height (ILS Category I). It is keyed to transmit alternate dots and dashes, with the alternate dots and dashes keyed at the rate of 95 dot/dash combinations per minute on a 1300 Hz tone, which is received aurally and visually by compatible airborne equipment. (Refer to AIM.)

minimum crossing altitude/MCA The lowest altitude at certain fixes at which an aircraft must cross when proceeding in the direction of a higher minimum en route IFR altitude (MEA).

minimum descent altitude/MDA The lowest altitude, expressed in feet above mean sea level, to which descent is authorized on final approach or during circle-to-land maneuvering in execution of a standard instrument approach procedure where no electronic glideslope is provided.

minimum en route IFR altitude/MEA The lowest published altitude between radio fixes which assures acceptable navigational signal coverage and meets obstacle clearance requirements between those fixes. The MEA prescribed for a Federal airway or segment thereof, area navigation low or high route, or other direct route applies to the entire width of the airway, segment, or route between the radio fixes defining the airway, segment, or route. (Refer to FAR Parts 91 and 95; AIM.)

minimum fuel Indicates that an aircraft's fuel supply has reached a state where, upon reaching the destination, it can accept little or no delay. This is not an emergency situation but merely indicates an emergency situation is possible should any undue delay occur. (Refer to AIM.)

minimum holding altitude/MHA The lowest altitude prescribed for a holding pattern that assures navigational signal coverage, communications, and meets obstacle clearance requirements.

minimum IFR altitudes/MIA Minimum altitudes for IFR operations as prescribed in FAR Part 91. These altitudes are published on aeronautical charts and prescribed in FAR Part 95 for airways and routes, and in FAR Part 97 for standard instrument approach procedures. If no applicable minimum altitude is prescribed in FAR Parts 95 or 97, the following minimum IFR altitude applies:

   a. In designated mountainous areas, 2000 feet above the highest obstacle within a horizontal distance of 4 nautical miles from the course to be flown; or

   b. Other than mountainous areas, 1000 feet above the highest obstacle within a horizontal distance of 4 nautical miles from the course to be flown; or

   c. As otherwise authorized by the Administrator or assigned by ATC. (Refer to FAR Part 91.)

minimum obstruction clearance altitude/MOCA The lowest published altitude in effect between radio fixes on VOR airways, off-airway routes, or route segments which meets obstacle clearance requirements for the entire route segment and which assures acceptable navigational signal coverage only within 25 statute (22 nautical) miles of a VOR. (Refer to FAR Part 91 and 95.)

minimum reception altitude/MRA The lowest altitude at which an intersection can be determined. (Refer to FAR Part 95.)

minimum safe altitude/MSA

   a. The minimum altitude specified in FAR Part 91 for various aircraft operations.

   b. Altitudes depicted on approach charts which provide at least 1,000 feet of obstacle clearance for emergency use within a specified distance from the navigation facility upon which a procedure is predicated. These altitudes will be identified as minimum sector altitudes or emergency safe altitudes and are established as follows:

1. Minimum Sector Altitudes. Altitudes depicted on approach charts that provide at least 1,000 feet of obstacle clearance within a 25-mile radius of the navigation facility upon which the procedure is predicated. Sectors depicted on approach charts must be at least 90 degrees in scope. These altitudes are for emergency use only and do not necessarily assure acceptable navigational signal coverage.
2. Emergency Safe Altitudes. Altitudes depicted on approach charts that provide at least 1,000 feet of clearance in nonmountainous areas and 2,000 feet of obstacle clearance in designated mountainous areas within a 100-mile radius of the navigation facility upon which the procedure is predicated and normally used only in military procedures. These altitudes are identified on published procedures as "Emergency Safe Altitudes."

minimum safe altitude warning/MSAW A function of the ARTS III computer that aids the controller by alerting when a tracked Mode C-equipped aircraft is below or is predicted by the computer to go below a predetermined minimum safe altitude. (Refer to AIM.)

minimums/minima Weather condition requirements established for a particular operation or type of operation: for example, IFR takeoff or landing, alternate airport for IFR flight plans, VFR flight, etc. (Refer to FAR Part 91, AIM.)

minimum vectoring altitude/MVA The lowest MSL altitude at which an IFR aircraft will be vectored by a radar controller, except as otherwise authorized for radar approaches, departures, and missed approaches. The altitude meets IFR obstacle clearance criteria. It may be lower than the published MEA along an airway or J-route segment. It may be utilized for radar vectoring only upon the controller's determination that an adequate radar return is being received from the aircraft being controlled. Charts depicting minimum vectoring altitudes are normally available only to the controllers and not to pilots. (Refer to AIM.)

missed approach

a. A maneuver conducted by a pilot when an instrument approach cannot be completed to a landing. The route of flight and altitude are shown on instrument approach procedure charts. A pilot executing a missed approach prior to the missed approach point (MAP) must continue along the final approach to the MAP. The pilot may climb immediately to the altitude specified in the missed approach procedure.
b. A term used by the pilot to inform ATC that the missed approach is being executed.
c. At locations where ATC radar service is provided, the pilot should conform to radar vectors when provided by ATC in lieu of the published missed approach procedure. (*See* missed approach point.) (Refer to AIM.)

missed approach point/MAP A point prescribed in each instrument approach procedure at which a missed approach procedure shall be executed if the required visual reference does not exist.

moving target indicator/MTI An electronic device that will permit radar scope presentation only from targets that are in motion. A partial remedy for ground clutter.

National Airspace System/NAS The common network of U.S. airspace; air navigation facilities, equipment and services, airports or landing areas; aeronautical charts, information and services; rules, regulations and procedures, technical information, and manpower and material. Included are system components shared jointly with the military.

navaid classes VOR, VORTAC, and TACAN aids are classified according to their operational use. The three classes of navaids are:

a. T—terminal.

b. L—low altitude.

c. H—high altitude.

The normal service range for T, L, and H class aids is found in the AIM. Certain operational requirements make it necessary to use some of these aids at greater service ranges than specified. Extended range is made possible through flight inspection determinations. Some aids also have lesser service range due to location, terrain, frequency protection, etc. Restrictions to service range are listed in Airport/Facility Directory.

negative "No," or "permission not granted," or "that is not correct."

negative contact Used by pilots to inform ATC that:

a. Previously issued traffic is not in sight. It may be followed by the pilot's request for the controller to provide assistance in avoiding the traffic.

b. They were unable to contact ATC on a particular frequency.

night The time between the end of evening civil twilight and the beginning of morning civil twilight, as published in the *American Air Almanac,* converted to local time.

no gyro approach/vector A radar approach/vector provided in case of a malfunctioning gyro-compass or directional gyro. Instead of providing the pilot with headings to be flown, the controller observes the radar track and issues control instructions "turn right/left" or "stop turn" as appropriate. (Refer to AIM.)

nondirectional beacon/radio beacon/NDB An L/MF or UHF radio beacon transmitting nondirectional signals whereby the pilot of an aircraft equipped with direction finding equipment can determine the bearing to or from the radio beacon and "home" on or track to or from the station. When the radio beacon is installed in conjunction with the instrument landing system marker, it is normally called a compass locator.

nonprecision approach procedure/nonprecision approach A standard instrument approach procedure in which no electronic glideslope is provided: for example, VOR, TACAN, NDB, GPS, LOC, ASR, LDA, or SDF approaches.

nonradar Precedes other terms and generally means without the use of radar, such as:

a. Nonradar Approach. Used to describe instrument approaches for which course guidance on final approach is not provided by ground-based precision or surveillance radar. Radar vectors to the final approach course may or may not be provided by ATC. Examples of nonradar approaches are VOR, NDB, GPS, TACAN, and ILS approaches.

b. Nonradar Approach Control. An ATC facility providing approach control service without the use of radar.

c. Nonradar Arrival. An aircraft arriving at an airport without radar service, or at an airport served by a radar facility and radar contact has not been established or has been terminated due to a lack of radar service at the airport.

d. Nonradar Route. A flight path or route over which pilots perform their own navigation. The pilot may be receiving radar separation, radar monitoring, or other ATC services while on a nonradar route.

e. Nonradar Separation. The spacing of aircraft in accordance with established minima without the use of radar: for example, vertical, lateral or longitudinal separation.

notice to airmen/NOTAM A notice containing information (not known sufficiently in advance to publicize by other means) concerning the establishment, condition, or change in any component (facility, service, or procedure of, or hazard in the National Airspace System) the timely knowledge of which is essential to personnel concerned with flight operations.

a. NOTAM(D). A NOTAM given (in addition to local dissemination) distant dissemination beyond the area of responsibility of the Flight Service Station. These notams will be stored and available until canceled.

b. NOTAM(L). A NOTAM given local dissemination by voice and other means, such as teleautograph and telephone, to satisfy local user requirements.

c. FDC NOTAM. A NOTAM regulatory in nature, transmitted by USNOF and given system wide dissemination.

notices to airmen publication A publication issued every 14 days, designed primarily for the pilot, which contains current NOTAM information considered essential to the safety of flight as well as supplemental data to other aeronautical publications. The contraction NTAP is used in NOTAM text.

on course

a. Used to indicate that an aircraft is established on the route centerline.

b. Used by ATC to advise a pilot making a radar approach that the aircraft is lined up on the final approach course.

on-course indication An indication on an instrument, which provides the pilot a visual means of determining that the aircraft is located on the centerline of a given navigational track, or an indication on a radarscope that an aircraft is on a given track.

option approach An approach requested and conducted by a pilot that will result in either a touch-and-go, missed approach, stop-and-go, or full stop landing. (Refer to AIM.)

outer marker/OM A marker beacon at or near the glideslope intercept altitude of an ILS approach. It is keyed to transmit two dashes per second on a 400 Hz tone, which is received aurally and visually by compatible airborne equipment. The OM is normally located 4 to 7 miles from the runway threshold on the extended centerline of the runway. (Refer to AIM.)

parallel ILS approaches Approaches to parallel runways by IFR aircraft which, when established inbound toward the airport on the adjacent final approach courses, are radar-separated by at least 2 miles.

parallel runways Two or more runways at the same airport whose centerlines are parallel. In addition to runway number, parallel runways are designated as L (left) and R (right) or, if three parallel runways exist, L (left), C (center), and R (right).

pilot weather report/PIREP A report of meteorological phenomena encountered by aircraft in flight. (Refer to AIM.)

practice instrument approach An instrument approach procedure conducted by a VFR or an IFR aircraft for the purpose of pilot training or proficiency demonstrations.

precision approach procedure/precision approach A standard instrument approach procedure in which an electronic glideslope/glidepath is provided: for example, ILS and PAR.

precision approach radar/PAR Radar equipment in some ATC facilities operated by the FAA and/or the military services at joint-use civil/military locations and separate military installations to detect and display azimuth, elevation, and range of aircraft on the final approach course to a runway. This equipment may be used to monitor certain nonradar approaches, but is primarily used to conduct a precision instrument approach (PAR) wherein the controller issues guidance instructions to the pilot based on the aircraft's position in relation to the final approach course (azimuth), the glidepath (elevation), and the distance (range) from the touchdown point on the runway as displayed on the radarscope. (Refer to AIM.)

preferential routes Preferential routes (PDRs, PARs, and PDARs) are adapted in ARTCC computers to accomplish inter/intrafacility controller coordination and to assure that flight data is posted at the proper control positions. Locations having a need for these specific inbound and outbound routes normally publish such routes in local facility bulletins, and their use by pilots minimizes flight plan route amendments. When the workload or traffic situation permits, controllers normally provide radar vectors or assign requested routes to minimize circuitous routing. Preferential routes are usually confined to one ARTCC's area and are referred to by the following names or acronyms:

   a. Preferential departure route/PDR. A specific departure route from an airport or terminal area to an en route point where there is no further need for flow control. It may be included in a standard instrument departure (SID) or a preferred IFR route.

   b. Preferential arrival route/PAR. A specific arrival route from an appropriate en route point to an airport or terminal area. It may be included in a standard terminal arrival (STAR) or a preferred IFR route. The abbreviation PAR is used primarily within the ARTCC and should not be confused with the abbreviation for precision approach radar.

   c. Preferential departure and arrival route/PDAR. A route between two terminals which are within or immediately adjacent to one ARTCC's area. PDARs are not synonymous with preferred IFR routes but may be listed as such as they do accomplish essentially the same purpose.

preferred IFR routes Routes established between busier airports to increase system efficiency and capacity. They normally extend through one or more ARTCC areas and are designed to achieve balanced traffic flows among high density terminals. IFR clearances are issued on the basis of these routes except when severe weather avoidance procedures or other factors dictate otherwise. Preferred IFR routes are listed in the *Airport/Facility Directory*. If a flight is planned to or from an area having such routes but the departure or arrival point is not listed in the *Airport/Facility Directory*, pilots may use that part of a preferred IFR route which is appropriate for the departure or arrival point that is listed. Preferred IFR routes are correlated with SIDs and STARs and may be defined by airways, jet routes, direct routes between navaids, waypoints, NAVAID radials/DME, or any combination thereof. (Refer to *Airport/Facility Directory*.)

procedure turn inbound That point of a procedure turn maneuver where course reversal has been completed and an aircraft is established inbound on the intermediate approach segment or final approach course. A report of "procedure turn inbound" is normally used by ATC as a position report for separation purposes.

procedure turn/PT The maneuver prescribed when it is necessary to reverse direction to establish an aircraft on the intermediate approach segment or final approach course. The outbound course, direction of turn, distance within which the turn must be completed, and minimum altitude are specified in the procedure; however, unless otherwise restricted, the point at which the turn may be commenced and the type and rate of turn are left to the discretion of the pilot.

radar advisory The provision of advice and information based on radar observations.

radar approach An instrument approach procedure which utilizes precision approach radar (PAR) or airport surveillance radar (ASR). (Refer to AIM.)

radar approach control facility A terminal ATC facility that uses radar and nonradar capabilities to provide approach control services to aircraft arriving, departing, or transiting airspace controlled by the facility. Provides radar ATC services to aircraft operating in the vicinity or one of more civil and/or military airports in a terminal area. The facility may provide services of a ground controlled approach (GCA); i.e., ASR and PAR approaches. A radar approach control facility may be operated by FAA, USAF, U.S. Army, USN, USMC, or jointly by FAA and a military service.

radar contact
  a. Used by ATC to inform an aircraft that it is identified on the radar display and radar flight following will be provided until radar identification is terminated. Radar service may also be provided within the limits of necessity and capability. When a pilot is informed of "radar contact," the pilot automatically discontinues reporting over compulsory reporting points. (Refer to AIM.)
  b. The term used to inform the controller that the aircraft is identified and approval is granted for the aircraft to enter the receiving controller's airspace.

radar contact lost Used by ATC to inform a pilot that radar data used to determine the aircraft's position is no longer being received, or is no longer reliable and radar service is no longer being provided. The loss may be attributed to several factors including the aircraft merging with weather or ground clutter, the aircraft operating below radar line of sight coverage, the aircraft entering an area of poor radar return, failure of the aircraft transponder, or failure of the ground radar equipment.

radar environment An area in which radar service may be provided.

radar flight following The observation of the progress of radar identified aircraft, whose primary navigation is being provided by the pilot, wherein the controller retains and correlates the aircraft identity with the appropriate target or target symbol displayed on the radarscope. (Refer to AIM.)

radar identification The process of ascertaining that an observed radar target is the radar return from a particular aircraft.

radar service A term that encompasses one or more of the following services based on the use of radar, which can be provided by a controller to a pilot of a radar-identified aircraft.

a. Radar monitoring. The radar flight-following of aircraft, whose primary navigation is being performed by the pilot, to observe and note deviations from its authorized flight path, airway, or route. When being applied specifically to radar monitoring of instrument approaches; i.e., with precision approach radar (PAR) or radar monitoring of simultaneous ILS approaches, it includes advice and instructions whenever an aircraft nears or exceeds the prescribed PAR safety limit or simultaneous ILS no transgression zone.

b. Radar navigational guidance. Vectoring aircraft to provide course guidance.

c. Radar separation. Radar spacing of aircraft in accordance with established minima.

radar service terminated Used by ATC to inform a pilot that any of the services that could be received while in radar contact will no longer be provided. Radar service is automatically terminated, and the pilot is not advised in the following cases:

a. An aircraft cancels its IFR flight plan, except within Class B airspace, Class C airspace, a TRSA, or where basic radar service is provided.

b. An aircraft conducting an instrument, visual, or contact approach has landed or has been instructed to change to advisory frequency.

c. An arriving VFR aircraft, receiving radar service to a tower-controlled airport within Class B airspace, Class C airspace, a TRSA, or where sequencing service is provided, has landed; or to all other airports, is instructed to change to tower or advisory frequency.

d. An aircraft completes a radar approach.

radar weather echo intensity levels Existing radar systems cannot detect turbulence; however, there is a direct correlation between the degree of turbulence and other weather features associated with thunderstorms and the radar weather echo intensity. The National Weather Service has categorized six levels of radar weather echo intensity. The levels are sometimes expressed during communications as "VIP LEVEL" 1 through 6 (derived from the weather radar's video integrator and processor, which produces the information = Video Integrator and Processor). The following list gives the VIP LEVELS in relation to the precipitation intensity within a thunderstorm:

a. Level 1. WEAK

b. Level 2. MODERATE

c. Level 3. STRONG

d. Level 4. VERY STRONG

e. Level 5. INTENSE

f. Level 6. EXTREME

radial A magnetic bearing extending from a VOR/VORTAC/TACAN navigation facility.

radio altimeter/radar altimeter Aircraft equipment that makes use of the reflection of radio waves from the ground to determine the height of the aircraft above the surface.

radio magnetic indicator/RMI An aircraft navigational instrument coupled with a gyro compass or similar compass that indicates the direction of a selected NAVAID and indicates bearing with respect to the heading of the aircraft.

remote communications air/ground facility/RCAG An unmanned VHF/UHF transmitter/ receiver facility which is used to expand ARTCC air/ground communications coverage and to facilitate direct contact between pilots and controllers. RCAG facilities are sometimes not equipped with emergency frequencies 121.5 MHz and 243.0 MHz. (Refer to AIM.)

remote communications outlet/RCO and remote transmitter/receiver/RTR An unmanned communications facility remotely controlled by air traffic personnel. RCOs serve FSSs. RTRs serve terminal ATC facilities. An RCO or RTR may be UHF or VHF and will extend the communication range of the air traffic facility. There are several classes of RCOs and RTRs. The class is determined by the number of transmitters or receivers. Classes A through G are used primarily for air/ground purposes. RCO and RTR class O facilities are nonprotected outlets subject to undetected and prolonged outages. RCO (Os) and RTR (Os) were established for the express purpose of providing ground-to-ground communications between air traffic control specialists and pilots located at a satellite airport for delivering en route clearances, issuing departure authorizations, and acknowledging instrument flight rules cancellations or departure/landing times. As a secondary function, they may be used for advisory purposes whenever the aircraft is below the coverage of the primary air/ground frequency.

request full route clearance/FRC Used by pilots to request that the entire route of flight be read verbatim in an ATC clearance. Such request should be made to preclude receiving an ATC clearance based on the original filed flight plan when a filed IFR flight plan has been revised by the pilot, company, or operations prior to departure.

resume own navigation Used by ATC to advise pilots to resume their own navigational responsibility. It is issued after completion of a radar vector or when radar contact is lost while the aircraft is being radar vectored.

roger I have received all of your last transmission. It should not be used to answer a question requiring a yes or a no answer.

route A defined path, consisting of one or more courses in a horizontal plane, which aircraft traverse over the surface of the earth. (*See* airway, jet route, published route, unpublished route.)

route segment As used in air traffic control, a part of a route that can be defined by two navigational fixes, two navaids, or a fix and a NAVAID.

say again Used to request a repeat of the last transmission. Usually specifies transmission or portion thereof not understood or received; e.g., "Say again all after Abram VOR."

say altitude Used by ATC to ascertain an aircraft's specific altitude/flight level. When the aircraft is climbing or descending, the pilot should state the indicated altitude rounded to the nearest 100 feet.

say heading Used by ATC to request an aircraft heading. The pilot should state the actual heading of the aircraft.

segments of an instrument approach procedure An instrument approach procedure may have as many as four separate segments depending on how the approach procedure is structured.

  a. Initial approach. The segment between the initial approach fix and the intermediate fix or the point where the aircraft is established on the intermediate course or final approach course.

  b. Intermediate approach. The segment between the intermediate fix or point and the final approach fix.

  c. Final approach. The segment between the final approach fix or point and the runway, airport, or missed approach point.

  d. Missed approach. The segment between the missed approach point or the point of arrival at decision height and the missed approach fix at the prescribed altitude. (Refer to FAR Part 97.)

separation In air traffic control, the spacing of aircraft to achieve their safe and orderly movement in flight and while landing and taking off.

severe weather avoidance plan/SWAP An approved plan to minimize the affect of severe weather on traffic flows in impacted terminal and/or ARTCC areas. SWAP is normally implemented to provide the least disruption to the ATC system when flight through portions of airspace is difficult or impossible due to severe weather.

severe weather forecast alerts/AWW Preliminary messages issued in order to alert users that a severe weather watch bulletin (WW) is being issued. These messages define areas of possible severe thunderstorms or tornado activity. The messages are unscheduled and issued as required by the National Severe Storm Forecast Center at Kansas City, Missouri.

sidestep maneuver A visual maneuver accomplished by a pilot at the completion of an instrument approach to permit a straight-in landing on a parallel runway not more than 1,200 feet to either side of the runway to which the instrument approach was conducted.

SIGMET/WS/significant meteorological information A weather advisory issued concerning weather significant to the safety of all aircraft. SIGMET advisories cover severe and extreme turbulence, severe icing, and widespread dust or sandstorms that reduce visibility to less than 3 miles. (Refer to AIM.)

simplified directional facility/SDF A NAVAID used for nonprecision instrument approaches. The final approach course is similar to that of an ILS localizer except that the SDF course may be offset from the runway, generally not more than 3°, and the course may be wider than the localizer, resulting in a lower degree of accuracy. (Refer to AIM.)

simultaneous ILS approaches An approach system permitting simultaneous ILS approaches to airports having parallel runways separated by at least 4300 feet between centerlines. Integral parts of a total system are ILS, radar, communications, ATC procedures, and appropriate airborne equipment. (Refer to AIM.)

special VFR conditions Meterological conditions that are less than those required for basic VFR flight in Class B, C, D, or E surface areas and in which some aircraft are permitted flight under visual flight rules.

special VFR operations Aircraft operating in accordance with clearances within Class B, C, D, and E surface areas in weather less than the basic VFR weather minima. Such operations must be requested by the pilot and approved by ATC.

standard instrument departure/SID A preplanned instrument flight rule (IFR) air traffic control departure procedure printed for pilot use in graphic and/or textual form. SIDs provide transition from the terminal to the appropriate en route structure. (Refer to AIM.)

standard rate turn A turn of 3 degrees per second.

standard terminal arrival/STAR A preplanned instrument flight rule (IFR) air traffic control arrival procedure published for pilot use in graphic and/or textual form. STARs provide transition from the en route structure to an outer fix or an instrument approach fix/arrival waypoint in the terminal area.

stand by Means the controller or pilot must pause for a few seconds, usually to attend to other duties of a higher priority. Also means to wait as in stand by for clearance. The caller should reestablish contact if a delay is lengthy. "Stand by" is not an approval or denial..

stepdown fix A fix permitting additional descent within a segment of an instrument approach procedure by identifying a point at which a controlling obstacle has been safely overflown.

stop altitude squawk Used by ATC to inform an aircraft to turn-off the automatic altitude reporting feature of its transponder. It is issued when the verbally reported altitude varies 300 feet or more from the automatic altitude report.

sunset and sunrise The mean solar times of sunset and sunrise as published in the *Nautical Almanac*, converted to local standard time for the locality concerned. Within Alaska, the end of evening civil twilight and the beginning of morning civil twilight, as defined for each locality.

surveillance approach An instrument approach wherein the air traffic controller issues instructions, for pilot compliance, based on aircraft position in relation to the final approach course (azimuth), and the distance (range) from the end of the runway as displayed on the controller's radarscope. The controller will provide recommended altitudes on final approach if requested by the pilot. (Refer to AIM.)

target The indication shown on a radar display resulting from a primary radar return or a radar beacon reply.

terminal VFR radar service A national program instituted to extend the terminal radar services provided instrument flight rules (IFR) aircraft to visual flight rules (VFR) aircraft. The program is divided into four types of service referred to as basic radar service, terminal radar service area (TRSA) service, Class B service, and Class C service. The type of service provided at a particular location is contained in the *Airport/Facility Directory*. (Refer to AIM.)

    a. Basic radar service: These services are provided are provided for VFR aircraft by all commissioned terminal radar facilities. Basic radar service includes safety alerts, traffic advisories, limited radar vectoring when requested by the pilot, and sequencing at locations where procedures have been established for this purpose and/or when covered by a letter of agreement. The purpose of this service is to adjust the flow of arriving IFR and VFR aircraft into the traffic pattern in a safe and orderly manner and to provide traffic advisories to departing VFR aircraft.

    b. TRSA Service: This service provides, in addition to basic radar service, sequencing of all IFR and participating VFR aircraft to the primary airport and separation between all participating VFR aircraft. The purpose of this service is to provide separation between all participating VFR aircraft and all IFR aircraft operating within the area defined as a TRSA.

    c. Class C Service: This service provides, in addition to basic radar service, approved separation between IFR and VFR aircraft, and sequencing of VFR arrivals to the primary airport.

    d. Class B Service: This service provides, in addition to basic radar service, approved separation of aircraft based on IFR, VFR, and/or weight, and sequencing of VFR arrivals to the primary airport.

touchdown zone The first 3,000 feet of the runway beginning at the threshold. The area is used for determination of touchdown zone elevation in the development of straight-in landing minimums for instrument approaches.

tower en route control service/tower to tower The control of IFR en route traffic within delegated airspace between two or more adjacent approach control facilities. This service is designed to expedite traffic and reduce control and pilot communication requirements.

track The actual flight path of an aircraft over the surface of the earth.

traffic advisories Advisories issued to alert pilots to other known or observed air traffic that may be in such proximity to the position of intended route of flight of their aircraft to warrant their attention. Such advisories may be based on:

    a. Visual observation.

    b. Observation of radar identified and nonidentified aircraft targets on an ATC radar display, or

    c. Verbal reports from pilots or other facilities.

Note 1: The word "traffic" followed by additional information, if known, is used to provide such advisories: e.g., "Traffic, two o'clock, one zero miles, southbound, eight thousand."

Note 2: Traffic advisory service will be provided to the extent possible depending on higher priority duties of the controller or other limitations: e.g., radar limitations, *volume of traffic, frequency congestion, or controller workload. Radar/nonradar traffic advisories do not relieve the pilot of the responsibility to see and avoid other aircraft. Pilots are cautioned that there are many times when the controller is not able to give traffic advisories concerning all traffic in the aircraft's proximity; in other words, <u>when a pilot requests or is receiving traffic advisories, the pilot should not assume that all traffic will be issued</u>.* (Refer to AIM.)

transfer of control That action whereby the responsibility for the separation of an aircraft is transferred from one controller to another.

transmissometer An apparatus used to determine visibility by measuring the transmission of light through the atmosphere. It is the measurement source for determining runway visual range (RVR) and runway visibility value (RVV). (*See* visibility.)

transmitting in the blind/blind transmission A transmission from one station to other stations in circumstances where two-way communication cannot be established, but where it is believed that the called stations may be able to receive the transmission.

transponder The airborne radar beacon receiver/transmitter portion of the air traffic control radar beacon system (ATCRBS) which automatically receives radio signals from interrogators on the ground, and selectively replies with a specific reply pulse or pulse group only to those interrogations being received on the mode to which it is set to respond. (Refer to AIM.)

T-VOR/terminal-very high frequency omnidirectional range station A very high frequency terminal omnirange station located on or near an airport and used as an approach aid.

unable Indicates inability to comply with a specific instruction, request, or clearance.

under the hood Indicates that the pilot is using a hood to restrict visibility outside the cockpit while simulating instrument flight. An appropriately rated pilot is required in the other control seat while this operation is being conducted. (Refer to FAR Part 91.)

UNICOM A nongovernment communication facility which may provide airport information at certain airports. Locations and frequencies of unicoms are shown on aeronautical charts and publications. (Refer to AIM, *Airport/Facility Directory*.)

vector A heading issued to an aircraft to provide navigational guidance by radar.

verify Request confirmation of information: e.g., "verify assigned altitude."

vertical separation Separation established by assignment of different altitudes or flight levels.

very high frequency/VHF The frequency band between 30 and 300 MHz. Portions of this band, 108 to 118 MHz, are used for certain navaids; 118 to 136 MHz are used for civil air/ground voice communications. Other frequencies in this band are used for purposes not related to air traffic control.

VFR-on-top ATC authorization for an IFR aircraft to operate in VFR conditions at any appropriate VFR altitude (as specified in FAR and as restricted by ATC). A pilot receiving this authorization must comply with the VFR visibility, distance from cloud criteria, and the minimum IFR altitudes specified in FAR Part 91. The use of this term does not relieve controllers of their responsibility to separate aircraft in TCAs as required by FAA Order 7110.65.

Appendix C

visibility The ability, as determined by atmospheric conditions and expressed in units of distance, to see and identify prominent unlighted objects by day and prominent lighted objects by night. Visibility is reported as statute miles, hundreds of feet or meters. (Refer to FAR Part 91, AIM.)

  a. Flight visibility—The average forward horizontal distance, from the cockpit of an aircraft in flight, at which prominent unlighted objects may be seen and identified by day and prominent lighted objects may be seen and identified by night.

  b. Ground visibility—Prevailing horizontal visibility near the earth's surface as reported by the United States National Weather Service or an accredited observer.

  c. Prevailing visibility—The greatest horizontal visibility equaled or exceeded throughout at least half the horizon circle which need not necessarily be continuous.

  d. Runway visibility value/RVV—The visibility determined for a particular runway by a transmissometer. A meter provides a continuous indication of the visibility (reported in miles or fractions of miles) for the runway. RVV is used in lieu of prevailing visibility in determining minimums for a particular runway.

  e. Runway visual range/RVR—An instrumentally derived value, based on standard calibrations, that represents the horizontal distance a pilot will see down the runway from the approach end. It is based on the sighting of either high intensity runway lights or on the visual contrast of other targets whichever yields the greater visual range. RVR, in contrast to prevailing or runway visibility, is based on what a pilot in a moving aircraft should see looking down the runway. RVR is horizontal visual range, not slant visual range. It is based on the measurement of a transmissometer made near the touchdown point of the instrument runway and is reported in hundreds of feet. RVR is used in lieu of RVV and/or prevailing visibility in determining minimums for a particular runway.

    1. Touchdown RVR—The RVR visibility readout values obtained from RVR equipment serving the runway touchdown zone.

    2. Mid-RVR—The RVR readout values obtained from RVR equipment located midfield of the runway.

    3. Rollout RVR—The RVR readout values obtained from RVR equipment located nearest the rollout end of the runway.

visual approach An approach conducted on an instrument fight rules (IFR) flight plan that authorizes the pilot to proceed visually and clear of clouds to the airport. The pilot must, at all times, have either the airport or the preceding aircraft in sight. This approach must be authorized and under the control of the appropriate air traffic facility. Reported weather at the airport must be ceiling at or above 1,000 feet and visibility of 3 miles or greater.

visual separation A means employed by ATC to separate aircraft in terminal areas. There are two ways to effect this separation:

  a. The tower controller sees the aircraft involved and issues instructions, as necessary, to ensure that the aircraft avoid each other.

  b. A pilot sees the other aircraft involved and upon instructions from the controller provides their own separation by maneuvering the aircraft as necessary to avoid it. This may involve following another aircraft or keeping it in sight until it is no longer a factor. (Refer to FAR Part 91.)

VORTAC/VHF omnidirectional range/tactical air navigation A navigation aid providing VOR azimuth, TACAN azimuth, and TACAN distance measuring equipment (DME) at one site. (Refer to AIM.)

vortices/wing tip vortices Circular patterns of air created by the movement of an airfoil through the air when generating lift. As an airfoil moves through the atmosphere in sustained flight, an area of low pressure is created above it. The air flowing from the high pressure area to the low pressure area around and about the tips of the airfoil tends to roll up into two rapidly rotating vortices, cylindrical in shape. These vortices are the most predominant parts of aircraft wake turbulence and their rotational force is dependent upon the wing loading, gross weight, and speed of the generating aircraft. The vortices from medium to heavy aircraft can be of extremely high velocity and hazardous to smaller aircraft. (Refer to AIM.)

VOT/VOR test signal A ground facility that emits a test signal to check VOR receiver accuracy. Some VOTs are available to the user while airborne, and others are limited to ground use only. (Refer to FAR Part 91, AIM, *Airport/Facility Directory.*)

wake turbulence Phenomena resulting from the passage of an aircraft through the atmosphere. The term includes vortices, thrust stream turbulence, jet blast, jet wash, propeller wash, and rotor wash both on the ground and in the air. (Refer to AIM.)

waypoint A predetermined geographical position used for route/instrument approach definition, or progress reporting purposes, that is defined relative to a VORTAC station or in terms of latitude/longitude coordinates.

wilco I have received your message, understand it, and will comply with it.

wind shear A change in wind speed and/or wind direction in a short distance resulting in a tearing or shearing effect. It can exist in a horizontal or vertical direction and occasionally in both.

Index

Illustrations appear in **boldface**.

INDEX

INDEX

INDEX

stress (*Cont.*):
    fear of unknown, 256–257
    flight test, 261
    flying stress, 256–258
    life change scale, **260**
    nonflying stress, 259–261
    physical factors, 258
supplementary exercises, 5

T

TAFs (Terminal Aerodrome Forecasts) 44–46
    METAR/TAF codes, **51–52**
takeoffs
    instrument, 230–231
    minimums, 34, 331–333
    runway checks, 79–80
    taxi checks, 179
TDZE (touch down zone elevation), 24
TEC (tower en route control), 17
temperature-dew point spread, 33–34
temperature gauges, 98, 278
TERPS (*United States Standard for Terminal Instrument Procedures*), 168–169, 318
thunderstorms, 28–31, 54
    embedded, 28–29
    forecasts/reports, 30–31
    hail, 28
    microbursts, 29–30, **29**
time, logging, 236, 241
    (*See also* flight log)
    pilot-in-command, 6–7
timed turns, 152–153
tips
    approach, 208–209
    cross-country, 243–244
    ILS, 217–218
    to reduce cockpit confusion, 72
total time en route, 63–64
tracking, 126, 160, 162–164, **162, 164**
transcribing weather, 49–52, **50, 51–52**
    weather shorthand, 50–52, **51–52**
transponder preflight check, 72
transition to higher performance, 274–275
trim in turns, 103

turbulence, 31–32
    high winds, 31–32
turn coordinator, 79, 98, 102–103, **91–92, 102–103, 150**
turns
    magnetic compass, 151–153, **153**
    oboe pattern, 103–104, **104**
    pattern A, 104–105, **105**
    steep, 143–144
    timed, 152–153
    trim in turns, 103
    vertigo in turns, 144–145
"two, two, and twenty," 86
two-way radio failure, 241, 335–336

U

uncontrolled airports, 234–235
unusual attitudes, 145–147

V

vertical S pattern, 113, **113**
vertigo, 144–145
visibility
    ground, 35
    importance of minimums, 35
    landing minimums, 176–177, 331–333, **177**
    prevailing, 35
    RVR, 35, 177, 362, **177**
    RVV, 35, 362
visual approaches, 228–230, **239**
void time clearances, 77, 235
VOR (VHF omnidirectional range)
    checks, 68, 70–71, **70**
    diagnostic test, **119**
    diagnostic test answers, **120**
VOR procedures, 117–129
    approaches, 195–198, **197, 199, 206**
    bearing/radial interception, 122–126, **123, 124, 127**
    receiver check, 68
    bracketing, 128
    chasing the needle, 128
    cleared direct, 126
    heading indicator errors, 117–118
    holding procedures, 131–138, **132, 135, 137**

VOR procedures (*Cont.*):
    interception mistakes, 125–126
    practice patterns, 129
    reference heading, 126
    16-point orientation, 118–120, **121**
    proficiency, 118
    station passage, 128
    time/distance check, 120–122, **122**
    wind corrections, 125–126
VOT (VOR test facility), 70–71, **70**
VSI (vertical speed indicator), 90, 98, 109, **91–92, 150**

W

weather, 27–40, 54–55
    (*See also* weather information)
    air stability/lapse rate, 29
    alternate airports, 37, **38**
    approach categories, 36–37
    departure minimums, 34, 55
    destination minimums, 34–37, **36**
    downbursts, 28
    embedded thunderstorms, 28–29
    fog, 33–34, 55
    freezing level, 32–33
    gust front, 28
    hail, 28
    high winds, 31–32
    icing, 32–33, 55
    microbursts, 29–30, **29**
    "one, two, three" rule, 37
    personal minimums, 37, 39
    reviewing weather factors, 39–40
    temperature-dew point spread, 33–34
    thunderstorms, 28–31, 54
    turbulence, 31–32, 55
    visibility, 35–37
    wind shear, 28
weather information, 41–64
    abbreviated briefings, 62
    AFSS, 41
    computer weather services, 42, 310–311
    DUATS, 43–44, 310–311
    EFAS, 236, 238
    faster service, 49
    forecast reliability, 46–47
    go or no-go?, 54–55
    HIWAS, 238, **239**

ABOUT THE AUTHORS

Henry Sollman is an award-winning flight instructor who has been flying for more than 50 years. He developed the highly successful 10-day instrument certification program on which this book is based.

Sherwood Harris is a flight instructor with 30 years' experience who is the author of several books and numerous articles on aviation.

Made in the USA
San Bernardino, CA
07 May 2013